DIVINE ACTION AND EMERGENCE

Divine Action and Emergence

An Alternative to Panentheism

MARIUSZ TABACZEK

University of Notre Dame Press

Notre Dame, Indiana

University of Notre Dame Press
Notre Dame, Indiana 46556
undpress.nd.edu
Copyright © 2021 by the University of Notre Dame

All Rights Reserved

Published in the United States of America

Library of Congress Control Number: 2020950412

ISBN: 978-0-268-10873-1 (Hardback)
ISBN: 978-0-268-10876-2 (WebPDF)
ISBN: 978-0-268-10875-5 (Epub)

To my Dominican brothers and sisters

Religion blushing veils her sacred fires,
And unawares Morality *expires.*
Nor public *Flame, nor* private, *dares to shine;*
Nor human *Spark is left, nor Glimpse* divine!
Lo! thy dread Empire, Chaos! is restor'd;
Light dies before thy uncreating word:
Thy hand, great Anarch! lets the curtain fall;
And Universal Darkness buries All.
—Alexander Pope, *The Dunciad* (1743)

Hope springs eternal in the human breast;
Man never Is, but always To be blest.
The soul, uneasy, and confin'd from home,
Rests and expatiates in a life to come.
—Alexander Pope, *Essay on Man*
(1733–34)

CONTENTS

	List of Figures	xi
	Preface	xiii
	Acknowledgments	xv
	Abbreviations	xvii
	Introduction	1
PART 1	**The Phenomenon of Emergence**	**11**
CHAPTER 1	Science and Metaphysics of Emergence	15
CHAPTER 2	Classical and New Aristotelianism and Emergence	61
PART 2	**God's Action in Emergence**	**105**
CHAPTER 3	Panentheism and Emergence in the Science-Theology Dialogue	109
CHAPTER 4	Problems of Panentheism and Theology Inspired by Emergentism	155
CHAPTER 5	Emergence and the Thomistic View of Divine Action	187
	Conclusion	233

Appendix 1	237
Appendix 2	239
Notes	241
Bibliography	301
Index	321

LIST OF FIGURES

FIGURE 1.1.	First-order emergent phenomenon: water surface tension	16
FIGURE 1.2.	Second-order emergent phenomena	18
FIGURE 1.3.	Third-order emergent phenomena	19
FIGURE 1.4.	Types of EM	34
FIGURE 1.5.	Kim's argument from causal exclusion applied to the concrete example of pain and escape reaction	37
FIGURE 1.6.	Two types of autogenesis	53
FIGURE 2.1.	The role of acetyl coenzyme A in the citric acid cycle	77
FIGURE 2.2.	A diamond lattice crystal structure	77
FIGURE 2.3.	Examples of isomerization	86
FIGURE 2.4.	Different views concerning ontology of properties	92
FIGURE 3.1.	Krause's diagram depicting the panentheistic notion of the relation between God and the world	119

PREFACE

The main objective of this book is to critically evaluate emergentist panentheism within the circles of the theology-science dialogue and to propose an alternative theological interpretation of emergence in terms of the classical and the new Aristotelianism and the Thomistic concept of the concurrent action of God in the universe.[1] In so doing, the book both builds upon and is a natural continuation of my earlier project entitled *Emergence: Towards a New Metaphysics and Philosophy of Science*, which should be accessible and insightful for philosophers and theologians alike.[2] Nonetheless, I decided to provide in the first part of the present project an extended summary of the research and the argumentation presented in my first book. I do this for at least three important reasons.

First, the theological argument presented here is deeply rooted in Aristotelian-Thomistic metaphysics, as well as the metaphysical evaluation of both the classical and Deacon's dynamical depth accounts of emergence. Hence, developing the argument without a more extended reference to its metaphysical grounding would make it lacking and unconvincing. Second, the length, depth, and highly detailed character of my first monograph may be discouraging for a theologian who, though acknowledging the importance of philosophy for the plausibility and coherence of any theological investigation, might search for a more concise and summary presentation of the philosophical foundations for the theological analysis developed in the second part of this project. Moreover, preparation of a more concise version of the argument from *Emergence* required revisiting and further clarifying some of its main objectives. This summary results in (1) the new approach in defining and describing the sources of scientific reductionism and nonreductionism, (2) a more comprehensible presentation of the classical account of emergence, (3) a clearer definition and evaluation of its three main versions, and (4) a more lucid introduction to the way in which Aristotle develops his fourfold typology of causes as an explanation of both stability and change in nature.

Finally, whereas my first book assumes basic knowledge concerning scientific description, analysis, and classification of emergent phenomena, the opening of the present project offers a short presentation of the natural-science approach to emergence, together with the scientific attempts at defining it. It should prove valuable and helpful for those theologians who approach the topic of emergence for the first time.

Hence, I believe that both parts of the research presented in this book form an organic whole and that the latter cannot go without the former. However, those who are already familiar with my first monograph may skip the first part and go directly to the second.

ACKNOWLEDGMENTS

I would like to express my deep gratitude to Fr. Michael Dodds, O.P., my teacher and mentor, whose monograph *Unlocking Divine Action: Contemporary Science and Thomas Aquinas* inspired me to study in depth scientific and philosophical aspects of the theory of emergence and that theory's implications for the theological notion of divine action. I am also grateful to Fr. Dodds for his invaluable help and guidance in my studies of Aristotle and Aquinas. Many thanks to Terrence Deacon for an invitation to study and discuss his dynamical depth model of emergence, and to Robert Russell, the director of the Francisco Ayala Center for Theology and the Natural Sciences in Berkeley, California, for his help in my exploration and engagement with the Protestant thinkers involved in the science/theology dialogue, as well as for his friendship and support.

This project has its origin in my doctoral research pursued in 2011–16 at the Graduate Theological Union in Berkeley and would not have been possible without the hospitality of the Dominican Community at the Priory of St. Albert the Great in Oakland, California, and the generous support and cordial friendship of many fellow Christians whom I was honored to serve as a priest during the time of my studies. I took another important step toward realizing the project during my fellowship at the Notre Dame Institute for Advanced Study in 2016/2017. I greatly appreciate the generosity of the Institute and the support and insights coming from its directors, staff members, seminar guest professors, and fellows. I am also deeply grateful to my family for their ever-steadfast love and support and to my Dominican brothers and sisters for their encouragement, trust, and confidence in me.

Many thanks to Ignacio Silva and the second anonymous reviewer of the book for the University of Notre Dame Press for their valuable comments and suggestions; to Fr. Reginald Martin, O.P., who made many helpful corrections; to Robert Banning, who copyedited the manuscript; and to Stephen Little, Matthew Dowd, and their coworkers from the University of Notre Dame Press, who helped to bring this project to its final realization.

ABBREVIATIONS

ABBREVIATIONS FOR THE WORKS OF ARISTOTLE

De gen. an.	*De generatione animalium* (*On the Generation of Animals*)
De gen. et corr.	*De generatione et corruptione* (*On Generation and Corruption*)
De part. an.	*De partibus animalium* (*On the Parts of Animals*)
Meta.	*Metaphysica* (*Metaphysics*)
Meteo.	*Meteorologica* (*Meteorology*)
Phys.	*Physica* (*Physics*)

ABBREVIATIONS FOR THE WORKS OF ST. THOMAS AQUINAS

De mixt. elem.	*De mixtione elementorum ad magistrum Philippum de Castro Caeli*
De prin. nat.	*De principiis naturae*
De sub. sep.	*De substantiis separatis*
In De div. nom.	*In librum beati Dionysii De divinis nominibus expositio*
In Meta.	*In Metaphysicam Aristotelis commentaria*
In Peri herm.	*Commentarium in Aristotelis libros Peri hermeneias*
In Phys.	*In octo libros Physicorum Aristotelis expositio*
In Sent.	*Scriptum super libros Sententiarum Magistri Petri Lombardi Episcopi Parisiensis*
Q. de pot.	*Quaestiones disputatae de potentia Dei*
Q. de ver.	*Quaestiones disputatae de veritate*
Quod.	*Quaestiones quodlibetales*

SCG	*Summa contra gentiles*
ST	*Summa theologiae*
Super De causis	*Super librum De causis expositio*
Super De Trin.	*Super Boetium De Trinitate*

OTHER ABBREVIATIONS

CVC	counterfactual view of causation
DC	downward causation
DVC	dispositional view of causation
EM	emergence
MDC	medium downward causation
MVC	manipulability view of causation
NP	nonreductionist physicalism
PA	panentheistic analogy
ProbVC	probability view of causation
ProcVC	process view of causation
RVC	regulatory view of causation
SDC	strong downward causation
SUP	supervenience
SVC	singularist view of causation
WDC	weak downward causation

Introduction

> *The West's sense of itself, its relation to its past, and its sense of its future were all profoundly altered as cognitive values generally came to be shaped around scientific ones. The issue is not just that science brought a new set of values to the task of understanding the world and our place in it, but rather that it completely transformed the task, redefining the goals of enquiry.*
> —Stephen Gaukroger, *The Emergence of a Scientific Culture*

> In the beginning, God created the heavens and the earth. *These first words of Genesis tell us two fundamental things about God: God is and God acts. If we believe in a God who acts, we can talk about God only if we speak about action, and to do that we need a language of causality.*
> —Michael J. Dodds, *Unlocking Divine Action*

1. THEOLOGY IN THE AGE OF SCIENCE

It has been commonly accepted in reflections on the history of the Western thought that when the scientific revolution began in the seventeenth century, religion and theology began to lose their importance for the common effort of explaining the phenomena occurring in nature. We are told that it was the growing secularization of the West that gave an autonomy to scientific endeavors, thus enabling scientists to develop a novel method of an adversarial discourse in which new theories about the physical world could be tested, demonstrated, and verified, to be eventually accepted across the world. In this context, theology is often perceived as taking a defensive position. Striving to face the challenges of mechanicism and physicalism, as well as the reductionist notions of stability, change, and causal dependencies in nature, it seems to be forced to gradually give up its territory to science. The God of the gaps—that is, God working in/through the phenomena that remain unexplained by proper divisions of natural science—tends to become ever smaller with the growing success of science.

Even if this picture is oversimplified and falls short in describing the complexity of the actual situation of theology in the age of science, we must acknowledge that the revival of atomism and the reduction of the rich Aristotelian notion of causal dependencies in nature to the realm of physical interactions—the effects of which are quantifiable and can be expressed in the language of mathematics—could not but significantly influence theology, and our understanding of divine action in particular. This change was so dramatic that it is noticed and described not only by theologians but also by many philosophers. Edwin A. Burtt says at one point that

> with final causality gone, God as Aristotelianism had conceived him was quite lost. . . . The only way to keep him in the universe was to invert the Aristotelian metaphysics and regard him as the First Efficient Cause or Creator of the atoms. . . . God thus ceases to be the Supreme Good in any important sense; he is a huge mechanical inventor, whose power is appealed to merely to account for the first appearance of the atoms, the tendency becoming more and more irresistible as time goes on to lodge all further causality for whatever effects in the atoms themselves.[1]

Gradual departure from the explanation involving the notion of teleology and formal causation resulted, in turn, in the dismissal of the Thomistic concept of the God-world relationship with its categories of God being the first formal, efficient, and final cause, and a source of primary matter, as well as the dismissal of that concept's distinctions between principal and instrumental causation and between primary and secondary causation, with Aquinas's emphasis on the importance of analogical predication in theology. These changes led to the idea of God's action conceived univocally as that of a physical agent, interfering (instead of concurring) with other physical forces.[2]

Although some of the prophets of the new science would gradually exclude divine action thus defined, considering it disruptive for the determined patterns of scientific laws—embracing a deistic, agnostic, or even atheistic stance—many theologians strove to find the "causal joint" between the causality of God and that of creatures (the category of "causal joint" is present today, e.g., in the writings of Austin Farrer, John Polkinghorne, Philip Clayton, and Arthur Peacocke). Another group of theologians chose to follow the suggestion of Schleiermacher, who claims that "we should abandon the idea of the absolutely supernatural because no single instance of it can be known by us, and we are nowhere required to recognize it."[3] Applying this general rule to the question of the nature of divine action, they seem to agree with Rudolf Bultmann and his opinion that God's action in the universe can only be perceived within the realm of a personal, existential encounter, and—just like other aspects of God talk—our notion of it has to be "demythologized."[4]

2. PARADIGM SHIFT

In this context, the current paradigm shift taking place in the life sciences and other branches of science studying complex phenomena becomes an important factor that radically changes the framework of our philosophical analysis and predication. Biologists are acknowledging that numerous properties of living organisms are irreducible, which motivates the advancement of the systems approach in life sciences. New research strategies are also developing for the study of dynamical aspects of structures and arrangements of mereological wholes. This new approach in science leads to the rediscovery and a new explication of the theory of emergence (EM) and downward causation (DC), as a necessary philosophical background for the new approach to complexity. It also questions the reductionist agenda, predominant in natural science over the last three centuries.

Moreover, moving to yet another level of explanation, the new framework of the scientific and philosophical analysis of complexity in nature provides a fresh inspiration for theology, which by definition approaches reality nonreductively. Taking account of the fact that nature reveals divine action of the Creator, theologians working in the theology-science dialogue find in the theory of EM and DC an inspiration for better and deeper understanding of God's presence and causal action in the universe.

3. EMERGENTIST PANENTHEISM

One of the most robust and thorough models of divine action—developed in reference to DC-based EM theory and grounded in panentheistic understanding of the God-world relationship—has been offered by Arthur Peacocke and supported by Philip Clayton. Defining emergentist panentheism, Peacocke speaks about the top-down causal influence of God on the totality of the world, understood as a flow of information—that is, a pattern-forming influence. He claims that this approach enables him to reconcile God's action with the current paradigms of physical, biological, human, and social sciences. According to his proposition, God's action on the world-as-a-whole does not abrogate the natural regularities of its processes, which are described by nonlinear non-equilibrium thermodynamics, the theory of chaos, relativity, and quantum mechanics.

Robert John Russell challenges this model scientifically, however, noting that according to the Big Bang cosmology the universe does not have a boundary, as needed by the concept of the "whole." Moreover, because it is rooted in panentheism (which holds that the world is in God, who is, nevertheless, more than the totality of the world), Peacocke's view of divine action seems to assume that the God-world relationship is real in both terms (i.e., both in the Creator and in creatures) and that both terms are therefore mutually affected by it. This view thus calls into question the classical understanding of God as immutable, omniscient, omnipotent, infinite, eternal, and impassible. I hold that the concept of God's self-limitation of the divine attributes—proposed by Peacocke and many other contemporary theologians—leads to an image of God as a superintelligent and a superpowerful agent, yet not a truly divine one.

4. VERSIONS OF EMERGENTISM

The classical version of the theory of EM itself raises an important metaphysical question concerning the nature of DC. It seems that the reduction of causality in modernity from the Aristotelian four causes (material, formal, final, efficient) to merely physical interactions has forced proponents of emergentism to think about top-down influence in terms of efficient causation alone, which tends to reduce top-down influence to efficient causes being posited on higher levels and acting on lower levels of complexity. In view of the manifest inadequacy of this model (Jaegwon Kim, Menno Hulswit), it has therefore been suggested that EM and DC can be possibly saved in the context of systemic causation, which goes beyond efficient causation, reaching toward Aristotle's concepts of formal and final causes (Claus Emmeche et al., Michael Silberstein, Charbel Niño El-Hani and Antonio Marcos Pereira, Alvaro Moreno and Jon Umerez).

This advice was followed by Terrence Deacon, who—in a number of publications, among which *Incomplete Nature* remains the most influential—offers a new version of EM theory, arguing in favor of a broader understanding of causation. Interestingly, developing his model of EM, Deacon rejects top-down mereological (whole-part) reasoning and suggests rethinking EM in dynamical terms. He introduces an intriguing notion of "constitutive absences" (constraints), understood as "possible features being excluded," as the core of his process view of EM, in which "what is absent is responsible for the causal power of organization and the asymmetric dynamics of a physical or living process."[5] In other words, the reduction of possibilities and options brings an increase in complexity and specialization, leading to the EM of the new features of inanimate and animate entities.

While fascinating and promising, Deacon's project raises important metaphysical questions. Although he explicitly rejects eliminative reductionism (assuming that everything reduces to physical particles), Deacon does not side with the classical antireductionist positions. Nor does he follow contemporary proponents of top-down causation. Rather, he suggests reinterpreting formal cause as a function and final cause as an emergent outcome of basic mechanical physico-dynamic processes—a position which is still compatible with some form of limited reductionism and departs from the Aristotelian understanding of these types of causation. Moreover, his idea of the causality of absences seems to be philosophically counterintuitive, as it assumes that "what is not" can act on "what is." Finally, following many proponents of the scientific notion of emergentism, Deacon rejects Aristotle's concept of hylomorphism (the view that things are composed of prime matter and substantial form). He is also critical of the process metaphysics of Whitehead, which he rejects as another version of panpsychism. At the same time, however, he does not seem to offer a fully developed alternative ontology for biological emergentism.

5. NEW AND CLASSICAL ARISTOTELIANISM AND EMERGENTISM

In the context of difficulties challenging both the top-down mereological and the dynamical depth versions of EM, I have proposed—in *Emergence: Towards a New Metaphysics and Philosophy of Science*—dispositional metaphysics and the corresponding view of causation as a possible solution to the ontological problems of emergentism. I will summarize the main points of my argumentation in the first part of the present project. Formed within the analytic philosophical tradition, dispositionalism defines powers in things and organisms as intrinsic properties characteristic for natural kinds and explains causation as a manifestation of these dispositions (Alexander Bird, Stephen Mumford, David Oderberg, Brian Ellis, George Molnar, and others). I argue that dispositional metaphysics opens the way to a more robust view of causation, while being attentive to cases of polygenic causation, where more than one cause is responsible for an effect, and to the problem of the causation of absences. Thus, I contend, this theory can serve as a metaphysical ground for emergentism, both in its classical DC-based account and in Deacon's dynamical depth model, based on the idea of causality of "constitutive absences."

Because it supports new essentialism (the view that beings have essences decisive for their natures) and involves a possible retrieval of formal and final causation, dispositional metaphysics can be regarded as a neo-Aristotelian position. It shows that Aristotle's explanation—employing scientific principles as well as the notion of causes that admittedly lie beyond the bounds of science (but are not facile explanations [homunculi], but legitimate [natural] principles of a philosophy of nature)—is still valid and applicable in the context of contemporary science.

6. NEW THEOLOGICAL INTERPRETATION OF EMERGENCE

A theory of EM based on dispositional metaphysics has a new explanatory potential. It not only reconciles Aristotelianism with emergentism but may also significantly affect the view of divine action developed in reference to the theory of EM. God's action would no longer be conceived panentheistically, as an influence on the totality of the world, a view that, metaphysically speaking, might assume that the causation of God and creatures is of the same kind and/or occurs on the same ontological level (univocal predication). Understood this way, emergentist panentheism might run the risk of collapsing into pantheism.[6] The recovery of the plural notion of causation allows us to recapture the classical understanding of divine action as proposed by Aquinas. God is regarded as the Creator of all substantial forms, acting through his divine ideas as exemplar causes of living and nonliving entities and acting as the ultimate source and aim of all teleology in nature. With regard to efficient causation, God's transcendence is protected by Aquinas's introduction of the categories of primary and principal causation of the Creator, who acts from the transcendent order of causation through the secondary and instrumental causality of his creatures, whose action takes place within the immanent order of causation. Therefore, God's immutability, omniscience, omnipotence, infinity, eternity, and impassibility are not challenged, while his immanent and constant presence in all worldly events is by no means undermined.

To develop such a model of divine action—built in reference to both the classical DC-based and Deacon's dynamic depth views of EM (reinterpreted in terms of the classical and new Aristotelianism) and the classical Aristotelian-Thomistic view of the God-world relationship—is the main goal of the project pursued in this volume. I want to propose it as an important alternative to the emergentist panentheism developed by Peacocke and supported by Clayton and by Niels Henrik Gregersen.

7. METHODOLOGICAL STRATEGY

The project has an interdisciplinary character and contributes to the ongoing dialogue between theology and the natural sciences. I believe, however, that the approach presented here is distinct due to the argument's strong emphasis on the role of metaphysics, philosophy of nature, and philosophy of science (philosophy of biology in particular) as foundational for the philosophical theology developed in the book's conclusion. It will prove crucial that I refer to all these disciplines of knowledge as I analyze and evaluate the two models of EM and that I apply the results of this inquiry to the theological understanding of divine action.

Concerning the distinction between natural theology and theology of nature, I will use the latter approach to the science and theology dialogue, incorporating the theory of EM into a theological reflection on the nature of divine action. I will also adopt the method of a realist critique of knowledge which acknowledges that concepts, models, and hypotheses in human cognition are never fully accurate because they do not give us an exact picture of the reality they describe, and therefore need to be revised. At the same time, however, I want to emphasize, following Étienne Gilson, that realism, even when understood critically, allows that our cognition gives us an access to the real world. This position—called "methodical" or "moderate" realism—prevents us from falling into the trap of Kantian epistemology in which we do not have access to *noumena* (things-in-themselves). For a true realist (accepting a realist theory of knowledge), being is always prior to knowing and a condition of knowing (knowledge is derived from being, not vice versa). Therefore, what is real is necessarily intelligible.[7]

Finally, one more methodological principle underlying the project states that any attempt thoroughly to describe the reality of the world has to take into account both its quantitative and qualitative aspects. The first group of properties belongs primarily to the domain of science, whereas the second opens the way to philosophical and theological investigation as well. Neglecting either one of them leads to reductionism or eliminativism.

8. PLAN OF THE PROJECT

The project is divided into two parts, each preceded by a short introduction. The first part concentrates on the phenomenon of EM and the metaphysical foundations of emergentism. It begins in chapter 1 with a series of examples of first-, second-, and third-order emergents and proceeds to analyze the origins of philosophical emergentism. The remaining part of chapter 1 presents and metaphysically evaluates the three main versions of the classical account of ontological EM (as distinguished from epistemological EM), as well as Deacon's dynamical depth model of ontological EM.

The second chapter applies classical and new Aristotelianism to EM theory. Beginning with Aristotle's response to ancient philosophical strategies in explaining stability and change in nature, it introduces his theory of four causes and their interrelatedness, as well as the Aristotelian view on the character of chance events in nature. Next, after exploring dispositionalism with its corresponding view of causation—unique among all other causal theories developed in analytic philosophy—I will analyze the neo-Aristotelian character of powers metaphysics. The chapter will conclude by summarizing the constructive proposal of a reinterpretation of both the classical DC-based and Deacon's dynamical depth accounts of ontological EM in terms of classical and new Aristotelianism, developed and presented in greater length in my *Emergence: Towards a New Metaphysics and Philosophy of Science*.

The second part of the project is dedicated to theological implications of EM theory for our understanding of God's action in the universe. In chapter 3 I study the most important historical facets of philosophical panentheism, which seems to inspire a number of theologians working in the science and theology dialogue. Having analyzed their application of panentheism in theological predication, I present the four available models of theological interpretation of emergentism, especially emphasizing Peacocke's emergentist panentheism.

Chapter 4 critically evaluates panentheism and examines the main shortcomings of all four ways of applying EM in the theology of divine action. Again, I attend especially to the difficulties and challenges of Peacocke's emergentist panentheism.

The book will conclude in chapter 5, which offers my constructive proposal of a new theological understanding of EM. In reference to the classical and neo-Aristotelian reinterpretation of both DC-based and Deacon's dynamical depth views of EM, it will entail Aquinas's theology of divine action and the God-world relationship, as well as a comparative analysis of Aquinas's and Deacon's notions of the ontological nature and the role of nonbeing (absences). A brief summary of advantages of my proposal over Peacocke's emergentist panentheism will complete the project.

PART 1

The Phenomenon of Emergence

The discussion of emergence has grown out of the successes and the failures of the scientific quest for reduction. Emergence theories presuppose that the once-popular project of complete explanatory reduction—that is, explaining all phenomena in the natural world in terms of the objects and laws of physics—is finally impossible.
—Philip Clayton, *Conceptual Foundations of Emergence Theory*

The phenomenon of emergence takes place at critical points of instability that arise from fluctuations in the environment, amplified by feedback loops. Emergence results in the creation of novelty, and this novelty is often qualitatively different from the phenomenon out of which it emerged.
—Fritjof Capra, *The Hidden Connections*

Philosophical reflection on the current status of natural science reveals its rather paradoxical character and nature. On the one hand, for more than three centuries, it has been driven by the reductionist agenda, developed and pursued by many practicing scientists and theorists of the scientific endeavor. The rapid development of biochemistry, molecular biology, and neuroscience in the twentieth century led many of them to believe that all phenomena investigated in the social sciences and psychology could be reduced to those studied in neuroscience, biology, and biochemistry, which could be, in turn, further reduced to chemistry, and eventually to physics. Questioning the reality of qualities, the advocates of the reductionist attitude in science explain them as a result of certain quantitative arrangements in matter, which can be expressed in mathematical language. Thus, the difference between red and green is for them just a matter of the difference of light wavelength, the melody of a song a variation of sound waves, while flavor, scent, hardness, and texture are simply functions of the constitution of elementary building blocks entering into the basic physical and chemical structures of things.

On the other hand—although it proved extremely successful for both describing and explaining an immense variety of natural phenomena, as well as for translating this knowledge into many practical and technological solutions and inventions—the reductively oriented science does not seem to provide an ultimate and an exhaustive explanation of the nature of material entities. Hence, its paradigm and quite radical aspirations have recently been objected to and criticized by a growing number of scientists and philosophers of science who claim it necessary to accept the ontological irreducibility of numerous phenomena, properties, and processes that characterize both inanimate and animate nature.

This irreducibility remains the object of study in the new, systems approach to biology, which originates in the growing availability of high computational power, developments in mathematical and algorithmic techniques, and the introduction of mass data production technologies (e.g., high-throughput data collecting). Applied to biological research, these technologically mediated methodological innovations questioned the reductionist and gene-centric strategies of investigating isolated molecular components or pathways. The "quantitative turn" of systems biology enabled supplementing such *in vitro* analysis and measurement of molecular properties and interactions with the *in vivo* study of their actual operation in living organisms.[1]

This new, nonreductionist approach in biology shows both that predictively accurate models for theoretical and practical purposes require a holistic approach and that qualitative properties of complex biological systems are not simply functions of quantitative aggregation of their physically simpler constituents. What is more, when analyzed from the philosophical point of view, the new paradigm in molecular biology opens the way to the rediscovery and further development of the theory of EM as a necessary background and grounding for biological theories.

The theory of EM is the main subject of my study presented in this part of the project. Starting by describing a number of emergent phenomena, the first chapter analyzes the origins of philosophical emergentism and presents the main objectives of its classical account. Among the main versions of ontological EM, I will especially attend to the one based on the concept of DC, that is, new, primitive, and top-down-oriented causal power, which is regarded by many as a decisive and the most characteristic trait of emergent systems. After discussing major shortcomings of the classical account of EM and showing the need and opening a way to its redefinition in terms of a more robust theory of causation, the latter part of chapter 1 will analyze the project developed by Terrence Deacon, in which he applies categories of causation related to those of Aristotle in his original dynamical depth model of EM.

The second chapter begins by introducing the legacy of Aristotle's classical theory of causation against the background of other views popular in his day. I then present and evaluate the neo-Aristotelian aspects of dispositional metaphysics and the corresponding theory of causation (inspired by the insufficiency of the six main views of causation offered in analytic philosophy). The latter part of chapter 2 proposes a reinterpretation of both the classical DC-based and Deacon's dynamical depth versions of ontological EM in terms of classical and new Aristotelianism.

CHAPTER 1

Science and Metaphysics of Emergence

Aristotle opens his study of metaphysics acknowledging that "it is owing to their wonder that men both now begin and at first began to philosophize."[1] If so, then we should begin our complex metaphysical investigation of emergentism with a deep breath filled with wonder and astonishment about the beauty and unique features of emergent phenomena.

1. THE WONDER OF EMERGENCE

Watching the surface of a lake or a river on a warm summer day, one can easily notice the so-called water striders or water skippers, that is, insects belonging to one of more than 1,700 species of Gerrids (*Gerridae*), known for their ability to walk on water. This unique skill of Gerrids is possible not only due to their light body weight but also because of the relatively high surface tension of water. The latter is regarded as an emergent phenomenon, that is, a phenomenon which shows, is realized, or arises out of some more fundamental phenomena and yet is novel and irreducible with respect to them. Thus, in the case of water surface tension, even if we can determine that this phenomenon is related to the characteristic V-shape of H_2O molecules, with an approximately 106° angle between the two O-H chemical bonds (which departs from the quantum equation for a system built of eighteen protons and

15

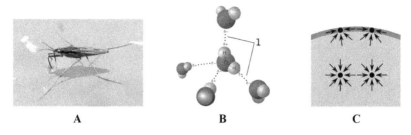

Figure 1.1. First-order emergent phenomenon: water surface tension
(A) Water strider of the genus *Gerris* walking on water. **(B)** The structure of water. The web of hydrogen bonds in a conglomerate of H_2O molecules is responsible for their mutual attraction. This attraction is higher (due to cohesion) than the attraction of H_2O to the molecules in the air (due to adhesion). The net outcome **(C)** is an inward force at the surface that causes the body of water to behave as if it was covered with a stretched elastic membrane. The relatively high attraction of water molecules effects a surface tension (72.8 millinewtons per meter at 20°C) higher than that of most other liquids.

Sources: **(A)** Webrunner, "Gerris," Wikipedia, May 25, 2010, https://en.wikipedia.org/wiki/File:Gerris_by_webrunner.JPG; **(B)** Qwerter, "3D Model Hydrogen Bonds in Water," Wikipedia, April 16, 2011, https://en.wikipedia.org/wiki/File:3D_model_hydrogen_bonds_in_water.svg; **(C)** User:Booyabazooka, "WassermoleküleInTröpfchen," Wikipedia, November 2008, https://en.wikipedia.org/wiki/File:Wassermolek%C3%BCleInTr%C3%B6pfchen.svg.

electrons and—typically—eight neutrons), the phenomenon in question occurs only in large conglomerates of water molecules and is irreducible to singular H_2O molecules. See figure 1.1.

Surface tension is just one example among a number of physical phenomena which result from basic characteristics and patterns of behavior typical of molecules or other building blocks of entities and substances when taken in bulks (conglomerates). These phenomena are classified as first-order emergent and include, among others, friction, viscosity, elasticity, tensile strength, temperature, and—according to a number of particle physicists—mass, space, and time (which are thought to be arising from Higgs bosons or strings).

Another set of examples of naturally occurring phenomena which both fascinate and puzzle human observers are (a) natural geometric patterns that develop through the interactions of shapes played out sequentially over time and (b) self-organizing (dissipative) systems. The former group includes polygonal and circular ground patterns, alternating stripes of stones and vegetation, water erosion formations, and so on.

Among the latter we find examples of water crystals forming on glass, formation of snowflakes, eddies forming in bodies of water, or convection cells (e.g., Bénard cells forming in heated liquids). These are all examples of the second-order emergent phenomena occurring in nature. See figure 1.2.

Finally, scientists distinguish third-order emergent phenomena as resulting from interactions sensitive to shape and time that show heritable and teleological features. As examples we can list the origin and organization of life (from subatomic level to the entire biosphere) and various cases of swarm behavior (ant and bee colonies, migrating insects, schooling fish, flocking birds, etc.). See figure 1.3.

However, the science of EM does not stop on this standard three-order classification of emergent phenomena. Going back to the bottom level of complexity, the physics of quantum mechanics seems to suggest that quantum entanglement and our perception of a deterministic reality—in which objects have definite qualitative features, positions, momenta, and so forth—are both emergent phenomena, grounded in the true state of matter described by a wave function, which does not allow us to assign to elementary particles definite positions or momenta. A similar situation obtains with the laws of physics as we know them today. It is believed that they all emerged from one fundamental law, which is yet to be found by science. Moving up on the scale of complexity, one may argue that the laws of chemistry emerged from those of physics and gave the origin to the laws of biology (including evolution). The laws of biology provided—in turn—a necessary foundation for the EM of the laws related to human mind, consciousness, and rationality—the object of study in psychology and social sciences.

Acceptance of this argument opens a way to consider as emergent a number of other familiar phenomena, such as spontaneous organizational tendencies characterizing groups of people, economic trends and the stock market, architectural and traffic patterns of modern cities, the World Wide Web, patterns of Internet traffic (including patterns in the social media), and so on.[2] These are all examples of decentralized complex occurrences exhibiting higher-order irreducible emergent properties. What characterizes these systems, according to Robert Laughlin, is broken symmetry, that is, the fact that the symmetry present on the lower level of complexity is not present on the higher level, due to phase transitions. This does not make the lower-order interactions irrelevant, but

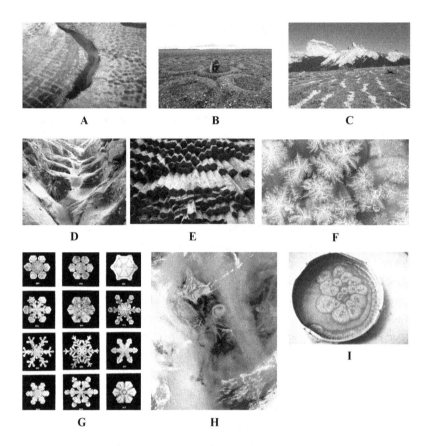

Figure 1.2. Second-order emergent phenomena

(A) polygons (so-called ice wedges in permafrost areas of arctic tundra); **(B)** circles (Svalbard Archipelago in Norway); **(C)** stripes of stones and vegetation (Glacier National Park in Montana); **(D)** water erosion steps (Red Rock Canyon in Oklahoma); **(E)** stone pattern (carved by the Goght River at Garni Gorge in Armenia); **(F)** water crystals forming on glass; **(G)** complex symmetrical and fractal patterns in snowflakes; **(H)** two currents (the Oyashio and Kuroshio currents) collide, creating eddies, and phytoplankton concentrates along the boundaries of these eddies, tracing out the motions of the water; **(I)** Bénard cells in a heated liquid.

Sources: **(A)** Dennis Cowals, "Alaska Patterned Ground 1973," Wikipedia, June 28, 2011, https://en.wikipedia.org/wiki/File:Alaska_patterned_ground_1973.jpg; **(B)** Hannes Grobe, "Permafrost Stone-Rings," Wikipedia, August 31, 2007, https://en.wikipedia.org/wiki/File:Permafrost_stone-rings_hg.jpg; **(C)** Walkswithgoats, "Patterned Ground in Glacier Park," Wikipedia, August 9, 2017, https://commons.wikimedia.org/wiki/File:Patterned_ground_in_Glacier_Park._.jpg; **(D)** Gina Dittmer, "Red Rock Erosion Steps," PublicDomainPictures.net, n.d., http://www.publicdomainpictures.net/view-image.php

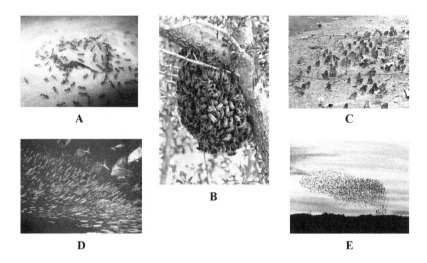

Figure 1.3. Third-order emergent phenomena
(A) a swarm of ants which have discovered a food source; **(B)** bee swarm on a tree; **(C)** Dark Blue Tigers (*Tirumala septentrionis*) migrating twice a year between Eastern and Western Ghats in India; **(D)** schooling anchovies; **(E)** flocking birds.

Sources: **(A)** Zainichi Gaikokujin, "Ants Eating Fruit," Wikipedia, October 9, 2009, https://en.wikipedia.org/wiki/File:Ants_eating_fruit.jpg; **(B)** Fir0002/Flagstaffotos, "Bee Swarm Feb08," Wikipedia, February 23, 2009, https://en.wikipedia.org/wiki/File:Bee_swarm_feb08.jpg, under the GFDL 1.2 license: https://www.gnu.org/licenses/old-licenses/fdl-1.2.en.html; **(C)** Kishen.das, "DarkBlueTigersCongregation," Wikipedia, February 4, 2010, https://en.wikipedia.org/wiki/File:DarkBlueTigersCongregation.jpg; **(D)** Bruno de Giusti, "Moofushi Kandu Fish," Wikipedia, January 27, 2007, https://en.wikipedia.org/wiki/File:Moofushi_Kandu_fish.jpg, under the Creative Commons Attribution-Share Alike 2.5 Italy license; **(E)** Christoffer A. Rasmussen, "Sort sol ved Ørnsø 2007," Wikipedia, September 26, 2007, https://en.wikipedia.org/wiki/File:Fugle,_%C3%B8rns%C3%B8_073.jpg.

?image=158104&picture=red-rock-erosion-steps; **(E)** "Garni Gorge," Wikipedia, May 19, 2007, https://en.wikipedia.org/wiki/File:Garni_Gorge3.jpg; **(F)** Rusfuture, "Water Crystals on Mercury 20Feb2010," Wikipedia, February 20, 2010, https://en.wikipedia.org/wiki/File:Water_Crystals_on_Mercury_20Feb2010_CU1.jpg; **(G)** Wilson Bentley, "SnowflakesWilsonBentley," December 9, 2010 [photos were taken in 1902 in Jericho, VT], Wikipedia, https://en.wikipedia.org/wiki/File:SnowflakesWilsonBentley.jpg; **(H)** Norman Kuring, "Spring Bloom Colors the Pacific near Hokkaido," Wikipedia, May 21, 2009, https://en.wikipedia.org/wiki/File:Spring_Bloom_Colors_the_Pacific_Near_Hokkaido.jpg; **(I)** WikiRigaou, "Bénard Cells Convection," Wikipedia, 2005, https://en.wikipedia.org/wiki/File:B%C3%A9nard_cells_convection.ogv.

simply tells us that their effects, observable in higher-order phenomena, have been renormalized. This assertion can be regarded as a more articulate and precise expression of the popular claim saying that EM simply tells us that the whole is more than the sum of its parts.[3]

If this is true, however, we need to acknowledge that the reductionist attitude in natural sciences interpreting all entities and phenomena in terms of their basal elementary constituents, predominant in the West after the scientific revolution of the seventeenth century, needs to be reevaluated and supplemented with a nonreductionist approach.[4] However, should we be willing to take this step so easily? Does natural science really need such radical reshaping of its own methodology, based on a number of natural phenomena that resist a reductionist explanation—an explanation that otherwise proved to be extremely efficient and successful? Will such phenomena show the same level of resistance in the future, taking into account the rapid progress and development of analytical devices and computing methods available to us? In other words, are we dealing here with phenomena that are irreducible on a merely epistemological level of analysis, or, rather, on an ontological level? And what is, after all, the exact meaning of EM? An attempt to answer these questions shifts our conversation about EM inevitably from a purely scientific ground toward the domains of philosophy of science, philosophy of nature, and metaphysics.[5] It is in this context that we will proceed with our investigation of emergentism in the remaining part of this chapter.

2. ORIGINS OF PHILOSOPHICAL EMERGENTISM

Tracing philosophical roots of emergentism, we need to go back to the origins of the reductionist agenda, which directs our attention toward a crucial distinction between (1) the study of mechanics, (2) the study of the physical aspects of matter, and (3) the study of deeper metaphysical questions concerning matter and the nature of being.

2.1. Mechanics, Physics, and Metaphysics

The study of mechanics, concerned with the behavior of physical bodies when subjected to forces or displacements, had been pursued since antiquity and became one of the central topics in natural philosophy.[6] It was construed as a form of practical mathematics and remained marginal to the physical theory of matter, regarded as another important yet separate division of natural philosophy. The theory of matter, in turn, was traditionally related to and followed by a metaphysical reflection construed in terms of essential features and asking the most fundamental questions of what there is (questions concerning being qua being) and of the ultimate causes of stability and change in nature.[7]

Although the great minds of the ancient and medieval West strove to build comprehensive systems that referred to all three aspects of knowledge about nature mentioned here, they would usually respect the division between mechanics on the one hand and both physical and metaphysical aspects of matter theory on the other. Moreover, looking at the protagonists of modern science, we realize that, for the most part, they seemed to respect the same distinction as well. Thinking about Isaac Newton's (1624–1727) laws of motion, which gave the foundation to classical mechanics, we realize that in formulating these laws, Newton remained agnostic about the ultimate character of matter. Similar was his approach in formulating the universal law of gravitation. Concentrating on empirical observation and inductive reasoning, he deliberately put aside the question about the ultimate causes of attraction between physical bodies and of gravitation. As a philosopher, he was naturally interested in these questions. Apart from assuming that gravity must be caused by "an Agent acting constantly according to certain laws," he considered a number of possible theories of gravity, with theories of matter related to them, including the following:

(a) matter as a subtle spirit pervading bodies and lying hidden in them;
(b) the efflux theory, which assumed that gravity was caused by a stream of particles moving rapidly towards the surface of the earth from above and pressing bodies;
(c) vitalistically conceived "fermental virtue" (crucial for transformation of species of matter);
(d) aether as a cause of gravity (made of particles smaller than air with almost no resistance to motion of things, and different from the Cartesian theory of dense aether dragging things around);

(e) matter defined as an impenetrable and opaque region of space;
(f) matter as force, which is in an important respect divine (close to occasionalism, the theory strives to accommodate forces in nature, but failing to find a place for them in the physical realm, it locates their source in God).

Once again, although Newton considered all these opinions as a philosopher, in his mechanics he distanced himself from any one of them and remained agnostic about a possible answer to ultimate questions concerning the theories of gravity and of matter.

A similar approach was adopted by Francis Bacon (1561–1626), who—when forging his famous empirical methodology, which valued eliminative induction more than logical syllogism—did not blur the distinction between mechanics on the one hand and physical and metaphysical matter theories on the other. Neither did he intend to dismiss any one of them—contrary to what is usually thought by those who call his agenda antimetaphysical or perceive Bacon as unconcerned with mechanics. It is true that he criticized the usefulness of the reference to formal and final causes in the theory of matter developed under the domain of natural philosophy. Nevertheless, that does not mean Bacon dismissed substantial form and teleology altogether. Rather, he thought they belonged to the domain of metaphysics (searching for axioms of the highest generality), and he differentiated them from material and efficient causal explanations, which he associated with the physical theory of matter (concerned with lower and middle axioms).

Speaking of mechanics, Bacon found it as indispensable as physics and metaphysics. For, whereas mechanics guided theoretical ascent from experiments to causes and axioms (the subject of physics and metaphysics), it was the task of mechanics to lead the operative descent to new works and new experiments. Hence, even if Bacon showed little interest in the emerging science of mathematically based mechanics (abstracting motions from material causes for the purpose of representing motions in geometrical or mathematical models), he still valued philosophical mechanics as sustaining the union of *res* and *mens* by its continual descent from axioms to experiments.[8]

2.2. From Mechanics to Mechanism and Reductionism

This division of labor in explanation between mechanics, natural philosophy (physics), and metaphysics was blurred with the advent of mechanism—an invention of the seventeenth-century philosophical reflection that accompanied the scientific revolution. The unique feature of mechanism was combining into a unified theory of explanation both matter-theoretical and mechanical considerations. Whereas the former provided an account of the behavior of physical phenomena in terms of their material constituents—supplemented with the metaphysical analysis of the highest axioms and causes—the latter investigated natural and forced occurrences in terms of motions and causes of these motions, mathematically quantified and modeled. The most conscious and explicit attempt at combining these approaches into an integrated theory of explanation, which has been classified as mechanism, was pursued by René Descartes (1596–1650). The main inspiration for this new strategy in natural philosophy was his bold assumption that we can name and specify one single, fundamental level of description of the material universe—that of inert microcorpuscles acting on one another through transfer of some mechanically specifiable and mathematically describable quantities, such as motion or momentum, on surface contact. Consequently, "the core claim of mechanism was that the behavior of macroscopic bodies could, by means of a programme of reduction, be accounted for fully in terms of such micro-corpuscularian interactions, and the very influential Cartesian programme of biomechanics extended the explanatory claims of mechanism to all physical phenomena, including the organic. In this way, mechanism offered an account of the physical realm that purported to be comprehensive."[9]

Further development of the Cartesian project, which dislodged both Aristotelianism and Neoplatonism, established its metaphysical and epistemological presuppositions as a general foundation for natural-philosophical inquiry. Once the metaphysically neutral project of mechanics was replaced with the ontologically loaded doctrine of mechanism, a new trajectory was established. It went gradually from questioning the reality of substantial forms, primary matter, and teleology toward the straightforwardly reductionist supposition that all causation in nature is of the efficient (physical) character. Consequently, having dismissed matter-theoretical questions, mechanism could be replaced once again by mechanics. This time, however, it was not mechanics understood as a form of practical mathematics. It was a rational mechanics, that is, a new natural philosophy that claimed to be able to give an ultimate explanation of all natural phenomena (e.g., electricity, magnetism, chemical reactions, physiological phenomena), based on purely experimental (phenomenal) data, without trying to couch these data in a single set of fundamental metaphysical principles.

Nonetheless, allegedly free from all matter-theoretical considerations, the new mechanics ended up grounding its assertions in a strong metaphysical conviction that goes back to the ancient atomism assuming that the universe is just a vast number of elementary particles, organized in different ways and influencing one another mechanically. In its more contemporary version, the same metaphysics argues that all there is is just mass, charge, and spin, located in the regions of space-time.

Inspired by the new mechanics, which saw causation as a necessary succession of phases in a physical (that is, mechanical) system, tantamount to the possibility of prediction, some thinkers—the followers of David Hume and a number of interpreters of quantum mechanics, coming from the logical-positivist movement in particular—went as far as to question the reality of cause/effect relationships in nature altogether. Humeans saw such relationships as a mere projection of human mind facing constant conjunctions of events in nature, while the proponents of the Copenhagen interpretation of quantum mechanics (e.g., Max Born) argued that when describing nature at its bottom level of complexity, we should ascribe a more fundamental role to chance than to causality.

Consequently, both groups of thinkers embraced a sort of causal nihilism, despite its undermining the pursuit of science as an explanation of stability and change in nature. What made them comply to this position was an agreement with the postulate that the nature of scientific endeavor should be rethought and redefined as (1) an analysis of human sense impressions and an attempt to organize them into a type of synthesis, enabling human beings to adapt themselves to natural conditions (phenomenalism of Ernst Mach); (2) a set of useful conventions, having their value in utility rather than pretending to discover ultimate truths about the reality (conventionalism of Henri Poincaré); or (3) a set of operations used for defining and measuring some basic physical concepts (operationalism of Percy Bridgman). Understood this way, science seemed to have a purely descriptional, practical, and pragmatic—rather than explanatory—character and nature.[10]

2.3. From Reductionism to Emergentism

Logical positivism proved to be the highest stage of development of the modern reductionist and empirical agenda. Its criticism marks what is oftentimes called a reemergence of EM in the second half of the twentieth century. This term becomes meaningful and informative once we realize that emergentism actually originated much earlier, among the protagonists of British empiricism in late-nineteenth-century Great Britain. Witnessing impressive advances in chemistry and biology at the time, John Stuart Mill (1806–73) and George Henry Lewes (1817–78) realized that some of the reductionist explanatory bridges between these sciences and physics became untenable.[11] They analyzed philosophical aspects of the "compositions of causes" and transition laws in cases investigated by chemistry and biology, and they emphasized that some laws and causal explanations of higher phenomena are irreducible. Following this analysis, Lewes introduced, for the first time—in both scientific and philosophical contexts—the concept of "emergent effects."[12]

As the struggle between reductionism and the growing skepticism about its validity in special sciences (beyond physics) continued, emergentism was reinvigorated once again at the beginning of the twentieth century. This time, it was proposed in a more specific context of philosophical issues arising in biology, as a middle ground between the extremes of vitalism and mechanistic reductionism. Its major proponents were Samuel Alexander (1859–1938), Conwy Lloyd Morgan (1852–1936), and Charlie Dunbar Broad (1887–1971).[13] Although the theory they offered seemed to show the hallmarks of a valid and intriguing proposition, it came to be seriously questioned and radically rejected by the antiphilosophically and antimetaphysically oriented agenda of logical positivism and thinkers who stood at the origins of the analytic tradition in philosophy.

The most recent revival of emergentism came with the critical evaluation of the program of logical empiricism in the second half of the last century, accompanied by a growing interest in and rediscovery of metaphysics in analytic philosophy. Both movements became crucial for the new methodological reflection in philosophy of science, inspired by the rapid changes in biology, making use of high computational powers, new mathematical and algorithmic techniques, and mass data production technologies (e.g., high-throughput data collecting). All these resources allowed molecular biologists to collect dense dynamic information from complex biological systems and study their constituents (biomolecules) *in vivo*. Such an approach became a viable alternative and a paradigm shift in the field of molecular biology, which used to measure molecular properties of biomolecules, discover interactions, and join them in causal pathways *in vitro*. The new techniques showed that an important part of the picture was often lost when molecules were analyzed with standard methods. Hence, the traditional reductionist approach in molecular biology was found in need of being enriched by a nonreductionist and "system-oriented" attitude, one that looks on an organism and investigates it holistically, as a dynamic system unifying a complex, causally related set of material constituents that are organized hierarchically and on several levels.

What becomes obvious with the acceptance of this new methodology is the reality of a number of new qualitative features and phenomena occurring on the higher levels of description, which show resistance to both the simple intertheoretical and the bridge-laws-based Nagelian types of ontological and causal reductionism. Both pathways of reduction mentioned here find their point of departure in Ernest Nagel's preliminary assertion that "a reduction is effected when the experimental laws of the secondary science (and if it has an adequate theory, its theory as well) are shown to be the logical consequences of the theoretical assumptions (inclusive of the coordinating definitions) of the primary science."[14] What differentiates the two pathways is that the former (simple intertheoretical reduction) covers homogenous cases, which do not require any reference to bridge laws, while the latter (bridge-laws-based reduction) covers nonhomogenous cases, which do need to implement the middle step engaging bridge laws. Nagel offers these strategies as a necessary theoretical base for the project of the ultimate reduction of sciences dealing with complex phenomena to physics. A systems approach in biology shows that, construed this way, the program of reduction fails, and—most importantly—it does so not merely on the epistemological ground, but primarily on the ontological. This fact makes the systems approach in biology one of the major arguments in support of rejecting the agenda of logical positivism and reductionist empiricism.[15]

As I have mentioned, this criticism is accompanied by a new appreciation for metaphysics and its role in providing a comprehensive account of the nature of reality that developed within the methodology of analytic philosophy. It originated from the conversation carried on by David Lewis and David Armstrong, which was later joined by Willard Quine and Saul Kripke. Their analysis brought back into consideration questions concerning possibility and necessity, universals, properties, composition, the possibility of essences, the nature of time, stability (permanence) and change, and the nature of causality. This new metaphysical debate provided a necessary theoretical base for the formulation of the theory of EM in philosophy of science.

Finally, we cannot ignore one more crucial contribution to the most recent revival of EM theory, which came from a growing interest and rapid development of mind/brain studies, on the level of science as well as in philosophy of science and metaphysics. The former developed into a separate discipline called neuroscience, and the latter into a separate division of philosophical studies called philosophy of mind. The contemporary mind/brain debate in philosophy was inspired by Mario Bunge (1919–2020), Jaegwon Kim (1934–2019), Karl Raimund Popper (1902–94), Roger Wolcott Sperry (1904–94), and John Jamieson Carswell Smart (1920–2012).[16] The conversation they started continues and engages a number of important thinkers representing both analytic and continental schools of thought. For those who support emergentism, mind/brain relation remains one of the flag examples in support of their argument. However, what is the exact formulation and what are the most important principles of the theory of EM?

3. CLASSICAL ACCOUNT OF EMERGENCE

We have referred already, in the first section of this chapter, to the scientific formulation of the theory of EM offered by the physicist Robert Laughlin. He defines it in terms of the broken symmetry of distribution and relations between entities constituting wholes, due to phase transitions between various levels of complexity distinguishable in those wholes. Analyzing the same phenomena from the biological perspective, Terrence Deacon defines EM as an "unprecedented global regularity generated within a composite system by virtue of the higher-order consequences of the interactions of composite parts."[17] What stands behind these definitions is an empirical and practical approach, typical for natural sciences trying to limit theoretical discussion to a minimum.

However, when analyzed from a more speculative, philosophical point of view, the theory of EM opens an interesting and thought-provoking conversation, addressing some crucial topics in metaphysics, including the theory of properties and composition, the reality of essences, and the nature of causality in complex dynamical systems and organisms. Consequently, providing a methodical definition of EM in philosophy of science requires taking into account a number of characteristic and unique aspects of phenomena classified as emergent.

In the most recent literature of the topic, we can distinguish two main philosophical accounts of EM. The first one goes back to British empiricism and its conclusions that certain rules of physical composition of causes and certain transition laws are inapplicable in chemistry and biology. These arguments were further developed in the second half of the last century into a mereologically grounded theory of EM that comes in three versions, classified as DC-based, whole-part constraint, and supervenience(SUP)-based EM. Its main objectives can be classified as (1) nonadditivity of causes; (2) novelty of emergent processes, entities, and properties; (3) ontological monism and qualitative difference of emergents; (4) necessity and insufficiency of parts for the existence of the emergent whole; (5) levels of organization of parts in emergent wholes; (6) new laws of nature characteristic of emergent wholes; (7) nondeducibility, nonpredictability, and irreducibility of emergents; and (8) radical difference between epistemological and ontological emergence. In what follows I will analyze briefly each one of these features of EM.[18]

3.1. Nonadditivity of Causes

The first formulation of the theory of EM comes with J. S. Mill's distinction of the physical (based on mechanics) composition of causes and transitional laws—referring to the vector or algebraic addition—from the chemical mode of combined action of causes and transition laws. Because products of chemical reactions are not simply algebraic sums of the effects of each reactant, Mill suggested calling them "heteropathic effects" (governed by "heteropathic laws"). He distinguished them from "homopathic effects" (governed by "homopathic laws") applicable in physical mechanics. Moreover, he claimed that heteropathic laws supersede the homopathic laws in showing that the laws of chemistry were not reducible to a small group of systematically organized laws from which all other laws of nature could be derived. If this is the case for the laws of chemistry, concludes Mill, then the laws of life must be nondeducible from the laws of its chemical ingredients all the more.

Mill's theory of homopathic and heteropathic effects and laws was followed by Lewes, who used a slightly different terminology of "resultant" and "emergent" effects. Distinguishing the latter as incommensurable and irreducible to the sum of their components, he used for the first time the very term EM in a philosophical context when stating that "there are two classes of effects markedly distinguishable as resultants and emergents. Thus, although each effect is the resultant of its components, the product of its factors, we cannot always trace the steps of the process, so as to see in the product the mode of operation of each factor. In this latter case, I propose to call the effect an emergent."[19]

3.2. Novelty of Emergent Processes, Entities, and Properties

An acknowledgment of the reality of heteropathic/emergent effects leads to an attempt to define the ontological nature of emergents more precisely. The three most popular ontological categories mentioned in this context are (1) emergent processes (events), (2) emergent entities (substances), and (3) emergent properties (qualities). Samuel Alexander, speaking of the "collocation of motions" possessing "a new quality distinctive of the higher complex" and "expressible without residue in terms of the processes proper to the level from which they emerge," seems to opt for defining emergents in terms of processes.[20] Referring to chemical examples, Mill, Lewes, and Morgan seem to concentrate more on entities (substances) as emergent. The most popular characterization of emergents, however, uses the language of emergent properties (qualities). Thus, Alexander speaks of emergent powers, dispositions, or capacities, whereas Broad mentions "ultimate characteristics," which he contrasts with "ordinally neutral" and "reducible characteristics."

3.3. Ontological Monism and Qualitative Difference: Nonreductionist Physicalism

The attempt to define emergents in terms of processes, entities, or properties requires some further clarification with regard to a still deeper metaphysical question concerning the ultimate character of reality. Here we encounter the first intrinsic tension within the theory of EM—the one between ontological monism and qualitative difference of emergents. For, on the one hand, contemporary emergentists seem to accept ontological monism, saying that "all individuals are constituted by, or identical to, micro-physical individuals, and all properties are realized by, or identical to, micro-physical properties."[21] On the other hand, they emphasize the novelty and qualitative difference of emergent properties, which cannot be instantiated simply by quantitative accumulation of the basic constituents of material entities and/or their properties. Their position is oftentimes classified as a nonreductionist version of physicalism (NP).

3.4. Emergence and Mereology

Understood in terms of the qualitative novelty of emergents, which, nonetheless, are realized by the most basic material constituents, the classical theory of EM classifies as mereological. Its main objectives are built around dependencies between the wholes and their parts. What is being emphasized by the proponents of this account of EM is both the necessity and insufficiency of parts for the existence of the whole. The claim of necessity strives to protect the emergentist position from dualism, while the claim of insufficiency contrasts it with reductionism.

Moreover, one of the main arguments in support of emergentism and NP builds on Hilary Putnam's idea of the "multiple realizability" or "compositional plasticity" of psychological phenomena. Extended to the realm of all emergents, the theory says that, although grounded in the most basic physical constituents (physical monism), emergents can be realized, instantiated, or implemented in endlessly diverse ways and thus cannot be identified with any particular kind of physical state.[22] This argument has clearly a mereological foundation, similar to another two contributions to the theory of EM, defining it in terms of DC and whole-part constraints. I will discuss both of them below. Here I just want to emphasize the mereological presuppositions standing behind the classical version of emergentism.

3.5. Levels of Organization

The fact that EM is grounded in ontological monism and mereology requires from it a theory and language describing the hierarchy of various stages of organization of the constituent parts of emergents. The language of levels of organization was first introduced by Alexander in the second volume of his *Space, Time and Deity*, where he says: "The higher-quality emerges from the lower level of existence and has its roots therein, but it emerges therefrom, and it does not belong to that lower level, but constitutes [in] its possessor a new order of existen[ce] with its special laws of behavior."[23]

More recently, a group of researchers led by Claus Emmeche proposed a more thorough ontological specification and classification of levels of complexity, emphasizing their physical interdependence and metaphysical "inclusivity." Following the argument from NP, they state that higher levels of complexity are materially related to lower levels of complexity in a way that does not violate the laws of nature typical of these lower levels. At the same time higher levels cannot be simply deduced from lower levels. This line of argumentation is set up, once again, to protect the theory of EM from falling into the pitfalls of ontological dualism and eliminativism.

In addition, Emmeche et al. emphasize that, ontologically speaking, the higher levels are as preeminent as the lower ones, even if they are presupposed by them. They propose a basic classification of levels that distinguishes four primary levels: (1) the physical, (2) the biological, (3) the psychological, and (4) the sociological. Each one of them can differentiate into various sublevels (e.g., the biological contains the cell, the organism, the population, the species, and the community). Finally, Emmeche et al. introduce the rule of the "nonhomomorphic" nature of emergent relations between particular levels and sublevels, which simply states that interlevel relations of dependence between different neighboring levels or sublevels always vary (e.g., the dependence between physical and biological differs from the one between biological and psychological).[24]

3.6. Emergent Laws

Another characteristic feature of emergent phenomena is that their scientific description leads to the discovery of new laws, unique and specific for higher levels of complexity. Acknowledging this fact, Alexander speaks about new laws ruling the emergents, whereas Mill—contrasting homopathic and heteropathic effects—concludes that the latter are governed by new, heteropathic laws. Broad analyzes emergents in terms of "trans-ordinal laws," which he defines as laws connecting "the properties of aggregates of adjacent orders" and which he differentiates from "intra-ordinal laws" connecting the properties of aggregates of the same order.[25]

When accepting this kind of argumentation, we have to remember that laws of nature are, strictly speaking, descriptive and not prescriptive; that is, they help us describe the way things and processes are, rather than making them be what they are. Nevertheless, the novelty of laws of nature describing higher levels of complexity remains an important indicator of EM.[26]

3.7. Nondeducibility, Nonpredictability, and Irreducibility of Emergents

Taking into account all aspects of EM discussed so far, it becomes apparent that it can be further characterized in terms of nondeducibility, nonpredictability, and irreducibility of emergent entities, properties, and laws.

First, deducibility—understood as deriving a conclusion from something already known—is not applicable to emergents. For the latter—for instance, emergent laws—are brute nomological and nonderivative facts which, according to Broad, must "simply be swallowed whole with that philosophical jam which Professor Alexander calls 'natural piety.'"[27]

Second, predictability—defined as envisaging future states of affairs—is also impossible in case of emergents. It is true not only because the occurrence of hierarchical levels or layers and relations between them depend on the fundamental indeterminism of the physical universe. The ontological nature of emergents is such that, even if we are able to form emergent laws inductively—that is, based on the former observation that a given property E_1 emerges from a certain lower-level property M_1 of a system S_1 in conditions C_1—we are not able to make the same kind of prediction about the EM of a possible new property of the kind E (E_2) theoretically. Such theoretical unpredictability is a fact due to either our not having the concept of E_2 before the property it refers to actually occurs, or some possible changes in the microstructure of M_2 or/and conditions C_2, which, transforming them into M_2^* or/and C_2^*, will cause the EM of E_2^*, instead of E_2.[28]

Third, emergents are resistant to reduction. They escape not only Nagelian simple intertheoretical and bridge-laws-based reductionist programs but also the functional model of reduction proposed by Kim. The latter position assumes that property E is reducible to the lower properties of the same system S iff E can be "functionalized," that is, construed or reconstrued as a property defined by its causal/nomic relations to the property M that instantiates it. In other words, property E is reducible iff it can be shown that it is a "function" of the lower property/ies $M(s)$ of the same system S. It is commonly acknowledged that ontologically emergent phenomena do not qualify for this type of reductionist description.[29]

3.8. Epistemological (Weak) and Ontological (Strong) Emergence

The classification of certain phenomena as ontologically emergent suggests that there might be some other phenomena whose irreducible and emergent character is predicated merely epistemologically. Moreover, it turns out that the ontological novelty of emergents can be defined in several different ways. These facts make us realize that EM is not an unambiguous term. Studied and analyzed from different perspectives, within different subdisciplines of science and philosophy, and in reference to a variety of phenomena on different levels of complexity, EM becomes, in a way, a term of family resemblance. Hence, our list of the main objectives of the classical formulation of EM theory closes with the classification of the main types of EM, depicted on figure 1.4. and discussed below.

The first distinction that needs to be mentioned in the classification of types of EM is the one between the weak (epistemological) and the

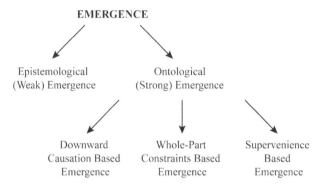

Figure 1.4. Types of EM

strong (ontological) versions of EM. The former, weak type of EM is perceived as an outcome of natural limitations in human analytic possibilities, knowledge, and understanding. Because these limitations are not unsurpassable, we can expect that, with the development of our cognitive skills and research methods, what today classifies as irreducible and emergent will find reductionist explanation in the future. This kind of EM is usually classified as epistemological and differentiated from ontological EM, which assumes that emergent phenomena belong to the fabric of the universe and thus are intrinsically irreducible. The latter version of EM, however, requires a more precise specification of the ontological novelty of emergents, which may put it in conflict with physical monism and physical causal closure. The literature on EM offers at least three ways of defining ontological novelty of emergents.

3.8.1. DC-Based Version of Ontological EM

The first and most popular way of defending ontological novelty of emergents refers to the concept of downward (top-down) causation. Because the mereologically oriented classical version of EM defines it in terms of dynamic causal relations between complex wholes and their constitutive parts, it seems plausible to argue that emergents show/exercise a novel and irreducible kind of causal influence. This new type of causation is usually given the name "downward causation" or "top-down causation" (some philosophers speak about downward- or macro-determination). Unlike upward (bottom-up) causation—that is, an instantiation of a higher-level property by a lower-level property—and same-level causation, DC occurs when an emergent higher-level property causes (has an influence on) the instantiation of a lower-level property.

A standard example of ontological EM defined in terms of DC is the phenomenon of human mind defined as a set of cognitive faculties—including consciousness, perception, thinking, judgment, and memory—that "arise" from the complex dependencies among neural cells of the brain. Because neurons become used to, and even develop a preference for, certain patterns of signal transmission, it is being argued that the brain can be trained to behave, and even gradually evolve, based on the activities of the mind. These activities can be thus regarded as exercising downward causal influence on the brain's neural networks from which it emerges.

The term DC was first defined in 1974 by Donald Thomas Campbell (1916–96). In his analysis of hierarchical levels of dependency in evolutionary processes, we find him saying that "all processes at the lower levels of a hierarchy are restrained by and act in conformity to the laws of the higher levels."[30] Those who support this claim find it decisive for the formulation of the general rule concerning qualitative difference of emergents from their basal constituents. They argue that despite emergents' physically monistic foundations, due to the novel and irreducible character of DC, emergents remain free from the charge of epiphenomenalism and resist reductionist programs in science and philosophy of science.

3.8.2. Difficulties of DC-Based Version of Ontological EM
Despite its explanatory potential, the DC-based version of ontological EM faces a number of critical challenges concerning its main objectives. As I argued in my study *Emergence: Towards a New Metaphysics and Philosophy of Science*, both proponents and more critical commentators of DC-based EM find it difficult to provide a unanimous answer to the three fundamental metaphysical questions related to different cases of DC.[31]

First, concerning the question of what is the cause in DC, we find them mentioning (1) general principles, laws, regularities, or wholes (Donald Campbell); (2) boundary conditions, context-sensitive constraints, or patterns of organization (Michael Polanyi, Arthur Peacocke, Nancey Murphy, George Ellis, Paul Davies, Alicia Juarrero, Terrence Deacon, Robert van Gulick, Emmeche et al.); (3) processes (Austin Farrer,

Terrence Deacon, Emmeche et al.); or (4) concrete entities (Nancey Murphy, Emmeche et al., Jaegwon Kim, Roger Sperry).

Second, concerning the question of what is being acted upon in DC, theorists of scientific and philosophical notions of EM mention (1) lower-level substances and structures, parts of a whole, constituents or units of a system, molecules in self-organizing processes (Donald Campbell, Paul Davies, Austin Farrer, Arthur Peacocke, Alicia Juarrero); (2) lower-level properties, properties of the lower-level basal constituents, lower-level conditions (Roger Sperry, Jaegwon Kim, Nancey Murphy); or (3) component constituent dynamics, lower-order interactions, lower-level processes, causal processes, action on the lower-levels, or microevents (Terrence Deacon, George Ellis, Alicia Juarrero, Robert van Gulick).

Finally, concerning the question of the ultimate character of DC, the followers of the DC-based EM face a critical dilemma, namely, the struggle between supporting the rule of physical causal closure and acknowledging the novelty and irreducibility of DC. The former presupposition, referred to as the causal inheritance principle, states that causal powers on higher levels of complexity are always a "product" of the causal powers of the basal properties on lower levels of complexity, and thus must belong to the same ontological category. The latter argument in favor of the ontological novelty of DC seems to suggest the opposite. This dilemma is closely related to the struggle between ontological monism and the qualitative difference of emergents mentioned above (section 3.3 of this chapter).[32] It was probably best expressed by Kim's famous argument from causal exclusion, in which he shows—on the example of mental causation arising from neural patterns—that the argument in favor of DC involves causal overdetermination, which consequently makes the whole project of EM fall into physicalism. See figure 1.5.[33]

Apart from his argument from causal exclusion, Kim rightly notes that the idea of DC of the whole influencing its own constituent parts implies a kind of self-causation or self-determination, which makes the whole explanation viciously circular. One may try to follow Kim in his suggestion to treat self-reflexive DC diachronically; however, such a strategy seems to be difficult to apply to dynamic systems in which the temporal framework of the relations of causal dependencies between wholes and their parts may be difficult to specify.[34]

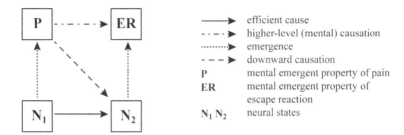

Figure 1.5. Kim's argument from causal exclusion applied to the concrete example of pain and escape reaction

An instance of emergent property of pain, P, causes another emergent property of escape reaction ER to instantiate, as an effect of the same-level causation. Both P and ER must have basal neural states (N_1 and N_2) from which they emerge. Consequently, as long as N_2 occurs, ER will be instantiated, whether or not ER's purported cause, P, occurs at all. The only way to save the claim that P caused ER appears to be to say that P caused ER by causing N_2. In this case, the same-level causation from P to ER entails DC from P to N_2. However, if causation is understood as nomological (law-based) sufficiency, N_1 is nomologically sufficient for N_2 and can be regarded its cause. Moreover, a causal chain from N_1 to N_2 with P as an intermediate causal link is questionable because the EM relation from N_1 to P is not, properly speaking, a causal one. Thus, if P is to be retained as a cause, we are faced with the highly implausible consequence that the case of DC (from P to N_2) involves causal overdetermination (since N_1 remains a cause of N_2 as well). But if DC goes—Kim concludes—DC-based ontological emergentism goes with it.

Kim's argument triggered a hot debate which still continues within the circles of philosophy of mind and within philosophy of science in general, since "if successful, [it] not only applies to the relation between mental and underlying physical events. It is a completely general argument-scheme, which can be applied to the relation between any two layers of reality, whenever one is held to be more fundamental or basic than the other, and the latter supervenes on the former."[35]

3.8.3. Whole-Part Constraints-Based Version of Ontological EM

Trying to answer the dilemmas of physical monism versus qualitative novelty of emergents and the dilemmas of physical causal closure versus novelty and irreducibility of DC, Carl Gillett suggests that it is possible to be an emergentist without abandoning the physicalist camp as long as one can specify the top-down determination in noncausal terms. In other words, physical monism and causal closure are not challenged if the emergent kind of determination "does not involve wholly distinct entities, and apparently involves no transfer of energy and/or mediation of force."[36]

Trying to provide an example of such determination, Gillett refers to the list of fundamental natural laws, which includes "higher laws that refer ineliminably to strongly emergent properties." He seems to think that defining emergents in terms of laws and properties enables him to avoid the problem of introducing ontologically new entities and the new type of causation that involves transfer of energy and/or mediation of force. He thus states that what can be called a "patchwork" version of physicalism "is committed to a mosaic of fundamentally determinative, and thus causally efficacious, entities, not just the microphysical properties, but also the strongly emergent realized properties with which the microphysical properties often share the determination of fundamental causal powers."[37]

As long as physical basal constituents contribute causally to complex systems that constrain, in turn, their activity and motions in some specifiable way—argues patchwork physicalism—nothing nonphysical is at work, nor does any new physical entity or causal force need to be introduced to explain the nature of emergents.

The idea of top-down noncausal determination of microphysical objects becomes a foundation for the whole-part constraints-based version of ontological (strong) EM. Building on the most basic formulation of this theory, Alicia Juarrero introduces the concept of "context-sensitive constraints" in complex dynamic systems, which connect objects with their environments (systems), strengthening their embeddedness. She redefines the concept of constraints, used formally in Newtonian mechanics, and presents them not only as reducing alternatives but also as a source of new possibilities. Understood this way, the theory of whole-part constraints serves as an argument in favor of an ontologically irreducible, and yet noncausal, character of emergents.[38]

The same idea of whole-part constraints inspires an original theory of EM offered by Terrence Deacon. What differentiates his position from the one offered by Gillett is that he seems to ascribe to constraints (absences) a causal character. We will analyze Deacon's model of EM more carefully in a separate section below.

3.8.4. Difficulties of the Whole-Part Constraints-Based Version of Ontological EM

Although the whole-part constraints-based version of EM may sound promising, it raises at least two important questions. First, the fact that the patchwork type of physicalism remains committed to "strongly emergent realized properties" seems to suggest the reality of emergent wholes understood as entities of a new kind. For the properties in question—like any kind of properties—cannot exist by themselves, that is, separate from entities that instantiate them.[39] If this is the case, however, the objective rules of physicalism and ontological monism seem to be in jeopardy.

Likewise, the idea of "constraining the activity of microphysical objects" needs a further qualification. Because the action of microphysical objects in question is commonly understood in terms of their exercising efficient (physical) causation, it is difficult to envision an influence constraining such activity that is not causal and does not involve any "transfer of energy and/or mediation of force."

3.8.5. SUP-Based Version of Ontological Emergence

The same struggle to reconcile physicalism and physical causal closure with qualitative novelty of emergents and their ability to exercise DC inspires one more—probably the weakest—version of ontological EM, which is grounded in the theory of supervenience (SUP). "Supervenience," providing one of the most popular explanations of mind/brain dependency, is defined in terms of covariance between two types of properties: "A set of properties A supervenes upon another set B just in case no two things can differ with respect to A-properties without also differing with respect to their B-properties. In slogan form, 'there cannot be an A-difference without a B-difference.'"[40]

However, in the context of emergentism, it has been argued that SUP entails not only metaphysically neutral covariance between physical and emergent but also their nonidentity, dependence, and directionality, all three having an ontological dimension. If this is the case—the argument continues—SUP becomes a possible metaphysical foundation for ontological EM. Without introducing physically novel entities and causes (DC), the fact of nonidentity, dependence, and directionality of supervenient properties on their subvenient basal property counterparts suffices to build upon it a SUP-based theory of ontological EM.

3.8.6. Difficulties of the SUP-Based Version of Ontological EM
Similar to DC-based and whole-part constraints versions of ontological EM, the SUP-based theory of ontological EM raises some important questions. First, since the covariation principle that defines SUP does not entail by itself an ontological dependency between supervenient and subvenient entities/properties, we are left with the question of what actually does the ontological "work" in the SUP-based theory of EM. In other words, the claim that SUP entails ontological nonidentity, dependence, and directionality cannot be simply treated as another—supplementary to covariance—aspect of SUP, because this claim exceeds the limits of the SUP theory. Indeed, says Kim, the question of what must be added to covariation to yield dependence remains an interesting and deep metaphysical query for which there is no easy answer.[41]

Consequently, a number of emergentists emphasize a radical difference between SUP and EM, which puts in question the plausibility of the SUP-based version of ontological EM. Paul Humphreys suggests that SUP is acceptable merely as a consistency condition, enabling attribution of concepts concerning properties characteristic for different levels of complexity, whereas EM provides an explanation of ontological relationships between them.[42] Hong Yu Wong, differentiating emergent (nonstructural) from resultant (structural) properties, claims that in the case of the former kind of properties SUP is sui generis and insufficient for establishing their ontological status: "Emergent properties are *non-structural* properties, in contrast to resultant complex properties which are *structural* properties; but both supervene on basal properties. Since emergent properties are non-structural, supervenience in the case of emergent properties must be *sui generis*; it is not a matter of constitution, identity, realization, causation, or any of the usual relations that ground supervenience; it is a matter of fundamental, non-derivative emergent laws."[43]

Finally, assuming SUP does yield ontological dependence—notes Kim—it seems that reconciling it with the requirement of irreducibility may cause a problem. Knowing that the relation between subvenient and supervenient properties needs to be weak enough to avoid their identity—which would entail reducibility—and strong enough to provide for their dependence, he states: "The main difficulty has been this: if a relation is weak enough to be nonreductive, it tends to be too weak to serve as a dependence relation; conversely, when a relation is strong enough to give us dependence, it tends to be too strong—strong enough to imply reducibility."[44]

It seems that this difficulty becomes a serious challenge for those who support the SUP-based ontological version of EM.

4. PROBLEMS OF THE CLASSICAL ACCOUNT OF EMERGENCE

Our critical evaluation of all three versions of ontological EM brings us to a more general assessment of the classical account of emergentism. It seems that one of its main challenges is the persistent dilemma of reconciling physical monism with ontological novelty of emergents and reconciling physical causal closure with the reality and irreducibility of DC of emergents. We have seen questions related to both difficulties coming back repeatedly throughout the entire section dedicated to the presentation of the classical account of EM.

These difficulties remain unresolved as long as we do not realize that emergentism stands at the intersection of natural science and philosophy. For, on the one hand, EM is clearly defined in the context of recent discoveries in physics, chemistry, and systems biology in particular—discoveries that have led us to become aware of complexities and properties that escape the reductionist description in terms of basic-level constituents of any given natural system. On the other hand, the theory of EM is sensu stricto philosophical because it is developed in reference to the language of causation, entities, and properties, accompanied by the question concerning their ontological status.

Once we acknowledge and distinguish the scientific and the philosophical aspects of EM theory, the difficulties we are trying to resolve seem to be less of a problem. It may be true that, based on the outcomes of detailed empirical research, a scientific account of the physical aspects of reality finds all perceivable entities to be composed of some basic elementary particles. Moreover, analyzed metaphysically, this fact (physical monism) reveals an important part of the ontological truth about the way things are. Nevertheless, this truth is only partial and does not necessarily make invalid the argument about ontological novelty and irreducibility of entities/properties on higher levels of complexity of matter. Even if their unique character escapes scientific description, it may find a proper expression in the language of philosophy of nature and of metaphysics. Moreover, the same philosophical disciplines can contribute importantly to the explanation of the nature of matter on the lower levels of complexity as well.[45]

Speaking of the second dilemma of emergentism, once we acknowledge that physical (efficient) description of causal relations does not exhaust all possible aspects of cause-effect dependencies in nature, physical causal closure on some levels of complexity, or characteristic of some types of explanation, does not exclude the possibility of metaphysically new and irreducible DC exercised by emergents. This fact is of special importance and requires some more comment, since the DC-based version of ontological EM is the most influential and popular among the followers of classical emergentism.

4.1. Downward Causation and the Fallacy of Causal Reductionism

As mentioned above, the DC-based version of EM faces the challenge of the argument from physical causal closure, as well as Kim's charge of overdetermination and the collapse of DC into physical explanation of higher-level phenomena in terms of causal operations characteristic for the lower levels in which they find their basal constituency. I acknowledge that Kim's critical argument poses a real problem for the classical account of EM. It needs to be noticed, however, that the main objective of Kim's critique resides in the metaphysical presupposition that all causation in nature is of the efficient character, that is, that it is based on the physical principle of mass and energy (ex)change and is amenable to a mathematical description. Those among emergentists who follow such metaphysics face the dilemma of reconciling physical causal closure and the ontological novelty of DC. This same dilemma calls into question the plausibility of NP, embraced by many ontological emergentists. Since physicalism assumes, *ex definitione*, either reductionism or eliminativism concerning all complex phenomena with respect to their basal physical constituents, it is difficult to explain metaphysically what the nonreductionist aspect of NP could be about. Hence, Kim might be right in his critical comment on the popularity of NP in which he points out that its ontology is physicalistic, while its "ideology" is dualistic.[46]

Trying to avoid the dilemma of reconciling physical causal closure and the ontological novelty of DC, van Gulick proposes a new explanation of EM in which he reinterprets DC in terms of higher-order patterns being capable of the "selective activation" of lower-order causal powers. However, the question of the ultimate nature of DC in his theory remains unanswered. If selective activation operates as physical efficient cause, the whole argument seems to be circular and vulnerable to reductionism. If it is not a physical cause, the principle of physical causal closure is violated.[47]

Nancey Murphy struggles with the same difficulty. Thinking about ontological aspects of EM and trying to define unique causal factors of emergents, she avoids describing DC in terms of new causal forces operating over and above those known to physics. Postulating them would violate the causal closure of physics. She suggests instead approving the idea of new causal powers that cannot be reduced to the summary of lower-level processes.[48] The status of these causal powers, however, remains as problematic as van Gulick's idea of "selective activation." For how are the new emergent powers realized? If their manifestation is not of a physical nature, how can they have an impact on physical constituents of lower levels? What is the ultimate "causal joint" between high and low levels of organization of matter? The nonphysical aspect of the causal powers in question seems to contradict, once again, the principle of physical causal closure.

The failure of these reinterpretations of the nature of DC leads Menno Hulswit to state that "the concept of 'downward causation' is muddled with regard to the meaning of causation and fuzzy with regard to what it is that respectively causes and is caused in downward causation."[49] Taken to its extreme, this criticism seems to suggest that the ontological status of DC is impossible to determine and that the very concept of ontological EM, which relies heavily on it, is consequently incoherent and self-contradictory.

However, the status of DC-based ontological EM turns out to be radically different once we give up trying to fit DC into the framework of efficient (physical) explanation of causal dependencies in nature. We need to realize that what an explanation in terms of DC actually shows is that our narrow understanding of causation is insufficient and needs to be extended to other types of causes, which are not necessarily physical and may not be describable in the language of mathematics. In other words, DC-based EM becomes a strong argument against the fallacy of causal reductionism, which has gradually dominated both science and philosophy since the beginning of modernity. This fact has been emphasized by a number of thinkers supporting the theory of EM.

4.2. Downward Causation and the Extended Typology of Causes

One of the first suggestions to broaden the notion of causation in the context of emergent studies came with the trifold division of DC proposed by Emmeche et al. In their article published in 2000, Emmeche, Køppe, and Stjernfelt present an original classification of strong, medium, and weak DC (SDC, MDC, and WDC, respectively), which argues in favor of reintroducing the distinction between efficient, formal, and final types of causation in philosophical analysis of natural phenomena.

Defined as an efficient type of influence of the higher-level entities on lower-level entities that ground them, SDC is not—according to Emmeche at al.—an appropriate option for defending DC-based ontological EM. They find it erroneous not only because it defines DC in terms of efficient causation, located in space and time, and vulnerable to the reductionist and the overdetermination arguments discussed above. Because SDC assumes that higher-level entities are substantially different from the lower-level entities, it seems to entail substance dualism, which violates the principle of interdependence and of metaphysical inclusivity of ontological levels (see section 3.5 above). Moreover, it is not entirely clear what the character and nature of the exchange of mass and energy between substantially (ontologically) different levels of complexity of matter is—an exchange presupposed by any type of efficient causal relations.[50]

The WDC version of the theory is not a proper argument in defense of DC-based ontological EM either. Defined by Emmeche et al. in terms of the phase-space physical theory, it sees higher levels of complexity as forms into which the constituents of lower levels are arranged. Treating the higher levels as organizational phenomena, rather than functions of new emergent substances, WDC can be compared—through analogy—with the concept of an "attractor" in a dynamical system, that is, a steady state toward which the system may evolve. And yet, just as an attractor in the phase-space theory does not, sensu stricto, exercise causation, the question of the reality and nature of causation in WDC remains without a clear answer. This fact seems to disqualify WDC as a candidate for a metaphysical grounding of DC-based ontological EM.[51]

At the same time, however, the concept of WDC refers to the principle of formal causation, which extends the typology of causes and thus opens the way to the formulation of MDC. In reference to the language of boundary (or constraining) conditions, Emmeche et al. define this version of DC in terms of higher-level entities as constraining conditions for the emergent activity of lower levels. They suggest that "the higher level is characterized by *organizational principles*—lawlike regularities—that have an effect ('downward,' as it were) on the distribution of lower level events and substances."[52] Most importantly, departing from the reductionist understanding of causation, Emmeche et al. claim that "medium DC does not involve the idea of a strict 'efficient' temporal causality from an independent higher level to a lower one, rather, the entities at various levels may enter part-whole relations (e.g., mental phenomena control their component neural and biophysical sub-elements), in which the control of the part by the whole can be seen as a kind of functional (teleological) causation, which is based on efficient, material as well as formal causation in a multinested system of constraints."[53] Again, criticizing the constraint of causation to efficient causes, they state "there is a place for a rational concept of downward causation (in some version) in science and philosophy, but only with a broader framework of causal explanation. Very often 'causality' is implicitly equated with the usual notion of efficient causality, but if downward causation is regarded as an instance of efficient causality it will form a 'strong version' of the concept, which . . . is not a plausible one. The notion of causality should therefore be enlarged to make sense of downward causation."[54]

The argument of Emmeche et al., who ground the DC-based version of ontological EM in the extended typology of causes, clearly refers to Aristotle's philosophy of causation and his fourfold division of causal dependencies. Interestingly, similar suggestions can be found in writings of other contemporary emergentists:

1. Searching for a middle ground between reductionism and metaphysical dualism, C. N. El-Hani and A. M. Pereira emphasize the importance of the entanglement of matter and form, the role of higher-order structures constraining lower ones, and the role of functional causation.[55]
2. In his defense of ontological EM, Michael Silberstein states that "systemic causation means admitting types of causation that go beyond efficient causation to include causation as global constraints, teleological causation akin to Aristotle's final and formal causes, and the like."[56]
3. Speaking about biological systems, Alvaro Moreno and Jon Umerez suggest introducing a new type of causation which "is 'formal' in the sense that it infuses forms, i.e., it *materially restructures matter according to a form.*"[57]
4. An authority in the science of nonlinear phenomena, Alwyn Scott, refers to all four Aristotelian causes, suggesting—at the same time—that in modern terms Aristotle's material and formal causes should be put together and classified as "distal causes," that his efficient cause should be referred to as a "proximal cause," and that the final cause should be simply given its traditional name "teleological cause."[58]
5. Although he does not explicitly refer to Aristotle, George Ellis speaks about causally effective "teleonomic" goals, central for feedback control systems, and based on the information about these systems' desired ways of behavior. He thinks such goals can be either in-built (e.g., homeostasis), learned, or consciously chosen. The analogy of his argument to Aristotle's teleology is apparent.[59]

These preliminary suggestions of the followers of the classical mereological DC-based version of ontological EM to expand the typology of causes in reference to Aristotle's philosophy of causation found a particularly intriguing implementation and development in the work of Terrence Deacon. His new and original account of EM becomes an important alternative to the classical one, and thus it requires a separate analysis, to which we shall now turn.

5. DYNAMICAL DEPTH ACCOUNT OF EMERGENCE

A specialist in biological anthropology—turning most recently toward philosophy of biology—Terrence Deacon puts a strong emphasis on the dynamical aspects of living systems, which inspires him to offer a new theoretical model of EM. He defines it in terms of the three dynamic modes or levels of what he calls, together with Spyridon Koutroufinis, a "dynamical depth"[60] and describes in reference to an idea of the importance of "constraints" ("specifically absent features" and "unrealized potentials") for the causal activities in complex systems. Interestingly, while developing this new account of ontological EM, in opposition to the classical mereological and DC-based EM, Deacon seems to rediscover Aristotelian categories of causation, which he applies to a concrete and thorough model of the growing complexity of nonliving systems, the origin of life, and the origin of consciousness. He finds consciousness to be inseparably related to the phenomena of meaning, purpose, significance, value, and so on—the phenomena which can be classified under a general category of "ententional" (the term Deacon coins).[61] I have offered a very careful presentation and critical metaphysical evaluation of Deacon's project in the third chapter of my first monograph, *Emergence: Towards a New Metaphysics and Philosophy of Science*. Here I will limit my account to a short summary of that research, which will prove both necessary and sufficient for the main objectives of the present study.

5.1. The Engine of Emergence

Deacon's interest in persistently far-from-equilibrium self-organizing systems and his appreciation for some aspects of the theories of life's origin that have been developed in reference to the unique features of these systems do not prevent him from criticizing a major shortcoming of the argument these theories are based upon. For a self-organizing process—says Deacon—has "no capacity for persistence beyond the extrinsically imposed gradients that it develops in response to. When such external perturbation ceases, the acquired regularity of the local system dissipates and the system re-approaches equilibrium following the Second Law of thermodynamics."[62] Consequently, what is necessary for the phenomenon of life (a primitive biological "self" characterized by "self-directedness") to emerge is the presence of an inherent, self-propagating, and teleological formative power.

Once in operation, such power is not susceptible to a mechanistic reduction. The way it comes into existence, however, requires a clarification that can and should be developed in terms of the standard reductionist scientific method of explanation. In other words, Deacon finds that the question concerning the origin of the ontologically irreducible phenomenon of life boils down to "an explanation of how a biological self can emerge from non-self components."[63] This dialectic of reductionism and nonreductionism is brought into a synthesis in Deacon's dynamical depth model of EM, which is based on the principle that

> *emergent phenomena grow out of an amplification dynamic that can spontaneously develop in very large ensembles of interacting elements by virtue of the continuing circulation of interaction constraints and biases, which become expressed as system-wide characteristics.*[64]

This principle is unique for two reasons. First, because of its concentration on the dynamics of the "very large ensembles of interacting elements" rather than on mereological (whole-part) organization, supervenience, and causal inflection, it seems to do well without introducing the concept of DC. Indeed, Deacon sees one of the main virtues of his version of ontological EM in avoiding the language of DC as well as the ideas of "abstract formal properties" or "abstract ideal forms" when categorizing higher levels of organization.[65]

Second, the same interest in the dynamics of large conglomerates of causally active entities directs our attention to "interaction constraints and biases," which define the "engine" of Deacon's view of EM. It is expressed in his suggestion that "it's not so much what *was* determined to happen that is most relevant for future states of the system, but rather what *was not* cancelled or eliminated. It is the negative aspect that becomes most prominent. This is the most general sense of *constitutive absence*: something that is produced by virtue of determinate processes that eliminate most or all of the alternative forms. It is this, more than anything else, which accounts for the curious 'time-reversed' appearance of such phenomena."[66]

Indeed, the notion of "constitutive absences"—that is, "possible features being excluded"—becomes the core of Deacon's concept of EM. His departure from the descriptive notion of form is aimed at showing "how what is absent is responsible for the causal power of organization and the asymmetric dynamics of a physical or living process."[67] In other words, "emergent properties are not something added, but rather a reflection of something restricted and hidden via ascent in scale due to constraints propagated from lower-level dynamical processes."[68] In his article written with Koutroufinis, Deacon describes the nature of constraints that characterize particular levels of complexity scale as follows:

> The term "constraints" refers to all factors that reduce the number of the possible states of a system, so that its behavior resides only in a limited part of its state space. Since entropy increases with the number of a system's possible states, it follows that constrained systems are not in the state of their maximum possible entropy and therefore are able to perform work. Constraints can be imposed from outside on a system or can be generated internally, i.e., by a system's own dynamics. We describe the former as extrinsic and the latter as intrinsic constraints.[69]

Because of its strong emphasis on the role of constraints, Deacon's theory is sometimes classified as an "increased constraints"-based model of ontological EM. As such, it is both related to and differentiated from the whole-part constraints-based version of ontological—that is, strong—EM (described in sections 3.8.3 and 3.8.4 above), because the latter does not assign causal powers to constraints.[70]

To better explain the phenomenon of an emergent transition, Deacon introduces a distinction between "orthograde" and "contragrade" changes. He defines the former as natural and spontaneous changes in the state of a system (regardless of external interference), and the latter as extrinsically forced changes, ones that run counter to orthograde tendencies. He claims the fundamental reversal of orthograde processes in contragrade changes is a "defining attribute of an emergent transition."[71]

5.2. Three Stages of Dynamical Depth

Differences in the ways processes "eliminate, introduce, or preserve constraints" inspire Deacon to list at least three main levels of dynamical depth, which involve "the nesting of stochastic dynamical processes within one another."[72] These levels are partly reminiscent of the classification of the three orders of emergent phenomena listed at the beginning of this chapter.

5.2.1. Homeodynamics

The first level of dynamical depth includes higher-order linear thermodynamic phenomena that show a spontaneous tendency toward reaching thermodynamic equilibrium through elimination of constraints and increasing global entropy. In other words, homeodynamic change occurs when the features responsible for orthograde changes characteristic of a given entity cancel one another in contragrade changes proper for a dynamic aggregate of these entities. The process effects an instantiation of a higher-order state characterized by higher-order properties.[73]

Deacon notes that homeodynamics explains higher-order properties of stochastic systems such as liquid properties, including laminar flow, surface tension, and viscosity.[74] Other examples of the first-level EM include chemical reactions, simple nonidealized mechanical systems (e.g., a harmonic oscillator or a pendulum with friction), compound mechanical systems (e.g., clockwork), and computational devices.[75]

5.2.2. Morphodynamics

The second level of dynamical depth explains characteristic properties of systems showing a spontaneous self-regularizing tendency to become more organized and globally constrained over time. Such a tendency is an effect of intrinsic constraints that are produced to dissipate extrinsically imposed constraints. The amplification and propagation of the former occurs due to constant perturbation brought by the latter, which results in contragrade reversal of the typical thermodynamic orthograde tendency.

Morphodynamic systems are usually classified as "self-organizing." However, Deacon suggests they should be rather described as "self-simplifying," since the internal dynamics of their constituents diminishes in comparison to "being a relatively isolated system at or near thermodynamic equilibrium."[76] This terminological nuance becomes even more meaningful and explanatory once we take into account the fact (mentioned already in section 5.1) that morphodynamic systems depend on the persistent presence of extrinsic constraints. Because all that these systems can do is degrade these extrinsic constraints through the production and propagation of intrinsic constraints, they cannot, in fact, contribute to their own persistence in a particular thermodynamic state. That is why Koutroufinis describes morphodynamic EM as "self-organization without self."[77]

Deacon describes several examples of morphodynamic systems. His list is similar to the catalogue of the second-order emergent phenomena mentioned in section 1 in this chapter and includes the formation of regular spiral whorls of plant structures (called spiral phylotaxis), the formation of geological polygons, the formation of an eddy in a stream, the formation of Bénard cells in a heated liquid, the growth of snow crystals, the generation of laser light, and autocatalytic reactions.[78] In each of these examples "we find a tangled hierarchy of causality, where micro-configurational particularities can be amplified to determine macro-configurational regularities and where these in turn further constrain and/or amplify subsequent micro-configurational regularities."[79]

5.2.3. Teleodynamics
The third level of dynamical depth is called teleodynamics because of its end-directedness and consequence-organized features. Unlike morphodynamic processes, which generate order but lack representation of the external environment and functional organization, teleodynamic systems show both of these features. These characteristics enable them not only to resist entropy increase but also persistently to decrease it within themselves and their progeny over the course of evolution.

Deacon claims that systems of this kind emerge from the contragrade change involving two or more complementary and interdependent orthograde morphodynamic systems. Together with Koutroufinis, he attributes to them the category of "selfhood," based on two crucial facts. First, unlike the lower-order intrinsic constraints of a morphodynamic system, channeling an externally imposed gradient of energy, the higher-order intrinsic constraints of teleodynamic systems prevent the disruption of the synergy between the complement morphodynamic systems. Thus, these intrinsic constraints perpetuate the intrinsic teleodynamic target state, which can be described as a "higher order attractor." At the same time—and this is the second important fact about teleodynamic systems—it all happens in relation to the external environment and in a way that sustains supportive external influences and compensates for unsupportive or destructive ones (by self-reparation or reproduction). Hence, we may say that teleodynamic systems have an *Umwelt* ("self-centered world"), which represents the external environment and reacts in response to it.[80]

Emphasizing the interdependency of all three basic levels of dynamics in a three-stage nested hierarchy of emergent ascent, Deacon and Koutroufinis conclude:

> What we term *dynamical depth* then, is this hierarchic complexity and irreducibility of constraint-generating dynamics, such as distinguishes teleodynamics from morphodynamics and morphodynamics from homeodynamics. Each of these transitions is characterized by the *generation of intrinsic constraints* on the relationships between processes at lower levels and as a result with increasing autonomy from extrinsically imposed constraints. Since constraints are a prerequisite for producing physical work, the increasing autonomy of constraint generation with dynamical depth also corresponds to an increasing diversity of the capacity to do work. Thus the flexibility with which a dynamical system can interact with its environment also increases with dynamical depth.[81]

5.2.4. Autogenesis

Deacon offers a theoretical exemplary model of teleodynamics, which he calls "autogenesis."[82] He defines it in terms of a specific form of reciprocal coupling between morphodynamic processes of autocatalysis and an enclosure-generating process called "self-assembly." He thinks it is possible that one or more of the side products of catalysis show a tendency to self-assemble, forming thus an enclosure for the autocatalytic activity, similar in its shape to a polyhedron (e.g., a virus shell) or a hollow tube (e.g., a microtubule formation). Such a process gives an origin to an "autogen" or an "autocell"—a self-generating system. Whereas normally the rate of autocatalysis must be equal to or exceed the diffusion rate (otherwise the process will cease), in autogenesis diffusion is impeded by physical barriers to molecular movement generated in the process of self-assembly. This produces what Deacon calls a "negentropy ratchet effect" in which full dissipation of constraints is never completed—"*a unit structure that contains within itself the very set of constraints that are necessary and sufficient to re-create these same constraints in a new system in a supportive environment.*"[83] See figure 1.6.

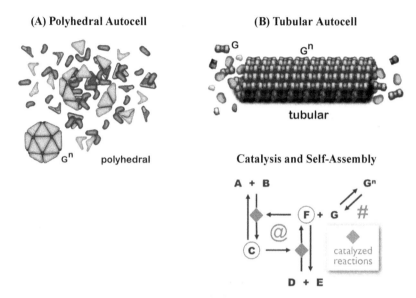

Figure 1.6. Two types of autogenesis

An autocell produced by polyhedral containment is depicted in (**A**), and an autocell produced by spirally elongated tubular containment is depicted in (**B**). Both are minimal autocells, products of a reciprocal catalytic cycle @, depicted with an arrow diagram. The cycle engages two catalysts C and F. One of the molecular side-products (G) tends to self-assemble (#) into a closed structure G^n, encapsulating the ensemble of reciprocal catalysts. Both catalysts are depicted as synthesized from two substrate molecules in each case (A and B, D and E). A polyhedral autocell completely encloses the complementary catalyst, achieving structural closure and allowing no further growth. A tubular autocell does not completely enclose its interior, but contained molecules are retained by viscosity of van der Waals interactions with the inner walls. The tubular autocell also retains the ability for continual elongation. Reproduction in both cases depends on extrinsic forces to break containment.

Source: Courtesy of Terrence Deacon. See Terrence W. Deacon, "Reciprocal Linkage between Self-Organizing Processes Is Sufficient for Self-Reproduction and Evolvability," *Biological Theory* 1 (2006): 141.

Deacon asserts that autogenesis, which demarcates the boundary and an emergent transition between teleodynamic and morphodynamic processes, becomes the threshold zone between living and nonliving systems. The self-reparation and self-replication tendency of an autogen in response to physical damage of its shell makes it a potential candidate for the operation of natural selection. Hence, an evolving autogen becomes not only a "negentropy ratchet" but also a "ratchet of life."[84] Most importantly, because maintenance, regulation, self-integrity, and reproduction of an autocell are embodied holistically (not in specialized molecules such as DNA or RNA), "autocell dynamics demonstrate that these fundamental attributes of life can emerge without the presumptive critical and ubiquitous role of separate information-bearing molecules; for example, nucleic acid polymers functioning as templates."[85] Deacon maintains that this fact distinguishes his theory of life's origin from all other models and helps to specify the foundation, as well as the ultimate locus, of primitive biological "selfhood."[86]

5.3. Dynamical Depth and the Extended Typology of Causes

One of the main philosophical objectives inspiring Deacon's dynamical depth account of ontological EM is his strong argument in favor of causal nonreductionism. Deacon emphasizes the negative outcomes of the philosophy of nature at the beginning of the Renaissance, when Bacon, Descartes, and Spinoza questioned and then rejected final causality. He attributes to them the origin of the process that led to "restricting the conception of causal influence to the immediate pushes and pulls of physical interaction" and that, in effect, made scientists "replace these black boxes and their end-directed explanations of function, design, or purposive action with mechanistic accounts."[87]

Moreover, Deacon criticizes eliminative materialism for its assuming that concepts like information, representation, and function are ultimately reducible to mechanistic accounts. He disagrees with the claim that, even if an attempt to explain goal-directed concepts in the language of basic physical properties and interactions of things is clumsy and arduous, it does not change the fact that such things are nothing more than just a sum of these properties and interactions. Contrary to this assertion, Deacon finds concepts such as information, function, purpose, meaning, intention, significance, consciousness, and value intrinsically irreducible.[88]

What makes Deacon's argument particularly intriguing is that, when criticizing causal reductionism in science, he recommends as a remedy the plural notion of causality in Aristotle. Deacon lists Aristotle's four causes (material, formal, efficient, and final), finding in his philosophy "the most sophisticated early recognition of a distinction between ... different modes of causality."[89] He then implements formal, efficient, and final types of causation in his explanation of the dynamical depth model of EM, adding to the Aristotelian list his emphasis on the importance of absences (constraints) for the overall picture of causal dependencies in dynamical systems.

5.3.1. Formal and Efficient Causes
The importance of the distinction between orthograde and contragrade changes for an emergent transition in Deacon's view of EM requires grounding it in a theory of causation, explaining both the source and the character of these changes. Trying to provide such a theoretical base of dynamical depth, Deacon refers, respectively, to formal and efficient types of causal dependency in nature. He states, first, that natural and spontaneous orthograde changes are expressions of formal causation. In reference to systems dynamics, Deacon defines those changes as functions of the geometric properties of a probability space that come about spontaneously, irrespective of any extrinsic influence on a given system. Their occurrence is "an unperturbed reflection of the space of possible trajectories of change" characteristic and unique for a dynamic process in question. He then concludes: "I take this to be a reasonable way to reinterpret Aristotle's notion of a formal cause in a modern scientific framework, because the source of the asymmetry is ultimately a formal or geometric principle."[90]

Deacon's new interpretation of formal causation plays an even more significant role in one of his more recent publications, coauthored with Alok Srivastava and Augustus Bacigalupi, in which he introduces terms such as "*forms* of the constituent dynamical processes," "formal reciprocity constraint," "formal 'semiotic' constraints," "formal source of regulation," "'formal' disequilibrium," and "formal relationship" of morphodynamic processes. What is being emphasized in this terminology is the idea of the formative aspects of absences (constraints). According to Deacon, it is not only what is present and analyzable in terms of parts of the system depicted in the probability space, but also what is absent, that provides for the formal identity of dynamic processes. I will say more about the role of absences in Deacon's view of EM below.[91]

Speaking of the other, contragrade type of changes, as extrinsically forced and running counter to orthograde tendencies, Deacon associates them with the efficient type of causation. "Forcing change away from what is stable and resistant to modification," efficient cause is defined in systems dynamics as "the juxtaposition of different orthograde processes," which "can produce complex forms of constraint."[92]

Summing up, it becomes apparent that Deacon finds both formal and efficient types of causation necessary in defining the causal grounding of the "engine" of his dynamical depth account of ontological EM. The intrinsic interdependency of orthograde and contragrade changes in an emergent transition finds an analogy and reflection in the suggested intrinsic interrelation of formal and efficient causation, which provide a causal explanation for these changes. This view departs significantly from the reductionist account of causation that is typical for scientific accounts of the natural phenomena.

5.3.2. Teleology

Deacon further expands his typology of causes with reference to a whole range of phenomena which he associates with the third level of dynamical depth. His suggestion that the irreducibility of information, function, purpose, meaning, intention, significance, consciousness, and value is a function of their intrinsic incompleteness inspires him to classify these phenomena as "ententional." Deacon acknowledges that the term "ententional," which he himself coins, requires a reference to another type of causal dependencies, one that proves indispensable for his theory of the origin of life. This type of causation has been traditionally known as teleological or final. Following Aristotle, Deacon defines teleology as that for the sake of which something is done, and he stresses that the concept of teleology does not assume future goals have a mysterious causal influence on the present: "Being organized for the sake of achieving a specific end is implicit in Aristotle's phrase 'final cause.' Of course, there cannot be a literal ends-causing-the-means process involved, nor did Aristotle imply that there was."[93] Again, referring to teleological aspects of systems dynamics and the role of absences (constraints) in the character and direction of development of these systems, Deacon asserts:

> This physical disposition to develop toward some target state of order merely by persisting and replicating better than neighbouring alternatives is what justifies calling this class of physical processes *teleodynamic*, even if it is not directly and literally a "pull" from the future. . . . The "constitutive absences" characteristic of both life and mind are the sources of this apparent "pull of yet unrealized possibility" that constitutes function in biology and purposive action in psychology. The point is that absent form can indeed be efficacious, in the very real sense that it can serve as an organizer of thermodynamic processes.[94]

While reintroducing the notion of teleological explanation and emphasizing its importance for the dynamical depth model of ontological EM, Deacon asserts that to remain consistent with the scientific understanding of natural phenomena, it is necessary to explain the way in which teleological properties emerge from nonteleological ones. He thus claims that teleodynamics—a dynamical realization of final causality—can be described in quasi-mechanistic terms.[95] Arguing in favor of the emergent character of teleology, Deacon states:

> Until we explain the transformation by which this one mode of causality [physical] becomes the other, our sciences will remain dualistically divided, with natural science in one realm and the human sciences in the other.[96]

> Hierarchically organized constraint relationship provides a plausibility proof for the emergence of *telos* (a form of Aristotelian "final causality") from mere physico-dynamic processes (Aristotelian "efficient causality").[97]

Searching for an explanation of the ways in which the causal dynamics of teleological processes emerges from simpler blind mechanistic systems, Deacon notes once again that ententional phenomena are defined by their fundamental incompleteness; they are reaching out toward something they are not.[98] This makes him suggest that absences have causal efficacy and that the development (or EM) of teleology occurs due to the growing limitation of the degrees of freedom, which is an outcome of the operation of absences in constraint-generating dynamical systems. Thus, Deacon concludes that the traditional motto of complex systems biology, which states the whole is more than the sum of its parts, must be replaced by an opposite assertion, namely, that the whole is actually less than the sum of its parts.[99]

5.3.3. Absences (Constraints)
Deacon's emphasis on the importance of absences—that is, ways in which dynamical systems are constrained—for the causal activity of each stage of complexity of emergents inspires him to ascribe to absences a causal activity. He thus lists constraints along with other types of causation necessary for elucidating the basic mechanism of all emergent transitions.

In his writings we find Deacon referring, on numerous occasions, to "absence-based causality," "absential influence," and "efficacy of absence," or emphasizing that "absent form can indeed be efficacious."[100] To better explain his argument, Deacon refers to the examples of what he calls "a specific absence, rather than, well, just nothing," given by the sixth-century-BC philosopher and father of Taoism, Lao-tzu, in the *Tao Te Ching*. He speaks about a wheel's hub as an empty space that makes it useful, a clay being shaped into a vessel to take advantage of the emptiness it surrounds, or doors and windows being cut into walls of a room to make it serve some function.[101] He sees absences as meaningful and thus essential for an explanation of why things are what they are and why processes progress the way they do.

Trying to explain the causation of absences still more specifically, Deacon says in *Incomplete Nature* that "to argue that constraint is critical to causal explanation does not in any way advocate some mystical notion of causality."[102] He sees constraint simply as a reduction of options for change in one dynamic process, a reduction that can account for a further reduction of options in another dynamic process which depends on the first. Understood this way, specific absences "shape" the reality and nature of these processes.

6. PROBLEMS OF THE DYNAMICAL DEPTH ACCOUNT OF EMERGENCE

The typology of causes offered by Deacon as a foundation and explanation of the "engine" of his dynamical depth model of ontological EM becomes an intriguing and original development of the preliminary suggestion to expand the notion of causation in the analysis of emergent phenomena. Raised by the followers of the classical mereological (DC-based) account of ontological EM, and realized more thoroughly in Deacon's project, the recommendation of bringing back into consideration formal and final causes seems to subscribe and contribute to a recent revival of Aristotle's philosophy of nature.

At the same time, however, Deacon's reinterpretation of the nature of formal causation in terms of "functions of the geometric properties of a probability space" and his argument in favor of the EM of teleological phenomena from nonteleological ones may raise doubts and questions among Aristotelian scholars. They might react similarly to Deacon's introduction of the idea of causation by absences. Moreover, an apparent ignorance of the importance of the material cause and its relation to the formal causation in Aristotle's account may classify Deacon as remaining still further away from Aristotle's philosophy of causation and his account of stability and change in nature. Indeed, aware of these differences, Deacon himself acknowledges and states:

> In many ways, I see this analysis of causal topologies as a modern reaffirmation of the original Aristotelian insight about categories of causality. Whereas Aristotle simply treated his four modes of causality as categorically independent, however, I have tried to demonstrate how at least three of them—efficient (thermodynamic), formal (morphodynamic), and final (teleodynamic) causality—are hierarchically and internally related to one another by virtue of their nested topological forms. Of course there is so much else to distinguish this analysis from that of Aristotle (including ignoring his material causes) that the reader would be justified in seeing this as little more than a loose analogy. The similarities are nonetheless striking, especially considering that it was not the intention to revive Aristotelian physics.[103]

This careful and modest evaluation of his own project, with respect to the possibility of its bringing about a retrieval of Aristotle's philosophy of nature, seems to speak in Deacon's favor. However, we need to realize that the question that needs to be asked is not limited only to whether Deacon offers in his works a truly "modern reaffirmation of the original Aristotelian insight about categories of causality" or whether it offers, instead, just "a loose analogy" to it. I think that a critical analysis of the metaphysical basis of Deacon's project with reference to both classical and new Aristotelianism may call into question the very success of his project. At the same time, such analysis might trigger another query concerning a possible reinterpretation of both the classical and Deacon's accounts of EM in reference to a more thorough account of Aristotle's philosophy of nature, and of causation in particular. I will address all these questions in the next chapter.

CHAPTER 2

Classical and New Aristotelianism and Emergence

The metaphysical investigation of the classical and the dynamical depth accounts of EM presented in the previous chapter shows that both theories make important references to Aristotle's fourfold division of causes. However, the fact of their advocating in favor of Aristotelian philosophy of causation requires some further metaphysical qualifications. We have seen that the classical account of EM seems to make the preliminary suggestion that all four causes must be taken into account in emergent studies, without specifying how exactly particular types of causation should be understood and applied in EM theory. Deacon's reinterpretation of formal and final causes, on the other hand, followed by his suggestion to attribute causal character/importance to constraints, leaves us with the pending question about the extent to which his dynamical depth model of EM actually follows Aristotle's philosophy of nature, rather than just loosely alludes to it.

An attempt to meet and answer these challenges and questions requires from us a careful inquiry into Aristotelian teaching on causation, both in its classical and contemporary analytic versions. I will pursue it in the first two sections of the present chapter.[1]

1. ARISTOTLE'S PHILOSOPHY OF CAUSATION

Of all ancient Greek philosophers, Aristotle (384–322 BC) is thought of as offering the most advanced theory of causation. His typology of causes, however, was naturally both a development of and a response to other theories of causal dependencies in nature offered at the time. Hence, our analysis of Aristotle's view of causation needs to be preceded by a short explanation of its contextual background.

1.1. Ancient Theories of Change and Stability

All theories of causation that were developed in ancient Greece find their origin in the most fundamental question about the source of change and stability of entities and processes observed in nature. The fathers of Western philosophy realized the answer to this question was necessary for the explanation of the obvious coherence of the world-system and its phenomena, as given in our sense experience. Trying to situate Aristotle's opinion on change and stability among the whole range of views developed and circulating among the members of different schools of Greek philosophy at the time, we must note first that his proposition was an attempt to avoid the two extreme positions of Heraclitus and Parmenides.[2]

Heraclitus (fl. 500 BC), struck by the continual change of material things, speculated that change should be considered the main feature of the reality and of matter. For him, the essence of any material being was change, that is, a continuous becoming. He thus saw all things as ephemeral patterns of continuity in the perpetual flux of the world, made of changing primary matter—fire, which seems to remain always the same, even though we know that, in fact, it is subject to continuous change: "The changed states of Fire are, first, sea; half of sea is earth, and half is stormcloud. All things are exchanged for Fire and Fire for all things, as goods for gold and gold for goods."[3]

On the other extreme of the same conversation about change and stability in nature we find Parmenides (fl. 500 BC), who assumes that nothing really changes in nature. Departing from the basic assertion that "only being exists" and "nonbeing does not exist," Parmenides proceeds and says that things either "are" or "are not." Since there is no other option, he concludes, nothing can change. For a new thing could come into being only from nonbeing (which is impossible), or from being (which would mean it existed before). Therefore, change must be an illusion of the human sense experience.[4]

The difficulty of both answers to the problem of the origin of change and stability in nature, offered by Heraclitus and Parmenides, is that each of them treated one of the phenomena in question as illusory. The Heraclitean view of being as endless becoming led to the denial of the reality of stability and permanence in nature, and thus found it difficult to give an account and an explanation of unity and perseverance of entities through time.[5] The Parmenidean view of being as one and unchangeable, on the other hand, led to the denial of the reality of multiplicity and change, which contradicts our common experience of the character and nature of the universe.

Acknowledging the weaknesses of both theories, Aristotle strove to find a middle ground between them. In fact, he was not the first among philosophers who acknowledged the reality of both change and stability in nature. Those who preceded Heraclitus in the search for the most elementary (primitive) matter thought of it as a source of the common origin of things, both changing and remaining temporarily permanent, rather than the principle of constant change with mere appearance of permanency. Thales (620–550 BC) claimed that the fundamental constituent grounding the stability of all entities and explaining all change was water (existing as solid, liquid, or vapor), whereas Anaximenes (570–500 BC) saw air as the basic "stuff" and thought that everything was really air in different forms of rarity or density. Anaximander (610–525 BC) took an important step by emphasizing that none of the basic elements (earth, air, fire, water) can be the first principle, for it would destroy other elements. He thus argued for considering the most fundamental principle as undetermined and called it ἄπειρον (*apeiron*, "the infinite"). He thought it was by the joining and separating of opposite qualities present in "the infinite" that all things in the universe came into being and changed.

The whole conversation took a new direction with Anaxagoras (ca. 510–428 BC), who spoke about cosmic intelligence (νοῦς, *nous*, "mind") as the cause of everything. Because intelligence seeks what it values, his argument seems to exceed the interest in material cause and introduce a primitive notion of final causation (teleology).[6] Empedocles (ca. 495–435 BC) added yet another aspect to the reflection on change and stability when he claimed that besides the elements of earth, air, fire, and water, two further principles, "love" and "strife," were needed to combine or keep apart the basic elements. His point can be regarded as the primitive notion of efficient cause.[7] Lastly, Plato (428/427 or 424/423 to 348/347 BC), who regarded matter as the stable and eternal receptacle for transcendent Ideas (Forms)—responsible for an ephemeral and temporal stability of entities that reflected them—became the first proponent of the formal type of causation.

This variety of opinions, offered as an explanation of both the reality and the origin of change and stability in nature, introduced preliminary versions of all four types of causes that were brought together and listed repeatedly in the works of Aristotle. At the same time, his philosophy of causation proved to be an important alternative to one more causal theory, one developed in ancient Greece and predominant in natural science since modernity: Democritus's (ca. 460–ca. 370 BC) atomic view, which strove to follow the Parmenidean conclusion that being is unchangeable without giving up the reality of change. Convinced that quantitative differences, subject to mathematical analysis, remained within the realm of the intelligible, whereas qualitative differences were not intelligible and therefore impossible, Democritus saw being as divided by space (the void) into many beings, that is, atoms. He thought his theory preserved the unity of being by assuming that all atoms were of the same kind (nature), while the fact that they differed in size and figure (shape) provided for the infinite variety of entities in nature. He also thought that the ability of atoms to aggregate and permeate in aggregates explained stability of things, while loosening atomic connections and disengaging from an aggregate explained physical change (evaporation, melting, motion, etc.).

Although it shared the same objective of treating as real and explaining both change and stability in nature, Democritus's theory departed from the views of other philosophers of his time, because it did not refer to formal or final aspects of causal dependencies in nature. At the same time, many found and still find it attractive because of its metaphysical sparsity and its explaining causation purely mechanically. Nevertheless, Aristotle thought that the alleged simplicity of Democritean atomism came at the price of its insufficiency for explaining the whole variety of entities with their intrinsic changes given in sense experience. He found it particularity limited in dealing with qualitative changes, which he perceived as real and resistant to an explanation in terms of quantitative alterations in atomic aggregates.

Summing up our short analysis of the contextual background of Aristotle's theory of causation, we can say that he developed it in response to the fallacy of the extreme views of Heraclitus and Parmenides as well as the insufficiency of the mechanical and reductionist view of Democritus. At the same time, he built on the seeds of truth found in all three theories and in a number of other opinions that preceded them and contained preliminary accounts of primary matter, substantial form, efficient cause, and teleology.

1.2. Aristotle on Types of Being and of Change

Answering the controversial thesis about unchangeability and unity of being, Aristotle criticizes Parmenides for taking being in a univocal sense, whereas it really has several meanings. He is convinced that we need to distinguish, first, between things, which have in themselves their own "to be," and the properties of things, such as their size, shape, color, and so on, which do not have their own "to be." Whereas in the first instance "to be" refers to something that exists in itself (a substance), in the other it refers to something which is "in" another or is predicated "of" another (an accident). Whiteness, for example, will be different from what has whiteness. Although we have a good reason to use just one concept "to be" with respect to both whiteness in itself and an entity that has it, we need to acknowledge different ways in which concrete realities show the realization of "to be."[8]

Second, Aristotle introduces one more fundamental distinction of being: that between being-in-potency (being-in-capacity) and being-in-act (being-in-perfection). When we think about an acorn—to follow his favorite example—it would be equally right and wrong to say that it is and is not an oak tree. We need to acknowledge that an acorn has a certain capacity or potency to become an oak. The reality of potency (δύναμις, *dynamis*) and act (ενέργεια, *energeia*) drives a wedge into Parmenides' assertion that "only being exists" and "nonbeing does not exist." For a capacity-of-being is neither simply nonbeing nor simply being. It is being-in-potency, distinct, on the one hand, from absolute nonbeing and, on the other hand, from actualized being-in-act.

Moreover, speaking of potency, Aristotle defines it as "the source, in general, of change or movement [ἀρχή μεταβολῆς ἢ κινήσεως, *archē metabolēs ē kinēseōs*] in another thing or in the same thing qua other, and also the source of a thing's being moved by another thing or by itself qua other."⁹ This basic definition proposed by Aristotle introduces an important distinction between active and passive potency. The former means a capacity to bring about an effect in another thing (e.g., fire burning a wooden log) or in the same thing as if it were the other ("qua other," e.g., a doctor healing herself). The latter means a capacity to be affected by another thing (e.g., a wooden log's being burned by fire) or by the same thing as if it were the other ("qua other," e.g., a doctor's being healed by her own action).

Finally, it is important to note that Aristotle defines potency in terms of a general category of "change" (μεταβολή, *metabolē*) and does not limit it to physical "motion" (κίνησις, *kinēsis*). In the opening paragraph of book 9 of the *Metaphysics* he states, "Potency and actuality [δύναμις καὶ ἡ ἐνέργεια, *dynamis kai hē energeia*] extend beyond the cases that involve a reference to motion."¹⁰ Moreover, he also distinguishes between nonrational and rational active powers, showing thus that potency is truly one of the most general characteristics, applicable to all material beings: "Since some such originative sources are present in soulless things, and others in things possessed of soul, and in soul, and in the rational part of the soul, clearly some potencies will be non-rational and some will be accompanied by a rational formula."¹¹

The distinction between different kinds of being allows Aristotle to address the controversy concerning the nature of change. On one hand, in response to the Parmenidean argument from unintelligibility and impossibility of change, Aristotle states that a new being does not have to come into existence from either absolute nonbeing or an actualized being. It can come "to be" from something which is not yet in some respect. To go back to Aristotle's example, from an acorn which is not yet a tree an oak tree can come to be, and from burning a wooden log which is yet nothing more or else than just a wooden log, a pile of ashes may come into being. Consequently, we may say that defining "coming to be" as the transition from being-in-potency to being-in-act enables Aristotle to explain the reality and the origin of change.

On the other hand, in response to the Heraclitean argument from the constant flux and the impossibility of stability, permanence, and unity of being, the distinction between substance (existing in itself) and accident (existing in/predicated of another) allows Aristotle to distinguish between the two general types of change: substantial and accidental. Describing first the latter, he classifies it in *De gen. et corr.* by a term ἀλλοίωσις (*alloiōsis*), commonly translated as "alteration," referring to what occurs when a thing or being changes in its properties while remaining the same substance: "The body, e.g., although persisting as the same body, is now healthy and now ill; and the bronze is now spherical and at another time angular, and yet remains the same bronze." He then contrasts it with a situation "when nothing perceptible persists in its identity as a substratum, and the thing changes as a whole." He calls the latter "a coming-to-be of one substance and a passing-away of the other" (a substantial change), for example, a wooden log burning into ashes.[12]

The distinction between accidental and formal change enables Aristotle to explain how things that change continually in some respects can remain stable and temporally permanent in some other aspects of their nature. To give an example, the process of a puppy growing up and becoming a mature dog is extended in time and includes a constant flux of accidental formal changes concerning various features of its organism (the size of its bones and muscles, secretion of hormones, its vocal chords, etc.), while its substantial form of a particular dog remains the same. It is only at the moment of its death that—due to the substantial change—it will cease to be a dog (one substance: S_1) and will turn into a carcass (a new substance: S_2).

1.3. Material and Formal Causation

Aristotle grounds his distinctions of different kinds of being and of change in his fourfold division of causes. First, he proposes a doctrine of matter and form understood as two types of causality particularly related to each other. He thus defines them respectively in the *Physics* and the *Metaphysics*:

> In one sense, then, (1) that out of which a thing comes to be and which persists, is called "cause," e.g. the bronze of the statue, the silver of the bowl, and the genera of which the bronze and the silver are species. In another sense (2) the form or the archetype, i.e. the statement of the essence, and its genera, are called "causes" (e.g. of

the octave the relation of 2:1, and generally number), and the parts in the definition.[13]

"Cause" means (1) that from which, as immanent material, a thing comes into being, e.g. the bronze is the cause of the statue and the silver of the saucer, and so are the classes which include these. (2) The form or pattern, i.e. the definition of the essence, and the classes which include this (e.g. the ratio 2:1 and number in general are causes of the octave), and the parts included in the definition.[14]

What is crucial regarding the material cause is its not being reducible to basic chunks of physical stuff (elementary particles) out of which things are made. Although one may find it difficult to grasp it in the oft-cited quotations from the *Physics* and the *Metaphysics*, for Aristotle matter is the most basic principle of potentiality, that is, primary matter (πρώτη ὕλη, *prōtē hylē*), underlying nature (ὑποκείμενον φύσις, *hypokeimenon physis*), or primary substratum (πρῶτον ὑποκείμενον, *prōton hypokeimenon*) that persists through all changes that a given substance can be exposed to. As something that constitutes the very possibility of being a substance, it is called primary matter and should be distinguished from secondary (proximate) matter, which is perceptible to our senses and quantifiable. This becomes clear in the following passages from the *Physics* and the *Metaphysics*:

> The underlying nature [ὑποκείμενον φύσις, *hypokeimenon physis*] is an object of scientific knowledge, by an analogy. For as the bronze is to the statue, the wood to the bed, or the matter and the formless before receiving form to any thing which has form, so is the underlying nature to substance, i.e. the "this" or existent.[15]

> The matter comes to be and ceases to be in one sense, while in another it does not. As that which contains the privation, it ceases to be in its own nature, for what ceases to be—the privation—is contained within it. But as potentiality it does not cease to be in its own nature, but is necessarily outside the sphere of becoming and ceasing to be. . . . For my definition of matter is just this—the primary substratum [πρῶτον ὑποκείμενον, *prōton hypokeimenon*] of each

thing, from which it comes to be without qualification, and which persists in the result.[16]

By matter I mean that which in itself is neither a particular thing nor of a certain quantity nor assigned to any other of the categories by which being is determined. . . . The ultimate substratum is of itself neither a particular thing nor of a particular quantity nor otherwise positively characterized; nor yet is it the negations of these, for negations also will belong to it only by accident.[17]

And if there is a first thing, which is no longer, in reference to something else, called "thaten" [ἐκείνινον (*ekeininon*), e.g., box being wooden in respect to wood it is made of], this is prime matter.[18]

As a result of its unique metaphysical status, existence is not properly nor directly predicated of primary matter, but of substance (the composition of primary matter and substantial form).[19] And yet primary matter is real and exists, even if not with its own independent act of existence, but with existence of the substance. We can say primary matter, as pure being-in-potency, underlies each and every substance, remaining a principle of continuity in the process in which one substance (S_1) becomes another substance (S_2). Even if all physical aspects of S_1 change on the way to its becoming S_2, we are not dealing with a total annihilation of S_1 and coming to be out of nothing of S_2. Rather, due to primary matter as principle of potentiality underlying all existing substances, we observe the continuity of the process of S_1 changing into S_2. Moreover, it is due to primary matter that both S_1 and S_2 are characterized by the persistent passive potentiality for change, which is actualized by substantial form.

Concerning formal cause, Aristotle situates himself in a radical opposition to the transcendental character of Ideas in Plato. For him forms do not exist in a supernatural realm. Nor are they imitated imperfectly by the mundane reality. To the contrary, according to Aristotle forms must be in things, determining their actuality. This becomes clear from the quotations from the *Physics* and the *Metaphysics* cited above. In both passages, Aristotle, speaking of formal causality, uses the term ὁ λόγος τοῦ τί ἦν εἶναι (*ho logos tou ti ēn einai*), which R. K. Gaye translates as "the statement of the essence," and W. D. Ross as "the definition of the essence."[20] Form is for him a principle of each existing substance that makes it to be the particular kind of thing it is—a principle actualizing (determining) a pure possibility-of-being (primary matter) to be a concrete substance. As such, similar to primary matter, form is a simple metaphysical principle and not a thing that has the property of quantity or extension. For this reason, says Dodds, "we cannot make an imaginative picture of a substantial form. It is not imaginable, but it is intelligible."[21] Form cannot increase or decrease. Thomas Aquinas (1225?–74) will later say that it is "educed" from the potentiality of primary matter and remains present in the entire substance and its parts as a fundamental principle of operation.[22] It finds expression in essential qualities of a given substance, which classifies Aristotelian ontology as essentialist. In other words, form can be taken as an essence, that is, "the property of x that makes x the same subject in different predications—the property that x must retain to remain in existence."[23] It embodies the criterion of identity appropriate to a given entity and can be characterized as the ratio or proportion between the different elements that go through a substantial change in the process of the entity's formation. Thus form becomes an inner cause of a given entity's natural behavior.

Speaking of the formal principle, however, Aristotle uses also two other terms: μορφή (*morphē*) and εἶδος (*eidos*), which translate as "shape" and "appearance."[24] These may bring confusion leading to reduction of form to a geometrical shape (available for sensory perception and mathematical description), which flattens out Aristotle's original idea.[25] Trying to avoid this error, Terence Irwin rightly notes that "if the form of the statue is essential to it, then other features besides shape must constitute the form, and the reference to shape can at most give us a very rough first conception of form. If we turn from artifacts to organisms, it is even clearer that form cannot be just the same as shape."[26]

In addition to all these terms, we find Aristotle describing form, at several occasions, in terms of ἐντελέχεια (*entelecheia*), which relates formal to final causation (anticipating our analysis of it) and denotes form as actualized in the final state of a given entity. But even if we can speak about the final stage of the realization of form in an organism, we must not forget that the form has been fully present in that organism since the beginning of its existence, actualizing the primary matter that underlies it.

Contrary to primary matter, which is a principle of continuity and a passive principle of change (as pure potentiality), form is a principle of novelty and an active principle of change in causal processes. Hence, even if in a process of substantial change from S_1 to S_2 primary matter does not change, we distinguish S_1 and S_2 as separate substances due to different forms that inform primary matter in them and are educed from its potentiality. At the same time, however, if S_1 changes in a way that makes it different but does not lead to its transformation into a completely new substance S_2 (as in our example of a puppy growing up and becoming a mature dog), we are dealing with accidental, rather than substantial changes.[27]

1.4. Efficient and Final Causation

The complete explanation of the variety of different kinds of being and of change requires—according to Aristotle—an introduction of two other types of causality, which enable us to specify metaphysically the source of change and rest, as well as their goal/end. He thus defines the two causes in question in the *Physics*:

> Again (3) the primary source of the change or coming to rest; e.g. the man who gave advice is a cause, the father is cause of the child, and generally what makes of what is made and what causes change of what is changed.
>
> Again (4) in the sense of end or "that for the sake of which" a thing is done, e.g. health is the cause of walking about. ("Why is he walking about?" we say. "To be healthy," and, having said that, we think we have assigned the cause.) The same is true also of all the intermediate steps which are brought about through the action of something else as means towards the end, e.g. reduction of flesh, purging, drugs, or surgical instruments are means towards health. All these things are "for the sake of" the end, though they differ from one another in that some are activities, others instruments.[28]

Efficient cause, defined simply as an activity of an agent that is a source of change or coming to rest, is seen by Aristotle as "the fulfillment of ... potentiality ... by the action of that which has the power of causing motion."[29] As such, it is necessarily related to intrinsic features (natures) of entities that enter into a given causal relation. In other words, the possibility of an agent to act and of a patient to be acted upon is in each one of them a property dependent on substantial form. This fact helps us understand that the processual description of efficient causation in terms of causal events remains incomplete without a reference to the natures of its participants, analyzed in terms of actuality and potentiality.

Moreover, although we have seen Aristotle defining efficient causation in terms of "the action of that which has the power of causing motion," he realizes that it need not always be predicated in terms of some physical interaction and an exchange of energy. Indeed, in the definition quoted above, we find Aristotle saying that giving advice, which can hardly be associated with the realm of masses, forces, and physical action, is also an example of efficient causation.

Finally, it should not escape our attention that in his account of efficient causality Aristotle makes an important distinction, which would be later on classified as the one between primary and secondary causation. He alludes to it briefly in *Phys.* 2.2.194b13, saying that "man is begotten by man and by the sun as well." Putting aside its dependence on the ancient cosmology, what we learn from this short assertion is that one (primary) efficient cause can exercise its action through another (secondary) efficient cause. We will see this observation further developed by Aquinas.[30]

Speaking of the fourth, final cause, Aristotle defines it as "that for the sake of which" a thing is done, or a good that is proper for a being and that can be attained. It takes its other name, "teleology," from the Greek τέλος (*telos*), which translates as "end" or a "goal." Although he invokes necessity as an explanation of the availability of suitable matter, Aristotle acknowledges the need of an explanation in terms of purpose as a function of nature, to explain why given matter acquires the particular shape and structure it does.[31] It can be clearly seen in the case of animals, whose natures transmitted in the processes of generation function as causes, by way of being goals toward which animals develop and which are their proper good.[32] This observation inspires Aristotle to suggest a general proposition stating that "generation is a process from something to something, from a principle [ἀρχή, *archē*] to a principle—from the primary efficient cause, which is something already endowed with a certain nature, to some definite form or similar end [τέλος, *telos*]."[33]

This claim helps us realize that Aristotle extends teleology (goal-directedness)—which is usually associated with, and often wrongly limited to, conscious human decisions—to other living and nonliving entities. Indeed, as notes David Bostock, in *Meteo.* 4.12.389b25–390a21 "Aristotle does explicitly say that the elements, and the inorganic compounds that are formed from them, are 'for the sake of something,' equating this with the view that they have a 'function' (ἔργον [*ergon*]) which in turn is a power (δύναμις [*dynamis*]) to act or be acted upon."[34] Most importantly, when predicated about inanimate and animate but unconscious nature, teleology should not be understood as a mysterious, quasi-efficient, cause, directing things according to a preestablished harmony. Quite the contrary, it should be seen as a natural tendency of things to realize what is proper to their nature (e.g., a tree blossoming and bearing fruit, or sodium and chlorine acid reacting in a specific and definitive way). This tendency does not need to be necessarily known nor intended by a conscious agent. That is why Aristotle delineates in *Phys.* 2.8.199b26–27 that "it is absurd to suppose that purpose is not present because we do not observe the agent deliberating."[35] Obviously, an inorganic compound does not have a soul, but it does have a substantial form, which might fulfill the same task of holding and transforming elements into a composite entity. Hence, by analogy, just as the soul of a living organism can be identified with its goal and actions that realize it, so the form of an inorganic compound has a similar task to perform.[36]

1.5. Comprehensive Approach to Causation

Having described separately all four types of causation, Aristotle notes in *Phys.* 2.7.198a25–27 that "the last three [the form, the mover, and that for the sake of which] often coincide; for the 'what' and 'that for the sake of which' are one, while the primary source of motion is the same in species as these." This notion of the interrelatedness of causes becomes crucial for a proper understanding of the concrete situations of causal dependency.

1.5.1. Hylomorphism

The point of departure for Aristotle is his recognition of the fact that "of things that exist, some exist by nature [the animals and their parts, the plants and the simple bodies—earth, fire, air, water], some from other causes."[37] Acknowledging this basic truth, Aristotle states that "nature is a source or cause of being moved and of being at rest in that to which it belongs primarily, in virtue of itself and not in virtue of a concomitant attribute."[38] In the context of what I have said above, it becomes obvious that each particular "nature" is, first of all, a function of material and formal causes, intertwined such that their specification and distinction becomes possible only at the speculative level of our philosophical analysis. What we encounter in the world is always primary matter informed by different substantial forms, or—to put it the other way around—substantial forms actualizing different potentialities of primary matter.

Hence, it is the principle of hylomorphism (i.e., of ὕλη [*hylē*] intrinsically related to μορφή [*morphē*]) that lies at the very foundation of Aristotle's ontology. Understood this way, "nature" becomes a source of activity and reactivity, which links it to the other two causes listed by Aristotle. For he always sees efficient cause as exercised "by" or "over" concrete entities characterized by concrete "natures." Hence, it is only with reference to those entities that we can speak about natural teleology, that is, about entities realizing their natural possibilities and propensities.

1.5.2. Other Relations between Causes

Commenting on Aristotle's philosophical works, Aquinas mentions other cases of the interrelatedness of causes. Speaking of efficient and final causation, he notes that the efficient cause grounds the final as it begins motion toward it, whereas the final cause grounds the efficient insofar as it is the reason for the activity of the efficient cause.[39] Moreover, the interrelatedness of causes goes beyond material/formal and efficient/final causality tandems. At one point Aquinas shows the relation between formal and final causes, saying: "The form and the end coincide in the same thing," and "it must belong to the natural philosophy to consider the form not only insofar as it is form but also insofar as it is the end."[40]

All of these assertions make Aquinas agree with Aristotle in saying that "three causes can coincide in one thing, namely, the form, the end and the efficient cause."[41] It is true that noticing this interrelatedness of causes does not prevent Aristotle from attributing to teleology the primacy among causes: "Plainly, however, that cause is the first which we call that for the sake of which. For this is the account of the thing, and the account forms the starting-point, alike in the works of art and in works of nature."[42] This unique role of final causation, however, is not possible—in the context of what we have said above—without its reference and relation to material, formal, and efficient types of causality.

Finally, we can think about one more aspect of the interrelatedness of causes, in reference to the way in which different kinds of simple elements—with their active and passive potencies—are present in complex (mixed) substances. Commenting on Aristotle's *De gen. et corr.* 1.10.327b24–32 and trying to answer the challenging question of what happens with basic elements and their causal activities as they go through substantial changes that effect in the coming to be of complex substances, Aquinas develops a theory which is traditionally referred to as the doctrine of the virtual (*virtute*) presence of elements in mixed substances: "The powers of the substantial forms of simple bodies are preserved in mixed bodies. The forms of the elements, therefore, are in mixed bodies; not indeed actually, but virtually (by their power). And this is what the Philosopher says in book one of *On Generation*: 'Elements, therefore, do not remain in a mixed body actually, like a body and its whiteness. Nor are they corrupted, neither both nor either. For, what is preserved is their power.'"[43]

Despite its rejection by many followers of the contemporary version of atomism, the Aristotelian-Thomistic theory of virtual presence seems to offer a powerful and plausible argument against the Democritean view of matter. Leaving it up to the physicists to specify the most basic "primary components" that can be classified as physical objects, I assign to these entities the principles of primary matter and substantial form and claim that as such they can enter compounds and remain virtually present in them, with their powers retained yet (possibly) altered and with substantial forms not entirely corrupted away but instead retrievable in the processes of corruption of these "mixed" (composite) bodies or in the reclaiming of given elements from complex substances, which nevertheless "keep" their substantial form (e.g., an oxygen atom leaving my organism, which, nonetheless, remains the same organism).

On the physical, chemical, and biochemical level of observation, a given primary component or a more complex entity such as an atom, molecule, or chemical compound may be perfectly traceable in a composite being. This fact, however, does not prevent or invalidate a philosophical (metaphysical) reflection stating that the properties and causal powers (dispositions) of that primary component, although retained, are now properties of a given compound (which is informed by a new and separate substantial form). Moreover, due to the fact of being "a part of"—or, better to say "being now compounded" (e.g., a carbon atom consumed by me becomes me)—the set of properties and dispositions of a given elementary particle is usually altered; that is, we might attribute to it properties and dispositions it does not have when separated from the compound.

To take an example from contemporary science, a carbon atom double-bonded to an oxygen atom in the carbonyl group (C=O) of the acetyl group (CH_3CO) becomes crucial for the function of the acetyl coenzyme A. This molecule participates in biochemical reactions of protein, carbohydrate, and lipid metabolism. See figure 2.1.

The main function of acetyl-CoA (derived from carbohydrates, fats, and proteins) is to deliver the acetyl group to the citric acid cycle (Krebs cycle) to be oxidized to carbon dioxide (CO_2) and water (H_2O) for the production of chemical energy in the form of eleven molecules of ATP (adenosine triphosphate). Hence, we can say that in addition to the standard set of properties (dispositions) of a given carbon atom when separated, including its ability to become a "part of" ("be compounded in") the carbonyl group in the acetyl group, as the intrinsic "part" of acetyl-CoA engaged in the citric acid cycle, the same carbon atom has a crucial function in the production of the chemical energy. It thus participates in one of the most important series of biochemical reactions, typical for all aerobic organisms—an activity (or causal power) that cannot be ascribed to it when separated from a living organism. At the same time, some other reactivity properties of this particular carbon atom are suppressed in various ways on account of its being a part of a given organism. It cannot, for instance, be arranged with other carbon atoms in a variation of the face-centered cubic crystal structure called a diamond lattice (see figure 2.2). Such reaction of diamond growth, taking place at high temperature and pressure and occurring over periods from 1 billion

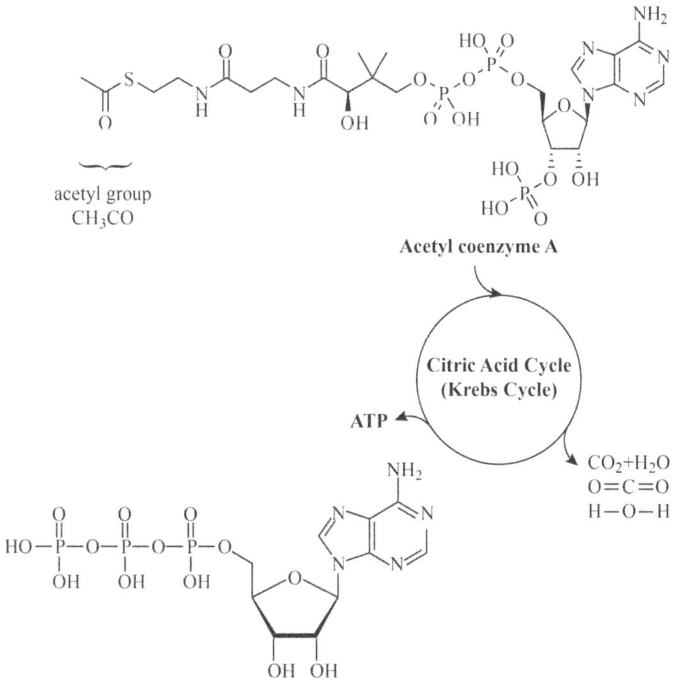

Figure 2.1. The role of acetyl coenzyme A in the citric acid cycle

Acetyl-CoA delivers the acetyl group (CH_3CO), which is oxidized to CO_2 and H_2O with the production of energy in the form of ATP.

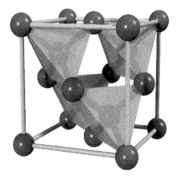

Figure 2.2. A diamond lattice crystal structure

Source: Pieter Kuiper, "Diamond Structure," January 15, 2011, https://commons.wikimedia.org/wiki/File:Diamond_structure.gif.

to 3.3 billion years at depths of 140 to 190 kilometers in Earth's mantle, requires as a substrate carbon in the form of carbon-containing minerals, and not the carbonyl group of the acetyl group in acetyl-CoA.

1.5.3. Consequences of the Plurality and Interrelatedness of Causes

The acknowledgment of the plurality of four causes and their interrelatedness in hylomorphism, as well as in efficient-final, formal-final, and formal-final-efficient causal interdependencies, and in the cases of basic elements and their causal activities virtually present in complex substances, enables us to offer a more accurate account and explanation of concrete situations of causal dependency (action and reaction) in nature.

Moreover, the awareness of the complex and multilevel character of causal dependencies proves crucial for properly understanding the methodology of natural science. It helps us realize that our use of scientific methods enables us to "identify the basic persistent subjects by reference to the properties that provide efficient causal explanations of change and stability. Some of these properties are material [i.e., defined in terms of secondary or proximate matter], but some are formal [i.e., defined in terms of substantial form and primary matter] and final [teleological]."[44]

Finally, an analysis acknowledging all types of causation needs a reference to, and becomes an argument in defense of, metaphysics, which

> inquires into the presuppositions of empirical science; for an empirical science assumes that it deals with an objective world, and with substances and their essential and coincidental properties. First philosophy [metaphysics] shows why we should accept these presuppositions, and what happens if we attempt to give them up.... While it would be quite wrong to claim that arguments in first philosophy are wholly non-empirical, it is still true that ... they are prior to empirical inquiry, in so far as they defend the assumptions taken for granted in empirical inquiry.[45]

1.6. Chance and Necessity

Before we move forward in our analysis, we need to allude to Aristotle's examination of the phenomenon of chance events in nature—especially in reference to those interpretations of quantum mechanics which tend to ascribe to chance a more fundamental role than to causality. The phenomenon of chance—which can refer to events happening both with and without a deliberate intention—has a rather peculiar character. In order to explain it, Aristotle uses the distinction between *per se* (καθ' αὐτὸ αἴτιον, *kath' hauto aition*) and incidental (κατὰ συμβεβηκὸς, *kata symbebēkos*), or *per accidens* causes. He sees *per se* causes as fundamental and essential efficient causes that come from nature (φύσις, *physis*) or intellect (νοῦς, *nous*). As such, they are naturally related to the formal and final causality of a given agent. In other words, an efficient cause is acting *per se* when its activity is performed by an agent, in accord with the agent's substantial form, to produce its proper effect.[46]

The character of the second, *per accidens* (accidental) type of causes, on the other hand, can be explained in reference to Aristotle's metaphysics of substance. Just as an accident (accidental formal feature) of an entity has no existence of its own, but is rooted in its substantial formal features, similarly an accidental cause must be related to a *per se* cause.[47] To give an example taken from Aristotle, the essential (*per se*) efficient cause of a statue is the sculptor. If he happens to be fair-skinned and musical as well, we may be justified in saying that a musician or a fair-skinned man made a statue. But his musical skills and the fact that he is fair-skinned are only incidental (coincidental, *per accidens*) causes related to the *per se* cause of his being a sculptor.[48]

Keeping this distinction in mind, Aristotle defines chance as an unusual incidental (*per accidens*) cause, which is inherently unpredictable, although it still falls in the category of events that "happen for the sake of something" (since it refers and is related to such occurrences). Thus, chance events are in a way posterior, since their occurrence and analysis require always a reference to *per se* causes of a given causal situation. Consequently, trying to specify its ultimate nature, Aristotle states that "chance is an incidental cause. But strictly it is not the cause—without qualification—of anything."[49] And yet, because it is distinguished as a unique type of occurrence which is not primary, and which yet is inherently related to nature (φύσις, *physis*) and intellect (νοῦς, *nous*), chance needs to be defined in reference to *per se* formal and final causality rather than blind material necessity.[50]

The Aristotelian account of the nature of chance events leads to a crucial assertion that their nature is not merely epistemological. They cannot be described simply as an unexpectedness due to the limitations of human understanding. For Aristotle, chance has primarily an ontological character. It is a kind of event that (ontologically) demands the agency/intentionality of nature or human will, since it appears to "happen for an end," but no such agency is involved in its occurrence.

Therefore, the need of reference to *per se* causes in the case of chance occurrences protects Aristotelian metaphysics not only from blind material necessity—defined as "tychism" (from the Greek τύχη, *tychē*),[51] that is, attributing everything to chance—but also from absolute determinism, which sees chance simply as lack of human knowledge of causes.[52] This fact has a significant influence on Aristotle's philosophy of nature and his understanding of necessary occurrences in the physical world. Necessity as such is for him never absolute but always suppositional. Things happen in accordance with causal patterns, but on the supposition that nothing interferes with given causal occurrences. In other words, what we observe in nature is a nomological necessity governing relations between metaphysically contingent entities and the processes they enter.

2. CAUSATION IN NEW ARISTOTELIANISM

The development of modern science and philosophy from the seventeenth to nineteenth centuries brought radical change and revision of Aristotelian theory of causation. Although not suddenly nor definitively, the course of both scientific and philosophical reflection on causality—influenced by the new appreciation of mathematics and its application in analysis and description of nature—shifted gradually toward dismissing teleology and formal causes.[53] Not amenable to empirical investigation and mathematical description, they seemed to be more and more obscure for those who followed the mainstream of modern thought. Aristotle's principle of primary matter as a part of causal explanation was rejected as well, which left the notion of efficient mass and energy (ex)changes as the only valid factors in the description of causal dependencies in nature.[54]

2.1. Insufficiency of the Neo-Humean Views on Causation

Interestingly, Aristotelian metaphysics and philosophical reflection on causation seem to come back in one of the more recent strains of analytic metaphysics, in the context of critical evaluation of neo-Humean perspectives on the question concerning cause/effect relationships.

Following modern criticism of the reality of substantial forms and natural teleology, and fostering a purely empirical approach in epistemology, Hume thought we are not justified in claiming access to the metaphysical basis of causation, defined in the past in terms of powers, forces, potentiality, actuality, and necessary connections. He thought the only reason for us to think about the reality of causal dependencies is our observation of similar events taking place one after another in time. He thus claimed that the tie of necessary connection "lies in ourselves, and is nothing but the determination of the mind, which is acquired by custom, and causes us to make a transition from an object to its usual attendant, and from the impression of one to the lively idea of the other."[55]

Although the neo-Humean philosophers of causation in the analytic tradition seem to agree with Hume that we do not have a direct access to the metaphysical basis of causation, they tend, nonetheless, to think that causation is ontologically real and not just a function of the constant conjunction of events producing an association of ideas in our mind. In order to defend their position, they offer at least six views of causation.[56] The main argument of each one of them can be summarized as follows:

1. Regulatory view of causation (RVC). Departing from Hume's assertion that causation can be defined in terms of "an object precedent and contiguous to another, and where all the objects resembling the former are placed in like relations of precedency and contiguity to those objects, that resemble the latter,"[57] the followers of RVC state that our observation of the regular succession of similar events suffices to claim they are ontologically causally related.
2. Counterfactual view of causation (CVC). Based on Hume's other definition of causation, which says that "if the first object had not been, the second never had existed,"[58] the proponents of CVC claim that our observation of the counterfactual dependency

between events allows us to conclude that, ontologically speaking, they are causally related.

3. Probability view of causation (ProbVC). Based on an observation that all relations among events in nature are usually accompanied by probabilistic dependencies, ProbVC states that our observation of an event A raising the probability of the occurrence of an event B suffices to claim that they are causally related (ontologically speaking).

4. Singularist view of causation (SVC). Critical of the analysis of causation in terms of laws and regularities between events or states of physical systems, the advocates of SVC insist that causal inquiry should be concerned with single occurrences. They claim that our ability to define a particular change (an event A) in the immediate environment in a particular situation that occurred just before an event B enables us to conclude A and B are ontologically causally related.

5. Manipulability view of causation (MVC). Based on our commonsense idea connecting causation with manipulation, MVC states that if any manipulation of an event A, which is usually followed by an event B, brings a relevant change in B, we are justified in regarding the relation between A and B as ontologically causal.

6. Process view of causation (ProcVC). Influenced by Russell's idea of "causal lines," defined as trajectories of things through time (replacing the primitive notion of causation in the scientific view of the world), the champions of ProcVC propose a theory of the causal world consisting in the nexus of processes and interactions. They define causes as processes transmitting a "mark" (understood as any local modification of a "characteristic," e.g., signal, information, energy), or a "conserved quantity" (e.g., mass-energy, linear momentum, charge, or spin). According to supporters of ProcVC, our ability to detect world-lines propagating marks or conserved quantities in space-time and to perceive their intersections that bring causal change allows us to specify the ontology of causal dependencies in nature.

Although each of the six views of causation mentioned here says something true about the nature of causal relations, they all face a number of critical doubts and questions concerning their metaphysical plausibility. The main objections can be classified as follows:

1. RVC: difficulty in identifying the ground of regularity; difficulty in specifying what makes a particular cause to cause a particular effect; problem of generalization of causal claims; treating accidental correlations as causal dependencies; difficulty in distinguishing causes from causal conditions; problem of specifying directionality of causal changes; interpreting as causal cases of joint effects of common causes (regularly co-occurring events A and B that are not causally related but are effects of another event C).
2. CVC: unspecified character (is CVC a folk concept, a philosophical concept, a scientific concept, or something in between?); dependency on possible worlds modality (in order to justify counterfactual statements which refer, *ex definitione*, to events that actually occurred in a given way); the problem of distinguishing between lawful and merely accidental generalizations of counterfactual dependencies; invalid character of counterfactuals concerning cases of late and simultaneous preemption (that is, causes canceling out, preceding in time, or trumping other causes, which would otherwise produce the same effect), over- and underdetermination (that is, cases of more than one cause bringing simultaneously the same event and cases of relations of counterfactual dependence that are not causal).
3. ProbVC: applicable in general accounts, ProbVC fails in particular cases; difficulty in forming lawful-kind generalizations based on individualized instances of probability rising; ProbVC requires an exclusion of all logically necessary conditions from the ranks of putative causes (otherwise causal conditions may be taken as causes); in many cases, an acknowledged cause of an event actually decreases the probability of its occurrence, instead of increasing it; difficulty in specifying the nature of probability (epistemological versus ontological).
4. SVC: difficulty in limiting description to a manageably small number of changes in the immediate environment of a singular causal change; the idea of regularity of causal events, which SVC opposes, sneaks back in with the requirement of the "commonality"

of causes in events of the same sort (proposed as a justification of treating a particular change as causal and not merely correlational).
5. MVC: confusing metaphysics with epistemology (MVC may suggest that agent interventions are constitutive for causal interactions); circularity (explaining causation in terms of "bringing about" a change by manipulation); anthropocentricity; reference to possible worlds modality in dealing with unmanipulative causes; the problem of forming lawful-kind generalizations with a certain level of invariance; difficult, if not impossible, to apply in cases related to fundamental physical theories that take into account the universe as a whole (due to impossibility of a manipulation on the universe as a whole).
6. ProcVC: difficulty in specifying the ultimate nature of processes and the ontological status of the "mark" or "conserved quantity" that is being transmitted in the process; the extent of persistence of a mark through time is unclear; the level of alteration within which a body (or wave) retains its self-identity is uncertain; a charge of reducing the metaphysical account into a watered-down version of a physical theory; challenged by issues concerning causal dependencies in branches of science other than physics.

2.2. Dispositional View of Causation

The number of critical difficulties concerning all six theories of causation in analytic metaphysics encouraged a number of scholars to depart from the Humean ontology of loose and separate entities or events which are related only externally and contingently (as Democritean atoms in the void) and to replace it with a metaphysics acknowledging the reality of powers and their manifestations in nature. In other words, while acknowledging the reality of regular occurrences and constant conjunctions, as well as counterfactual, probabilistic, manipulationist, and process aspects of dependencies among inanimate and animate beings—in various settings and combinations—the advocates of the dispositional view of causation (DVC) claim that the very reason they all occur is that there are real causal connections in nature. They also suggest grounding the latter in the metaphysics of dispositions and their manifestations.

2.2.1. Dispositions and Their Manifestations

According to dispositionalist metaphysicians, dispositions (or powers)[59] should not be treated as substance-like existents. They see them rather as mind-independent properties of a unique character. Alexander Bird defines powers as "properties with a certain kind of essence—an essence that can be characterized in dispositional terms,"[60] whereas Stephen Mumford suggests that we should treat them as "a distinct and basic ontological category in their own right, irreducible to any other ontological category."[61] Moreover, together with Rani Anjum, Mumford qualifies dispositionality as a primitive and unanalyzable modality, intermediate between pure possibility and necessity and providing a metaphysical base for both normativity (something ought to be the case but is not necessary) and intentionality (directness or aboutness, conscious and nonconscious). They suggest it can be understood as a "selection function" which picks up a limited number of possible outcomes that are manifestations toward which powers are directed.[62]

Classified as properties, dispositions can also be taken as universals having their instantiations in concrete substances/processes. As an example, we may think about the disposition of chemical substances to isomerize, that is, to go through the process in which one molecule is transformed into another molecule that has the exact same atoms, but the atoms have a different arrangement, either constitutionally (structural isomerism) or spatially (stereoisomerism). Both structural isomerization and stereoisomerization may occur spontaneously or be induced by a scientist (under specific conditions) and may effect production of substances that differ in some elementary qualities. For instance, heating of fuels such as diesel or pentane (C_5H_{12}) in the presence of a platinum catalyst (hydrocarbon cracking) leads to the production of straight- and branched-chain constitutional (structural) isomers that can be separated from the resulting mixture. As an example of spatial (stereo) isomerization we may think about the photochemical conversion (under exposure to ultraviolet radiation) of the *trans* isomer to the *cis* isomer of resveratrol (a substance produced by several plants in response to injury or when a plant is under attack by pathogens such as bacteria or fungi). These two cases of isomerization show the identity of the same disposition across various and distinct instantiations. See figure 2.3.

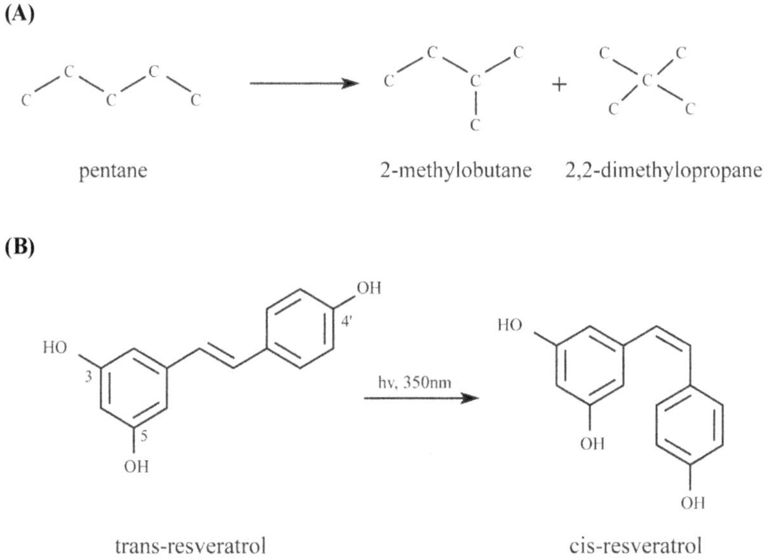

Figure 2.3. Examples of isomerization

(A) Constitutional (structural) isomerization: hydrocarbon cracking of pentane $CH_3(CH_2)_3CH_3$ into 2-methylobutane $CH_3CH_2CH(CH_3)_2$ and 2,2-dimethylopropane $C(CH_3)_4$. All three substances have the same molecular formula C_5H_{12}, but they differ in chemical structure. **(B)** Spatial (stereo) isomerization: photochemical conversion of the *trans* to the *cis* isomer of resveratrol (3,5,4'-trihydroxy-stilbene). Both stereoisomers have the same molecular formula $C_6H_4(OH)\text{-}CH_2\text{-}CH_2\text{-}C_6H_3(OH)_2$ but differ in spatial organization.

Dispositionalists note that the variety of powers ranges from the most basic physical properties (e.g., charge, mass, force) to very sophisticated ones, such as human agency.[63] What remains at the heart of their argumentation is an emphasis on each power being essentially, or necessarily, related to its manifestations of a specific kind, depending on the nature of an entity/process that has it. For example, the disposition of pentane to isomerization leads specifically to the production of 2-methylobutane and 2,2-dimethylopropane, whereas the same disposition in resveratrol leads to conversion of the *trans* to the *cis* isomer, both having a unique and specific shape. At the same time, however, powers and their manifestations are distinct existences, and even if not manifested, powers are real. This necessary relation of powers and their manifestations, even if they are not existent, is usually described as the natural directedness of powers toward their manifestations.[64]

2.2.2. Causal Powers

Powers ontology provides a metaphysical framework for a theory of causation that describes each event as an effect of powers manifesting themselves in a causal process. Although it may sound trivial at first, this assertion becomes profound in the context of all of the neo-Humean views of causation listed above. For it treats causation as a metaphysically real type and grounds it ontologically in the nature of concrete entities/processes, rather than in the reality of regular occurrences and constant conjunctions or in probabilistic, manipulationist, or process aspects of dependencies in nature. Following this line of reasoning, George Molnar characterizes causation as a manifestation of reciprocal powers,[65] whereas Brian Ellis clarifies that causal power is "a disposition to engage in a certain kind of process: a causal process."[66] This analysis is further developed by Mumford and Anjum, who state that "according to causal dispositionalism, causation involves an irreducible dispositional modality."[67]

DVC shows several advantages over other causal theories in analytic metaphysics. Because dispositions are decisive about intrinsic features of entities/processes that possess them, we can make general claims about their causal activity and reactivity, even if their powers are not manifested. Our judgment does not depend on an unspecified number of observations of particular regularities (RVC), counterfactual conditional dependencies between events (CVC), events raising the probability of an occurrence of other events (ProbVC), experimental manipulations of one kind of events changing other events (MVC), or particular world-lines propagating marks or conserved quantities in space-time and in their intersections (ProcVC).

The reality of dispositions and their manifestations enables us to distinguish between causes and causal conditions, solving thus one of the difficulties challenging RVC and CVC. Moreover, defining causation in terms of causal powers answers the problems of early, late, and simultaneous preemption, as well as over- and underdetermination, problems related to the same two theories of causation. It also helps us to be attentive to both singular (one cause bringing a particular effect) and plural cases of causal dependencies. The latter are characterized by polygeny of causes (more than one cause is responsible for an effect), which also opens the way to causal pleiotropy (one cause contributing to occurrence of more than one effect).[68]

The awareness of the polygeny of causal situations and pleiotropy of causal powers favors, in turn, the acknowledgment of the temporal simultaneity of causal relations. It leads Mumford and Anjum to suggest that causes and effects may occur without any time gap between them and may include reciprocity, as in the case of two books leaning against each other, or a kitchen magnet remaining attached to the front of the refrigerator. They claim we should see causation as the unfolding of a process (uniting cause and effect into a single, undivided whole), in which one thing affects/turns into another at the same time, whether it is instant or extended, often in a transparent and perceptible way. This process comes in temporally extended wholes.[69]

Taken together with polygeny of causal dependencies and pleiotropy of causes, the acknowledgment of their temporal simultaneity and reciprocity helps us to give an account of the puzzling and counterintuitive observation that sometimes causes are absences (e.g., my not wearing a winter hat in the cold weather made me sick). Because most of the cases of genuine causation involve manifestations of many dispositions, where an absence of action is invoked, we do not mean that it has a causal power, but rather that an absence of a particular power's manifestation changes the resultant causal effect which is brought by manifestation of other dispositions engaged in this particular causal situation. Hence, absences can be classified as enabling conditions rather than causes sensu stricto.[70]

The fact that manifestation of dispositions, though teleological in nature, depends usually on a set of specific conditions enables us to classify the necessity of causal relations as suppositional (avoiding both strict necessity and pure contingency). It also equips us with conceptual tools suitable in dealing with probabilistic cases in which causation is chancy yet probabilistically constrained. This aspect of DVC becomes a fitting alternative to ProbVC.[71]

Finally, paying attention to concrete individual cases of causal dependencies and yet being able to abstract from them to form general rules of causation, DVC avoids the evident limitations of SVC and the anthropocentricity of MVC. At the same time, it escapes the other extreme of CVC and its dependency on the possible worlds modality, which assumes the ontological reality of countless alternative worlds, the existence of which is not empirically verifiable.

2.3. Dispositionalism as New Aristotelianism

The emphasis on the importance of dispositions in causal explanation makes many commentators characterize dispositional metaphysics and DVC as neo-Aristotelian, which positions advocates of DVC in a radical contrast to the standard neo-Humean alternatives concerning causation. Nevertheless, the question whether dispositionalism truly follows Aristotle's original thought is rather complex and requires a closer examination.[72]

2.3.1. Potency and Act

In section 1.2 of this chapter I mentioned Aristotle's distinction between being-in-potency and being-in-act and its importance for explaining and defending the reality of change in nature. However, the same differentiation between potency and act and the additional distinction between active and passive potency have even more profound meaning for Aristotle's metaphysics and his causal theory. These distinctions not only explain why the change is possible at all but also decide about the character of the activity of an agent cause and its limitations to bring (produce) particular outcomes. In other words, these distinctions help us understand that the efficacy of a given causal occurrence is always limited primarily by the character of potentialities (active and passive) characteristic of concrete entities/processes participating in it (apart from limits imposed by circumstances and possible impediments).

Looking at the main principles of dispositional metaphysics and its emphasis on the role of powers, we may be sure it acknowledges the reality of being-in-potency. Nancy Cartwright and John Pemberton refer powers theory directly to Aristotle's metaphysics and say that, therefore, powers "are part of the basic ontology of nature—at least as nature is pictured through the lens of modern science."[73] Moreover, dispositionalists' emphasis on the relation of powers and their manifestations seems analogous to Aristotle's theory of potency and act, showing the same explanatory potential when it faces questions concerning the reality of change and the character of an agent cause and its efficacy in a given causal relation.

Nevertheless, dispositional metaphysicians do not refer openly to Aristotle's distinction between active and passive potency, which makes their account of causal occurrences one-sided. For the proper description of any causal change requires an actualization of both the ability to bring a change in an agent (an active potency) and the ability to receive a change in a patient, or in an agent when it acts on itself (a passive potency). Many dispositionalists concentrate only on the former, not paying enough attention to the latter aspect of potentiality in nature.

2.3.2. Teleology

Aristotle's theory of potency and act is closely related to his idea of intrinsic directedness toward an end in natural objects. Such a tendency is possible provided there is a real and objective potency in entities that can be actualized. As I emphasized above, Aristotle's notion of teleology is not limited to the realm of conscious and living beings, investigated by human and biological sciences. He extends it to inorganic entities and to inanimate dynamic systems, insofar as they are cyclical and show tendencies (potencies) toward certain end states (actualities).

The recognition of powers "pointing" or being "directed" toward their characteristic manifestations in dispositionalism certainly brings a revival of the classical Aristotelian notion of teleology. Hence, we find Molnar speaking about "physical intentionality," and John Heil about "natural intentionality," whereas Ullin Place describes dispositions as "intentional states" pointing beyond themselves.[74] What is at the core of their argumentation is, therefore, the fact of a stripped-down notion of finality, characteristic of inanimate objects as well as nonsentient, sentient, and conscious forms of living organisms. It leads David Oderberg to acknowledge that dispositionalism indeed brings back the notion of natural (inorganic and organic) teleology, which "fell by the wayside under the anti-Aristotelian assaults of empiricism and materialism and has not yet recovered."[75]

2.3.3. Hylomorphism and Grounding of Powers

However, it seems that the most important question concerning the neo-Aristotelian character of dispositionalism is related to the theory of hylomorphism. Looking once again on Aristotle's distinction between potency and act, we need to realize that—apart from explaining the possibility of change, the nature of efficacy of causal agents, and the phenomenon of their natural finality—it also refers us to the question concerning the ultimate nature of the possessors of active and passive potencies. For there needs to be something about the nature of inanimate and animate entities/processes that is a principle of their potentiality (a possibility to change), and something about their nature that is a principle of actuality (grounding their possibility to bring a change).

As we have seen in the first part of this chapter, Aristotle defines these principles as primary matter and substantial form. He states that they are characteristic features of both living and nonliving beings, constituting their essences (natures). Primary matter is the principle of passive potentiality, whereas substantial form determines actuality, realized and manifested in actual and perceivable characteristics of a nonliving being or an organism, as well as in its not yet realized (not yet manifested) active and passive powers and dispositions. The fact of the ultimate grounding of causal explanation in hylomorphism tells us, in turn, that for Aristotle, unlike for Hume, causes are not events but concrete nonliving beings and living organisms (substances). Even if they can be treated as processes, we cannot ignore the reality of their ontological unity, which goes beyond the neo-Humean theory of events as causes.

Recognizing the importance of hylomorphism for Aristotle's metaphysics and theory of causation, we need to ask to what extent dispositional metaphysics and DVC build upon the same foundation. The answer to this query is rather complex, because the very question of whether powers need to be grounded in any way is the subject of a vivid discussion among dispositionalists. Many of them state that all properties in nature are dispositional and that the whole of reality is constructed of powers localized in regiments of space-time (pan-dispositionalism).[76] Others distinguish between dispositional properties (responsible for things' modal characteristics, e.g., fragility, solubility, isomerization) and categorical properties (providing for things identity and underlying dispositions, e.g., macrostructures, molecular substructures, states, episodes, shapes, orientation) and claim that all properties, including powers, are or build upon categorical properties (pan-categoricalism).[77] Two additional versions of property monism are defined as neutral (dual-aspect) monism and identity theory. The former assumes there is only one kind of property, which is neither dispositional nor categorical, and this type of monism treats

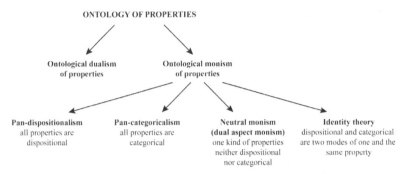

Figure 2.4. Different views concerning ontology of properties

the dispositional/categorical distinction as merely linguistic.[78] The latter states that the dispositional and the categorical are two modes of one and the same property, which is, thus, metaphysically simple (the advocates of this theory propose, as its model, a Necker cube, which can be seen now one way, now another, or the duck-rabbit figure).[79] Finally, the followers of the ontological dualism of properties claim that dispositional and categorical properties are two inherently distinct and yet interrelated kinds of qualities, necessary at least at some levels of explanation.[80] See figure 2.4.

How is the conversation about grounding of powers in categorical properties related to hylomorphism? On the one hand, it seems that the two theories are radically disconnected, if not opposed to each other. My analysis of different positions concerning ontology of powers shows that dispositionalism can be seen as accepting the neo-Humean "bundle" or "trope" theory of substance, defining its nature in terms of a collection or cluster of properties (here powers and/or categorical properties) and processes that a given substance enters (here manifestations of dispositions). Subscribing to this kind of metaphysics leads to a radical departure from Aristotle's hylomorphism.

On the other hand, however, we cannot ignore dispositionalists' emphasis on the fact that the range of powers characteristic for a given being is strictly defined and constrained. They seem to approve the reality of a more robust principle of unity, which goes beyond the mechanistic explanation of the similarity-generating constitution of the properties. They speak about "essential" and "intrinsic" properties and powers, which suggests they approve the reality of essences: "Dispositions are not relations to actual or possible manifestations, however. Objects possess dispositions by virtue of possessing particular intrinsic properties. The nature of these properties ensures that they will yield manifestations of particular sorts with reciprocal disposition partners of particular sorts."[81]

What is, then, the ultimate position of dispositionalists concerning the ontology of particulars (inanimate and animate entities/processes)? Even if they recognize the importance of the distinction between potency and act and refer (both indirectly and directly) to essentialism, they do not quite follow Aristotelian hylomorphism, because they do not mention primary matter and substantial form. Nor do they speak of the distinction between active and passive aspects of potency. Yet they do not follow the neo-Humean "bundle" theory of substance, popular in analytic metaphysics, either. Quite the contrary, we have seen many of them acknowledging the necessity of grounding powers in categorical properties, which ontologically means more than just forming bundles together with these properties.

Therefore, I want to argue that despite the considerable differences between powers metaphysics and classical Aristotelian ontology and philosophy of nature, one can still conclude that of all metaphysical theories concerning the nature of nonliving and living beings in analytic philosophy, powers metaphysics remains the closest to the orthodox notion of hylomorphism and essentialism. Its clear and undeniable reference to the classical theory of potency and act, as well as its emphasis on the kind-specific and intrinsic character of powers, situates it closer to Aristotelian hylomorphism and essentialism than to the Humean "bundle" theory of substance. Consequently, we may say that in this respect—in addition to its bringing a revival of teleology—dispositionalism can indeed be regarded as a neo-Aristotelian type of ontology.[82]

3. ARISTOTELIANISM AND THE CLASSICAL ACCOUNT OF EMERGENCE

After studying the classical and the new Aristotelian metaphysics and theory of causation, we can now follow the recommendation of a number of scholars listed in the section 4.2 of chapter 1 and apply Aristotle's thought in both the classical and the dynamical depth models of EM.

Beginning with the former, we will concentrate on the DC-based version of ontological EM and its core argument concerning the reality of the new and irreducible top-down causal influence characteristic for emergent phenomena.

My investigation offered in the first chapter proved that—when considered in reference to the neo-Humean ontology, which sees all entities in nature as loose, separate, and related only externally and contingently and defines causation in terms of regularity of causal events or in terms of counterfactuals—the concept of DC is incoherent and falls a victim to causal reductionism. Nonetheless, analyzed in the context of dispositional metaphysics and DVC, DC can be defined as a specific kind of manifestation of a new and unique set of dispositions, which are irreducible and kind specific, defined as intrinsic properties "with a certain kind of essence."

I argued in chapter 7 of *Emergence* that understood this way, DC has several complementary aspects to its character and nature, which sends us back to Aristotle and his robust theory of causation. I will now summarize my analysis of these aspects of DC when reinterpreted from the perspective of the new and the classical Aristotelian metaphysics and theory of causation.

3.1. Downward Causation and Hylomorphism

The first and the most fundamental aspect of DC can be described in terms of formal causation. Understood as the principle (logical predicate) of actuality, irreducible to efficient causation, decisive about the intrinsic nature of each entity and process ("making"/causing them to be what they are), substantial form becomes not only a powerful argument in defense of the irreducible character of DC but also a viable explanation of its nature.

I have said above that for Aristotle the formal type of causation is inextricably related to primary matter. These two principles form the first and the most basic ontological composition underlying the reality of all entities.[83] Applying Aristotle's hylomorphism to the DC-based ontological version of EM enables those who embrace it to solve the ambiguity of defining emergents in terms of properties. It suggests characterizing them in reference to new substantial forms of more complex/higher entities/organisms and new accidental forms of more complex dynamic processes engaging particular entities (e.g., molecules of H_2O in an eddy). The novelty of properties characteristic to emergent entities or processes finds its ground in the novelty of substantial and accidental forms proper to them. Thus, in reference to the formal aspect of DC, we can say that emergent properties describe and reveal the emergent character of higher (more complex) entities or processes, rather than being decisive about their emergent nature. But how do they do it? This question moves our analysis to the remaining aspects of DC.

3.2. Downward Causation and Efficient Causation

Because the new type of activity and reactivity proper to an emergent entity is inseparably related to its very nature, efficient causation becomes another indispensable explanatory aspect of DC. However, when defining DC in terms of efficient causation as it is understood by Aristotle, we have to remember that the latter does not stand on its own without a reference to the substantial form and accidental features of an entity that shows it. This fact relates the efficient aspect of DC necessarily to hylomorphic analysis of emergent entities and processes, even if it is otherwise a subject of scientific analysis and mathematical description.

Moreover, we have to remember that within the Aristotelian theory of causation, efficient causality has a broader meaning than the one defined within the domain of the natural sciences. We have seen that in his basic definition of efficient causation, Aristotle goes beyond physical efficacy and extends it to mental activity (see his example of a man giving advice in section 1.4 of this chapter). Interestingly, it is precisely causal activity of the mind that became one of the most common examples of DC in contemporary emergentism, which pays attention to causes that originate changes in the behavior of emergent wholes but are hardly measurable quantitatively and hardly describable in the mathematical language of natural sciences.[84]

Translating this argumentation into the language of dispositionalism, we can state that even if DC has an efficient character and is a subject of the scientific analysis and mathematical description, its irreducible nature is kept and is defensible within dispositional metaphysics because of its grounding of all cases of activity and reactivity in kind-specific powers, characteristic for an intrinsic nature (essence) of a given entity (organism/system). Moreover, advocates of dispositionalism emphasize the diversity and flexibility of powers ascriptions. We saw Mumford classifying the following as dispositions: fragility, belief, bravery, thermostats, and divisibility by 2 (see above, section 2.2.1, n. 63). If we want to analyze the efficient aspect of the manifestations of these dispositions, it becomes clear that our definition of efficient causation—similar to the one offered by Aristotle—must go beyond purely physical exchange of matter and energy, which also provides another argument in favor of the irreducibility of emergents and the DC they exercise.

3.3. Downward Causation and Teleology

In addition to the hylomorphic and efficient causal aspects of DC, the teleological character of dispositions proper for emergents—"pointing" or being "directed" toward their kind-specific manifestations and described as their "physical" or "natural intentionality"—becomes one more argument in favor of DC's irreducible character and one more crucial aspect explaining its nature.

In other words, we can analyze DC in terms of the new types/aspects of teleology, namely, tendencies to realize new potencies characteristic of a given emergent entity/organism or dynamic process. Because these tendencies are crucial for the future development of an organism (or future states of a nonliving entity or a dynamic process), final cause becomes an indispensable and irreducible aspect of DC.

3.4. Downward Causation and Chance

Finally, when trying to redefine the meaning of DC, we need to ask the question about the probability factor of causal occurrences in nature. Following Aristotle's teaching on this subject (summarized in section 1.6 above), although we find chance ontologically real, we see it merely as a *per accidens* cause, that is, strictly speaking, not a cause of anything by itself, unless it is related to the *per se* causes engaged in a given causal situation.

Consequently, I do not suggest regarding chance as one more crucial aspect of DC. Rather, I suggest treating it as an important feature of many causal situations engaging entities/properties that show DC, a feature that needs to be taken into account in both speculative investigations and practical applications of causal theories.

3.5. Downward Causation and Interrelatedness of Causes

Our reflection on the hylomorphic, efficient, and teleological causal aspects of emergents, together with a reference to the chance aspects of their nature, both shows the irreducibility and explains the very character of DC. However, one might legitimately ask whether such an explanation does not mean, in fact, explaining DC away. After all, why would we need a new and separate causal/descriptional category to express the reality explainable already within the Aristotelian plural notion of causation?

To answer this question, we need to go back to the phenomenon of the interrelatedness of causes, mentioned already in section 1.5 of this chapter. We have seen there, first and most importantly, both Aristotle and Aquinas noticing the intrinsic relationship between primary matter and substantial form. Both of them mentioned the relation among the formal, the efficient, and the final causes as well. In addition, Aquinas paid attention to the relation between formal and final causation. Moreover, both philosophers seemed to emphasize that in dealing with concrete examples of causal dependency, we need to refer to all four types of causation, even if we may find it difficult to specify one or more of them in particular situations. Finally, we found important one more level of causal interrelatedness expressed in the idea of the virtual presence of simple elements in complex (mixed) substances, introduced by Aristotle and further developed by Aquinas. It helps us understand how sets of causal dispositions proper for simple elements are both preserved and altered in substantial changes leading to the origin of new substantial wholes.

In this context, it becomes apparent that the analysis of DC in terms of Aristotle's four causes and his reflection on chance does not make the very term "DC" redundant nor spurious. Its distinctiveness and explanatory power is preserved in its emphasis on and expression of the many ways in which all four types of causation listed by Aristotle are actively interrelated. Because their interdependence is characteristic and unique for each particular emergent complex entity/system/organism (in a given environment), it is possible to argue that our use of the term "DC" is still legitimate as an expression of this uniqueness and particularity.

4. ARISTOTELIANISM AND THE DYNAMICAL DEPTH ACCOUNT OF EMERGENCE

As the last step taken in the first part of the project, I will now evaluate Deacon's account of EM in reference to both the classical and the new Aristotelianism. At the end of the first chapter, I mentioned several questions concerning Deacon's proposition of expanding the notion of causation in his dynamical depth model of EM, in reference to Aristotle's list of causes. The main objections seem to refer to his

(a) reinterpretation of the nature of formal causation in terms of "functions of the geometric properties of a probability space" and his apparent ignorance of the importance of the material cause and its relation to formal causation in Aristotle's account;
(b) introduction of the idea of causation of absences; and
(c) argument in favor of the EM of teleological phenomena from nonteleological ones.

In order to answer these objections, we must first pay attention to one of the most crucial aspects of Deacon's theory, which usually goes unnoticed among its commentators.

4.1. Spontaneity of Orthograde and Contragrade Changes

It seems that what matters most in the dynamical depth model of EM is an interplay between the regularity and spontaneity of orthograde (physical) changes and their juxtaposition in contragrade changes. It is precisely their reversal that brings new constraints, and thus becomes a decisive attribute of emergent transitions at various levels of EM (the "engine" of EM). Though appealing and promising, the whole theory begs the question concerning the very source of both spontaneous and forced changes and the behavior of entities and processes that are subject to these changes.

I want to suggest that this query finds an answer, first, in dispositional metaphysics and its emphasis on the reality of natural potencies and tendencies of entities, initiating causal changes and participating in dynamic processes. The behavior and activity of entities is a function of manifestations of their kind-specific dispositional properties. The source of all dynamics in the dynamical depth model of EM is, thus, located in natural potencies characteristic of all natural beings, including particles, molecules, substances, more complex inorganic structures, and living and conscious organisms. It cannot be attributed to or regarded as an outcome of the laws of thermodynamics and gradients of entropy, which themselves need an explanation (again, they are best explained in terms of dependencies between natural entities and processes).

4.2. Dynamical Depth and Hylomorphism

An acknowledgment of the reality of active and passive powers, characteristic for all entities/processes in nature and providing for the spontaneity of the orthograde and contragrade changes in Deacon's account of EM, sends us back to classical Aristotelianism and the question of hylomorphism, which we found to be the best grounding principle for dispositional properties. Interestingly, we have seen Deacon reintroducing a formal type of causation, while redefining it in terms of the geometric properties of a probability space, namely, as a function of the behavior of the constituents of dynamic systems. Moreover, I have also mentioned his description of the whole-part constraints, characteristic of those systems, as formative in their nature. Speaking about matter, Deacon appears to understand it simply as physical/tangible stuff out of which things are made.[85] Such a reinterpretation of the two foundational and complementary principles of Aristotelian metaphysics can be classified as a dynamic version of the mereological and structural neo-hylomorphism, which sees form as a dynamic structuralizing principle that organizes the whole and sees matter as consisting of building materials or elements that are being structured.[86]

Although it may look appealing from the point of view of natural science, concentrating primarily on the quantifiable aspects of dynamic and living systems, this new version of hylomorphism raises important doubts and questions when analyzed within the discourse of metaphysics and ontology. Its redefinition of form as the principle structuralizing material parts, entering into dynamical interactions, seems to suggest that an emergent substantial whole is just a temporal and spatial (re)organization of a given set of building blocks, whose efficient activity and reactivity is not radically changed but merely limited by absential constraints, proper to this particular dynamic whole. Indeed, Deacon's suggestion of the close tie between formal and efficient causation, which replaces the original Aristotelian emphasis on the complementarity of primary matter and substantial form, seems to support the mereological and structural reinterpretation of hylomorphism.[87] Consequently, lacking a clear identification of the principles of potentiality and actuality, the theory in question does not seem to offer a satisfactory explanation of the origin and the irreducible character of the properties of emergent entities/dynamic systems. Their occurrence seems to have a merely epiphenomenal or supervenient character.

In this context, while appreciating Deacon's recognition of the importance of the formal cause in explaining the nature of emergents, I want to argue in favor of Aristotle's classical hylomorphism and his understanding of the material and formal principles. I claim that his metaphysics provides a more suitable ground for an explanation of the reality of active and passive dispositions, which characterize all entities/processes in nature and provide for the spontaneous character of the orthograde and contragrade changes in Deacon's model of EM. The clear explanation Aristotle's hylomorphism gives of the principles of potentiality and actuality and the clear distinction he draws between them—calling them primary matter (with the idea of its proper disposition to receive a given form) and substantial form—together with the idea of virtual presence, allow us to speak about the reality of substantial changes on the ladder of the growing complexity of dynamic entities/processes, without denying the continuity of their parts. Hence, I think that by adopting the hylomorphic understanding of formal and material causation, Deacon's project will gain a solid metaphysical foundation for its otherwise adequate and intriguing scientific analysis of the growing complexity in nature.

4.3. Dynamical Depth and the Role of Absences

The biggest point of controversy in Deacon's account of EM is the role he attributes to absences (unrealized possibilities) of dynamic systems. I agree with his assertion that the reduced degrees of freedom that characterize each of these systems tell us something important about the ontology of the world around us. It is true that by observing dynamical/ processual occurrences in nature, we can infer much about their character in reference to the way in which they are limited and constrained. Besides, even the more stable objects of our sensory experience are constrained in many ways, which becomes crucial for our perception and description of their nature.

What is peculiar about Deacon's project, however, is not only his describing absences as related to material constituents of reality—constituents that are actual and extended in space and time (features depicted in reference to his rhetorical device of the hole at the wheel's hub)—but also attributing to them causal efficacy (see above, chapter 1, section 5.3.3). Such an argument obviously raises the question concerning the nature of the alleged causality of absences. It is not entirely clear whether it should be treated as an efficient, formal, final, or maybe a new and unique type of causation.[88]

Addressing this problem from the position of Aristotle's metaphysics and philosophy of nature, we need to mention that both in the *Physics* and in the *Metaphysics*—while explaining the fact that one and the same thing may be the cause of contrary results—he gives an example of the captain (pilot) whose absence becomes the cause of the sinking of the ship.[89] What we are dealing with in Aristotle's example, however, is a complex situation in which the structure of the ship and its crew was set up in such a way that in case of the absence of the captain he would be "missed," and thus might be described as a "missing" cause. Nevertheless, strictly speaking, his absence would not be a positive cause of anything. It may indeed contribute to the changed dynamics of the whole causal situation, that is, the operation of all other causes active in it. But it is precisely the changed activity of these causes that would bring a causal change, not the lack of the presence of the captain.[90]

Apart from the example of the ship and its captain, one may try to define incompleteness and absences as a privation of form. But although it is a principle of change for Aristotle—as the change proceeds from it—privation is never an actual cause, for nothing comes from nothing. "Whatever comes to be is always complex. There is, on the one hand, (a) something which comes into existence, and again (b) something which becomes that—the latter (b) in two senses, either the subject or the opposite."[91]

By "the opposite" Aristotle means a mode of nonbeing. Hence, when clay is shaped into a ball, the round comes from the nonround or the privation of roundness. Still, nothing comes from nothing, since the round comes from the nonround, not *per se*, but only *per accidens*, insofar as the nonround is not (simply) nonround but is clay. Clay in this case is the principle of potency from which the round comes to be *per se* (insofar as it is potential) and not merely *per accidens*. In other words, we can predicate causation about privations only analogically (*per accidens*), bearing in mind that as forms of nonbeing, they are defined in terms of deficiencies, gaps, imperfections, and so on, in actualities, which are *per se* sources of causal interactions.

Applied to Deacon's view of the role of absences, this analysis helps us realize that constraints, as such, are always functions of limitations introduced either by a mutual influence of the participants of two (or more) reciprocal processes (e.g., morphodynamic processes in autogenesis) or in a response of one and the same dynamic system to external energetic gradients. In other words, they are intrinsically related to concrete entities and processes, the nature of which determines the possible degrees of freedom of a given thermodynamic system in the first place. Consequently, any change of the degrees of freedom of the dynamic system in question follows the change (accidental or substantial) in the operation or in the very nature of entities that ground it.

Therefore, although I do agree with Deacon that what is not present matters and that constraints are both real and informative about the processes and concrete entities they are related to, I maintain that as such, they are effects of causal interactions rather than themselves causes.

4.4. Dynamical Depth and the Nature of Teleology

Another important point of controversy in Deacon's dynamical depth model of ontological EM is his treatment of final causation. On the one hand, we have seen him arguing in favor of the strategic meaning of teleology, which is the source of what he calls a primitive "self" of an autogen. Needless to say, Deacon's autogenesis theory of life's origin becomes an important development of Aristotle's philosophy of nature, which simply assumed eternity of life. Today we postulate that life emerged out of nonliving systems, and many theorists and experimentalists are trying to explain and reproduce this process. Deacon's great advantage over the alternative propositions of numerous theorists of life's origin is his emphasis on the role of teleology.

On the other hand, however, Deacon claims that teleology emerges from mere physico-dynamic processes, characterized by Aristotle's notion of efficient causation, and is specific only to the highest levels of dynamical depth. Hence, unlike Aristotle—for whom teleology is deeply related to formal and efficient causes of each natural entity and organism, but never emerges from them—Deacon strives to prove the reality and the irreducible character of "ententional phenomena" without introducing the classical concept of substantial form.[92]

Going back to the problem of the source of spontaneity of orthograde processes at the bottom level of each stage of Deacon's dynamical depth, discussed above in section 4.1, I have suggested accepting dispositional metaphysics with its emphasis on irreducible kind-specific dispositional properties, which are characteristic for every nonliving and living being and decisive for the nature of all causal processes they enter. We know by now that the same metaphysics acknowledges that each dispositional property is related to its proper manifestation under suppositional necessity. In other words, the property will be manifested in favorable and required circumstances, provided that no impediments prevent its occurrence. I have mentioned that dispositional metaphysicians see here a natural tendency of dispositional properties to be manifested. They call this tendency the "physical" or the "natural intentionality" and find it a universal characteristic of all beings, living and nonliving.

I want to argue that this neo-Aristotelian revival of teleology, discovered and described at all levels of complexity of matter, is relevant to Deacon's model of EM, in which he frequently emphasizes natural tendencies of processes at different levels of dynamical depth to act in a specific way. It becomes clear to me that teleology is present at all stages of dynamical depth and cannot be limited only to its highest level. Naturally, I agree with Deacon that with the EM of an autocell a new type of teleology can be traced and described, which is proper to autogenesis and is unlike the teleology characteristic of all lower dynamic systems. Hence, the highest level of dynamical depth can and should still be called teleodynamics, due to the novelty of its finality, providing for the primitive "self." But to say this unique type of teleology emerges from mere mechanical physico-dynamic systems is rather misleading. Contemporary dispositional metaphysics, following the classical teaching of Aristotle, finds teleology to be an irreducible type of feature (or type of causation), inextricably related to dispositions of concrete beings.

This suggestion concerning the character and the role of teleology in Deacon's dynamical depth model of ontological EM closes the investigation pursued in the present chapter and the first part of the entire project. The material gathered and discussed here provides a necessary scientific and philosophical background for the study of theological implications of the theory of EM for our understanding of the nature of divine action, to which we shall now turn.

PART 2

God's Action in Emergence

God, as Creator, endows nature from the beginning with existence and with capacities and dynamics to evolve the rich diversity of remarkable structures and organisms which have emerged in the course of cosmic history.

—William R. Stoeger, *Reduction and Emergence*

If there is reason to retrieve the notion of substantial form as an immanent principle in natural things to explain the existence and causality of whole entities as distinct from their parts, there is surely justification for theologians to use the idea of formal causality in their discussion of divine action. We might begin with the notion of form as an extrinsic exemplar cause.

—Michael J. Dodds, *Unlocking Divine Action*

Unlike social and political revolutions, usually marked by the radicalism of the views and actions of their participants and leaders, the panentheistic turn in theology of the twentieth century has been described as a "quiet revolution."[1] Rooted in some ancient Christian and non-Christian traditions, panentheism reappeared in German Idealism and gained much popularity in the last century, especially in the context of the process metaphysics of Whitehead and Hartshorne. Its core definition statement that the world is in God, who, nonetheless, is more than the world, is seen by its advocates as forging a middle path between classical theism and pantheism.² At first glance, this assertion does not seem to add much to the long tradition of theological reflection on divine immanence, the God-world relationship, and the nature of divine action. And yet today a host of theologians either openly identify themselves or are frequently identified by others as panentheists, which proves the "revolution" is not "small-scale." Quite the contrary, many followers of panentheism believe with Michael W. Brierley that panentheism "subverts the priorities of classical theism, and thereby undercuts its edifice and structure. It challenges classical theism's imperium, and places the doctrine of God in ferment."³ What they see at the heart of the panentheistic revolution is its emphasis on the reciprocal character of the God-world relationship, which has far-reaching consequences for the classical understanding of the attributes of God's nature.⁴

Among the many areas of theology which willingly apply the panentheistic metaphor, the dialogue between theology and natural science plays an important role. A number of its influential participants find the principles of panentheism fitting their understanding of the God-world relation and of divine action in the world. Moreover, panentheism serves as a background for the most developed analysis of the theological implications of the theory of EM, offered by Peacocke. In the first two parts of chapter 3 I will investigate the origin and some crucial stages of the historical development of philosophical panentheism, as well as an attempt of recent theology at providing a more precise explanation of panentheism's main objectives. The following sections of the same chapter will concentrate on the role of panentheism in the science/theology dialogue and the theological analysis of the theory of EM.

Despite the popularity of the contemporary version of panentheism, a more rigorous (scientifically and philosophically inspired) analysis of the category of divine containment of the world—that is, of the key preposition "in" (*en*) in pan*en*theism, reported in chapter 4—will show its ambiguity and lack of precision. This fact makes the definition of the core panentheistic metaphor one of a family resemblance, accommodating a variety of over a dozen interpretations. The inquiry pursued in chapter 4 shows that the contribution of panentheists working in the science/theology dialogue does not solve the problem of the precise definition of the term. Followed by a number of further aspects of the general critical evaluation of panentheism, my investigation, also offered in chapter 4, will conclude with a critique of emergentist panentheism, defined both in terms of the whole-part (mereological) interpretation of EM and in terms of the concept of DC.

My own constructive proposal of an understanding of the God-world relationship and of divine action is the main point of interest of chapter 5. It is based on EM as indicating the need for interpreting the reality of entities and processes in terms of a broader notion of causation, proposed by Aristotle. At this point, I will refer to both the classical DC-based and Deacon's dynamical depth accounts of ontological EM, analyzed in the context of both the classical and the new Aristotelianism of dispositional metaphysics. I will show how this interpretation of EM, referring to formal and final causation, fits with Aquinas's view of the primary and principal causality of God and the secondary and instrumental causation of creatures, as well as his concept of God as the final end (*telos*) and the Creator of all forms, acting through his divine ideas as exemplar causes of living and nonliving beings. I will also look at the role of nonbeing in Aquinas's theory of distinction and derivation of the many from the one and compare it to Deacon's causality of absences. This investigation—followed by the final list of advantages of the Thomistic interpretation of God's action in/through EM over emergentist panentheism—will conclude the project.

CHAPTER 3

Panentheism and Emergence in the Science-Theology Dialogue

The central point of my interest in this chapter is an exposition of the contemporary version of panentheism in the context of the science-theology dialogue and an analysis of emergentist panentheism. In order to better understand the main presuppositions of emergentist panentheism, however, I will explore at first (section 1) some crucial points of the historical development of philosophical panentheism. My investigation here is not intended to be complete nor comprehensive, because that would take us far away from the main topic of our study. I refer those who would like to research the history of panentheism in more detail to excellent works by Charles Hartshorne and William Reese and by John Cooper.[1] My goal is to show the Platonic and Neoplatonic roots of panentheism, as well as panentheistic implications of German idealism and of the contemporary philosophical theism of Whitehead and Hartshorne. I hope this will provide the necessary background for investigating some of the most recent theological clarifications of the meaning of the core panentheistic ontological claim (section 2), as well as for evaluating the panentheistic foundations of the writings of the chosen participants of the dialogue between theology and science (they usually refer—either directly or indirectly—to German idealism and/or process thought as an inspiration of their theistic positions) (section 3). The awareness of the Platonic roots of panentheism should also prove helpful in contrasting it, in the remaining chapters 4 and 5, with the classical theism of Aquinas, which is built predominantly on the philosophy of Aristotle (with several important references to Platonic concepts). The analysis offered in the present chapter will lead eventually (section 4) to exploration of the role of panentheism in theological reflection on the theory of EM, which corresponds to and further develops the main theme of my project.

1. HISTORICAL FACETS OF PHILOSOPHICAL PANENTHEISM

The doctrine of panentheism has a long tradition. It is true that a rich diversity of panentheistic views has developed in the last two and a half centuries, primarily in Christian traditions and in response to scientific thought. However, an attempt to conceive deity as neither radically isolated from the world nor identical with it has been popular across philosophical, theological, and practical religious traditions in the entire history of human thought.

1.1. Panentheistic Aspects of Ancient Religious Traditions

In their detailed study of various types of theism over the centuries and across various religious traditions, Hartshorne and Reese find quasi-panentheistic themes as early as in the hymns of Akhenaton, "the first monotheist" pharaoh in Egypt (ca. 1375–1358 BC). Although he saw the supreme reality of the sun god as not temporal, in the sense of earthly beings, Akhenaton and his subjects did not praise an abstract object (being, eternity, absoluteness, or infinity), but the supremely living, loving, wise, intelligent, and purposive maker of all—engaged in mundane reality.

These first signs of the two poles in a deity find a resemblance, according to Hartshorne and Reese, in the seemingly pantheistic God of Hindu scriptures (ca. 1000 BC), which is understood as devoid of temporality and spatiality. They find in the Atharva-Vedic *Hymn to Time* a primitive and mythical—yet clearly stated—account of the doctrine that becoming is the inclusive side of the primordial duality of being-becoming. This leads to a conclusion that both eternal and temporal components make up the reality of God, who—being the father of all beings (worlds)—became their son. Moreover, multiple intimations of panentheism are claimed to be found in the *Upanishads*, even if such an interpretation of their theological commitment is not predominant.[2]

The dipolar view of the Supreme Being is even clearer in the Taoist (fourth century BC) doctrine of truth, which is found not in one but in both sides of the ultimate contraries taken together. These are the contrarieties of the active and the passive, existence and nonexistence, the manifest and the unmanifest. Once again, Hartshorne and Reese find in the Taoist praise of the passive, responsive, receptive, undominating, yielding, tender, and pliant an anticipation of divine immanence, complementary to divine transcendence—a side of the Supreme Being which, in their opinion, has been overlooked in classical theism.[3]

1.2. Plato and Neoplatonism

The tendency of Taoism toward dialectics and emanationism finds its further development in the Neoplatonism of Plotinus and Proclus. Because their doctrine depends heavily on the philosophy and theology of Plato, we must refer to his thought first. Even if it does not qualify as a version of panentheism, his doctrine of the World-Soul has been considered by many to be an accurate analogy of the God-world relationship.

What remains crucial for Plato is that universal, eternal, and immutable Ideas are related to one another and form a coherent, unified system. This is possible because they are generated by and reflect the Good, or the Form of all Forms. Another expression of the uniting principle in Platonism, described in *Timaeus*, is a particular Form used by the Demiurge to make the cosmos as a whole—the Living Being, the prototype of all living things, including gods and humans. Thus, we can say with Plato that "the Cosmos . . . resembles most closely that Living Creature of which all other living creatures, severally and generically, are portions."[4] Cooper notes that the most important part of the cosmos as a living being in Platonism is its Soul, which the Demiurge "stretched throughout the whole of it, and therewith He enveloped also [with the Soul] the exterior of its body."[5]

Although this assertion may suggest that for Plato the physical world is "in" the Soul—a point conducive to panentheism—we need to remember that the theological status of the World-Soul in Plato's writings is highly ambiguous. Even if *Timaeus* 34a–c refers to the World-Soul as a god, this god was still planned and generated by the Demiurge and remains one of the many created beings Plato calls "gods." Moreover, the analysis of the creation story in *Timaeus* leaves no doubt that the material world, as such, does not abide in the Demiurge, the Good, or divine Reason. It is true that all things "participate" in ideal Forms, but these Forms are exemplary patterns reflected by earthly beings, which are not immanent in them.

Nevertheless, even if Plato himself cannot be classified as a panentheist, his thought has at least two important aspects opening the way to the principles of panentheism. The concept of the Good as the ground and source of all Ideas introduces a top-down emanationism of Forms, while the bottom-up way to the knowledge of the ultimate unity of all Forms—the Good itself—leads through dialectic, which enables us to distinguish and isolate the essential nature of Goodness from all other Forms. As we will see, both emanationism and dialectic are crucial for panentheism. They were further developed in Neoplatonism, to which we now turn.[6]

The first important point in Plotinus's (AD 204–70) works is his emphasis on the radical transcendence of Plato's Good, or the One Form—the single principle attained by dialectic reasoning. He sees it as utterly simple—with no parts, aspects, attributes, or powers—and absolutely infinite, not just in space, time, or number but also in its ultimate power of producing itself and everything else. At the same time, however, the One is not only totally transcendent, but also radically immanent, because it includes or contains all things that emanate from it—without diminishing its overflowing fullness and perfection.[7] In other words, creation is regarded as a natural phase of the divine life, and *creatio ex nihilo* amounts, in fact, to *creatio ex Deo*. Plotinus develops emanationism, formulating its hierarchy leading from the One, which generates the Intellect, which in turn produces the Soul, that is, the World-Soul animating the whole universe—the Life Principle of the physical cosmos. Consequently, we may state together with Cooper that for Plotinus "reality is a hierarchical order of different levels of being that extends from the One on top to the unformed matter of the physical universe on the bottom. Everything comes from the One and seeks to return to the One."[8]

Another emphasis in Plotinus is on the inevitable necessity of the divine emanation. This principle, together with the recognition of the strong ontological distinction between the One and the world, meets the requirements of panentheism and becomes, in fact, the foundation of its classical version. Plotinus's World-Soul is divine, includes in it the created world, and yet is distinct from and transcendent of it. Referring to the Platonic doctrine of the World-Soul discussed above, Plotinus states that "Plato rightly does not put the soul in the body when he is speaking of the universe, but the body in the soul." At the same time, he adds, we cannot ignore the fact that "there is a part of the soul in which the body is and part in which there is no body."[9] This is panentheism in a pure form. In his final remarks on Plotinus, Cooper claims that his panentheistic emanationism overcomes Plato's ultimate dualism, while his emphasis on the world being ontologically implicit in and a part of God radically changes Plato's doctrine of participation: "For Plato, things in the world participate in the divine Mind by reflecting or being patterned according to the Forms. But worldly things do not 'take part' in the ideal existence of the divine Mind. For Plotinus, *participation* means not only reflecting the Divine but also 'taking part' in it—directly in the World-Soul, mediately in the Intellect, and ultimately in the One. This is the meaning of *participation* that is characteristic of much panentheism: being part of God."[10]

An important supplement to Plotinus's emanationism is an elaboration and development of dialectics offered by another Neoplatonist philosopher, Proclus (410–485). He speaks about the two opposing principles—Infinity and Delimitation—proceeding from the One and interacting to constitute Being, Life, and Intelligence—the three "gods" or eternal principles generating subsequent emanations and constituting the universe. Beginning from Infinity and Delimitation, each level of reality is generated dialectically. Each is both the same as and different from its source, united with it and yet separated from it. This process, described in terms that anticipate the "thesis," "antithesis," and "synthesis" of the later proponents of dialectics, repeats itself down the Great Chain of Being and back. What remains crucial, however, is that for both Proclus and Plotinus the One itself is beyond the triunifying dynamics of dialectics. It was only in the philosophy of God developed by German idealists that dialectics was projected into the inner life of God.[11]

The Neoplatonic tradition found a particularly positive response in Pseudo-Dionysius's method of theology, based on the dialectical triad of the *via positiva*, the *via negativa*, and the *via eminentiae*, which was in turn embraced by many Christian philosophers, theologians, and mystics of the Middle Ages. Although the panentheistic overtones of his thought were less popular in the mainstream tradition of classical theism—represented by Anselm (1033–1109), Thomas Aquinas, and Duns Scotus (1266–1308)—they were readily accepted, strengthened, and transmitted by John Scotus Eriugena (810–77), Meister Eckhart (ca. 1260–1327), Nicolas of Cusa (1401–64), and Jakob Böhme (1575–1624). Böhme boldly declared—in contrast to the ancient Neoplatonists—that dialectics is at the very center of the intrinsic dynamics of God himself, eternally generating the Trinity and temporally generating the world. Böhme's ideas gave rise to the dialectical panentheism of Fichte, Krause, Schelling, and Hegel. Their thought is the subject of the following section.

1.3. German Idealism

The origin of German idealism is related to the "renaissance" of Spinozism at the end of the eighteenth century, when more and more radical changes were being introduced in all the sciences. As natural science and philosophy gradually assumed the place and role of theology, deistic, pantheistic, and materialistic doctrines became widely and strongly approved. Rationalism, empiricism, and sensualism became more prominent as well. Under these circumstances, Spinozism—regarded by many as the enemy of religion, morality, and civil order—was rediscovered as an important philosophical insight. Spinoza's pantheistic idea of *Deus sive Natura* was now found by many to be a necessary step on the way from substance to subject, from the God of classical theism—whom many saw as impersonal, self-contained, ontologically independent, and unrelated to creatures—to the concept of God as creative process, the Idea developing in existence and manifesting itself in Nature, God involving himself in the world and the world in himself. As thought shifted from abstraction and lack of concreteness to individualization and subjectivity, panentheism gradually replaced pantheism—leading to the panentheistic revolution in modern theology. We shall now analyze the thought of its founding fathers.[12]

1.3.1. Johann Fichte
As the first representative of German idealism whose thought has a panentheistic "flavor" we should mention Johann Gottlieb Fichte (1762–1814). Inspired by the Kantian debate concerning the possibility of

knowledge, the character of subject-object relation, and the nature and number of Transcendental Egos, Fichte developed the idea of finite subjects existing within the Infinite Subject—an "Absolute I," a ground of all experience, the very source of all reality, continuously creating it. He found irrelevant the postulate of soul and nature understood as things in themselves and beyond experience and claimed the "Absolute I" posits the world in relation to itself within experience.[13]

On the one hand, as notes Cooper, in "On the Foundation of Our Belief in a Divine Government of the World" (1798) Fichte argued that Ego, the Absolute, must also be Infinite, because the rational-moral activity that posits the world of individual egos and finite things cannot itself be relative and finite. Moreover, as Absolute Ego, God is not personal since this requires responsiveness and interrelatedness in reference to other persons, which would make him finite. On the other hand, God is identical with the eternal rational-moral activity that is immanent and manifests itself in human ego. He is not only the one infinite reason but also the one infinite will that necessarily manifests itself in the plurality of finite beings.[14] Recognition of this truth inspires Fichte to express his reflection which is clearly panentheistic in its understanding of the God-world relationship:

> This Will unites me with itself; it unites me with all finite beings like me and is the general mediator between all of us. That is the great secret of the invisible world and its fundamental law so far as it is *a world or a system of a number of individual wills: that union and direct interaction of a number of autonomous and independent wills with each other.* . . . All our life is its [Will's] life. We are in Its hand and remain there, and no one can tear us out of it. We are eternal because It is. Sublime living Will, which no one can name and no concept encompass, well may I elevate my mind to you; for you and I are not separate. Your voice sounds in me and mine resounds in you.[15]

Although it is not an entity, substance, or thing, but rather a rational-moral activity, the "Absolute I" ("Sublime living Will") resembles Spinoza's "substance." They can both be equated with God, who is immanently present in reality and remains in causal relation with Spinoza's modes of the one substance, or in Fichte's term, with each "relative I" (individual will). This causality is a dialectic process in which Ego posits itself as non-Ego to achieve a synthesis of Ego and non-Ego. Thus, notes Cooper, "Fichte is the first one to use the terms *thesis, antithesis,* and *synthesis*. He hereby appropriates traditional Neoplatonic dialectic and passes it to Schelling and Hegel." At the same time, Fichte modifies Spinoza when he ontologically distinguishes finite egos from the Absolute Ego. The former are more than just modes of the latter. They form an objective world of entities having their own metaphysical identity. The process of "Absolute I" positing multiple "relative I" becomes a creative act, a cause of existence of concrete beings. At the same time, however, because Fichte's "Absolute I" is a concrete subject, it cannot be truly universal. Thus, his idealism is regarded as "subjective idealism" and is opposed to Schelling's "objective idealism," which I shall analyze next. But first I must mention another German idealist, who is known to have coined the very term "panentheism."[16]

1.3.2. Karl Krause
In 1828 Karl Christian Friedrich Krause (1781–1832), inspired by Fichte's modification of the pantheism of Spinoza and Fichte's distinction between the "Absolute I" and "relative I," introduced the term "panentheism"— "Panentheismus, oder Allinngottlehre."[17] His point of departure was philosophy of science (scientific knowledge), where he wanted to establish a system of science-as-a-whole, that is, a monistic and universal approach that unites metaphysical grounding with panentheism in the philosophy of religion, with holism in philosophy of science, with ultimate explanation in epistemology, and panpsychism in philosophy of mind. He claimed that "science has unity of its principle only if the principle of recognition is nothing over and above the principle of being."[18] In other words, "If science is possible, then it has to entail intellectual intuition of the one principle, and everything which science recognizes has to be recognized through the principle."[19] Krause calls this principle "God," "Orwesen," or "the Absolute." How do we arrive at its recognition?

According to Krause we can distinguish two parts of science. The first—the "analytic-ascending" part of science—enables us, through transcendental reflection on the condition of the human subject, to discover the one principle of being and recognition. It is within the analytic-ascending part of science that we discern the set of basic transcendental categories according to which our cognition is structured. These are (1) material categories, which concern the fundamental properties of the object of our understanding (unity, wholeness, itselfness, and selfhood [when we reflect upon ourselves]), and (2) formal categories, which concern the fundamental mode of givenness of these objects (positivity, directness, composedness, and comprehension [when we reflect upon ourselves]). Because we perceive and understand objects in their unity (unity of properties), Krause introduces one more category, the category of unity of material and formal categories of an entity.[20] Moreover, he sees cognition not as a binary relation between the subject and the object, but rather a triadic relation which includes a principle that—respecting the mutual independence of subject and object—unites them. For Krause this principle is infinite.

The second—the "synthetic-descending" part of science—begins with the intellectual intuition and enables us to deduce the entire system of science stemming from it. It helps us understand science as explicating the categorical unity of the one and first principle of being and cognition. Consequently, just as the analytic-ascending part of science enables us to consider an entity "in itself"—that is, its internal constitution—the synthetic-descending part of science enables us to consider it as a unity (whole).[21] Most importantly, it also enables us to perceive the relation between God and the world as between the whole and its parts, such that the whole is more than the sum of its parts, whereas the parts are not external to the whole. This naturally opens the way to panentheism.[22]

Approaching the same crucial point from another perspective, we may conceive the Absolute "as such," which helps us realize that nothing can be "outside of" or "next to" or "in addition to" it, for otherwise it would not be the first principle, ultimate unity, and ground of everything. Hence, the world must be "in" this first principle, sharing its categorical essence in such a way that no finite entity becomes its limitation. At the same time, considered "in itself" the Absolute must be distinguished from the world as a whole is distinguished from its internal constitution and parts. As Göcke notes,

> Krause calls God insofar as God is distinguished from what he is principle of "Urwesen" and insofar as we consider the whole together with its internal constitution, that is, insofar as we consider

the essential unity between the principle and what it is principle of: "Orwesen." If we conceive of God as Orwesen, then the world is internal to God in the same way in which the parts are in the whole. If we conceive of God as Urwesen, God is outside of the world in the same way in which the whole is something over and above its parts."[23]

In Krause's own words, "Since in the intellectual intuition of God we find that Orwesen, as the One, also in itself and through itself, below itself is everything, also everything finite, we have to assert that the One in itself and through itself is the All, and since in the intuition of God we recognize that God is everything in Himself, below Himself and through Himself, it would not be false to call science panentheism."[24]

To enable his readers to better grasp his philosophical ideas, Krause developed a heuristic device, a diagram that strives to depict his notion of the dialectical moves of determination (unification), difference, and higher-level unity. Göcke uses it several times in his book in reference to various aspects of Krause's thought. I present it here solely in reference to Krause's notion of panentheism (with Göcke's description); see figure 3.1.

Göcke is right to note that Krause's distinction between *Orwesen* and *Urwesen* arguably is a predecessor of Hartshorne's dipolar concept of God, which I will discuss below.[25] At the same time, although it does refer to the logic of dialectics and applies it in philosophy of God, Krause's panentheism does not seem to put enough emphasis on the mutual influence of God on the world and vice versa, which becomes crucial for the more contemporary versions of this type of theism. Neither does it reflect upon the way in which God, the Absolute, changes in relation to the world. This aspect of the panentheistic position was developed in both the expressivist panentheism of Schelling and Hegel and in the more recent processual panentheism of Whitehead and Hartshorne. We will first investigate the former.

1.3.3. Friedrich Schelling
An important step in the progress of German idealism was made by Friedrich Wilhelm Josef von Schelling (1775–1854). Knowing and appreciating Fichte, he named and criticized some major shortcomings of his

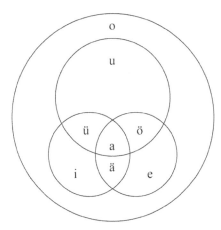

Figure 3.1. Krause's diagram depicting the panentheistic notion of the relation between God and the world

Göcke explains: "If we conceive of *Orwesen* as *o*, then the analysis of *Orwesen*, in itself, begins with the recognition that there is an opposition of essentialities *i* and *e*. Because *Orwesen* cannot be in any opposition, the opposition between *i* and *e* is an opposition in the essence of *Orwesen* itself. Since, however, because of the unity of the divine being, the *relata* of the opposition have to be united themselves, *ä*, and have to be united insofar as they constitute the intrinsic nature of *Orwesen*, there has to be a principle of unity, *u*, which Krause calls *Urwesen*, in virtue of which this is possible and actual. *Urwesen*, *u*, is both separated from *i*, *ä*, and *e* and united with each of them, *ü*, *a*, *ö*. Now because selfhood and wholeness are reason and nature [a conclusion from Krause's understanding of the transcendental constitution of the ego], because humanity is the union of reason and nature, and because the world is the union of nature, reason, and humanity, we can, in a first step, let *i* refer to reason, *e* to nature, and *ä* to humanity and then will be able to see that the world is an intrinsic determination of *Urwesen*. When, in a second step, we consider *Urwesen*, *u*, to be the principle in virtue of which the world exists, we immediately see that as the principle of the world *Urwesen* is both separated from and united with the world: *u*, *ü*, *a*, *ö*. All of this, even the distinction between *Urwesen* and the world, is an intrinsic and essential determination of the one divine being, *Orwesen*" (*Panentheism of Karl Christian Friedrich Krause*, 127–28 [I corrected some editorial mistakes in the quoted text]; the original diagram can be found in Krause, *Vorlesungen über das System*, 613).

system. Apart from the problem of the universality of the "Absolute I," Schelling notes that in Fichtean philosophy nature is seen merely as a field posited by the "Absolute I," passive rather than active. As a background of the human activity of positing that which is other and opposes us, nature becomes dependent on individual egos, which renders it inert and lifeless.

After studying mathematics, physics, and chemistry, Schelling developed a new concept of nature. He understood it as vital, dynamic, and polarized—existing apart from our consciousness of it and teleological in itself. He criticized the approaches of Kant and Newton as too mechanical and too heavily based on necessity. For Schelling the Absolute is Nature, a unifying principle defined in terms of an organism (a reason his idealism is classified as objective). It is a dynamic, dialectic process within which all things are products of an endless becoming and manifest the Absolute. Because both the subjective and the objective can be united in this flow of becoming, the problem of the relation between the ideal and the real can be solved. Nature becomes a self-reflecting organism. Thus, even if Schelling's system departs from orthodox Spinozism, it can be still classified as a type of monism. Copleston names it "a romantic and poetic spinozism."[26]

Schelling's theistic position can be classified as a modified panentheism, even if he himself defends it as a "true pantheism." In fact, according to Clayton, it was Schelling who in 1809—that is, almost two decades earlier than Krause—in his famous *Essay on Freedom* used for the first time the term "panentheism," although not as one word but as a composition of its three constitutive parts: "*Pan+en+theismus.*"[27] In reference to the Böhmian analysis of the essence of God, Schelling described three divine principles (potencies): (1) "the abyss," "emptiness," "non-being," "nothingness," "pure potentiality," "contingency," "the darkness in the heart of God," "the unconscious," "will," "the longing which the eternal One feels to give birth to itself," "the non-ground (*Ungrund*) which is the primal ground (*Urgrund*) of God's personal existence"; (2) "God's essence, a light of life shining in the dark depths," "the principle of being," "reason," "order," "positivity," "the principle of love"; and (3) actualizing love, which is responsible for the eternal, necessary, and self-generated synthesis of the other two principles.[28]

In reference to this dialectical sketch of the divine nature, Schelling emphasizes that God, dynamic and active in his essence, presents himself in human history. God and humans determine together the process of self-actualization. He goes further and says that God cannot exist without a free human action. At the same time, man exists not outside God but in God, and his activity belongs to God's life. To relieve the tension between the absolute identity of God on the one hand and divine and human freedom on the other, Schelling proposes a more personalist view of God. Eternal and immutable in his essence, God nevertheless grows, changes, and suffers in his existence. Personal in essence, he reveals himself in his actual existence and acquires personhood by developing with the world. The distinction between God's transcendent essence and his existence, understood as immanent actualization in human beings, becomes a dualist and dialectic reinterpretation of classical scholastic definitions. This is a new theism, that is, a modern and dynamic version of panentheism.[29]

However, although Schelling's philosophy of God was innovative and very forward-looking, it was Georg Wilhelm Friedrich Hegel (1770–1831) who won more popularity. He developed and improved Schelling's position in terms of explaining the relation of God and the world, and—what was his greatest achievement—he managed to unfold fully Fichte's and Schelling's still preliminary description and use of dialectics in philosophy of God.

1.3.4. Hegel's Dialectics

Copleston is right in saying that "if Fichte's emphasis on the Subject and Schelling's emphasis on the Object may be said to stand to one another as respectively thesis and antithesis, it was left to Hegel to attempt the synthesis."[30] Indeed, Hegel's philosophy brought a definite change from the philosophy of substance to the philosophy of subject. He understands reality as the dialectical self-actualization and manifestation of the Absolute Idea (Spirit, God) in and through the world, Nature, humanity, and philosophy. Moreover, in his thought the universe is no longer alien to a human being. We realize that the Absolute Idea becomes conscious of itself in us. Hegel regards this fact as an important achievement of his metaphysical system.[31]

Although radically new, Hegel's philosophy of the Idea was embedded in the thought of his predecessors. That he was aware of this we learn from his *Lectures on the History of Philosophy*, in which—beginning from antiquity—Hegel examines and criticizes great philosophers and philosophical schools. In the third volume he reflects on, among others, Spinoza, Fichte, and Schelling. Even if he paid tribute to Spinoza, saying, "To be a follower of Spinoza is the essential commencement of all Philosophy,"[32] and, "You are either a Spinozist or not a philosopher at all,"[33] Hegel did not hesitate to point out some essential shortcomings of Spinoza's philosophy:

1. *Method.* For Hegel the demonstrative method of geometry, applied in Spinoza's *Ethics*, is not proper for philosophy. He questions Spinoza's universal determinations such as substance, attribute, or mode. They are simply "assumed, not deduced, nor proved to be necessary."[34]
2. *Acosmism.* Unlike those who accused Spinoza of atheism, Hegel argues that in his system there is too much of God. If the world is merely a form of God, it has no true reality and is "cast into the abyss of the one identity."[35]
3. *Individuality.* There is no principle of subjectivity, individuality, and personality in Spinoza's philosophy, no explanation of how the modes of the one substance are produced, no place for self-consciousness in Being. Plurality and difference are an illusion.[36]

Hegel also recognizes the importance of the contribution of the Fichtean system. It emphasized that philosophy should be derived from one principle, in which all determinations are necessarily grounded. Nevertheless, this proposition also has its imperfections. Not only does the aforementioned subjective character of Fichte's "Absolute I" prevent it from being truly universal; his system does not reach the real unity of subject and object, of Ego and non-Ego.[37]

According to Hegel, it was Schelling—his classmate and former friend—who made an important advance upon the philosophy of Fichte. He "grasped the true as the concrete, as the unity of subjective and objective," and "pointed out in Nature the forms of Spirit."[38] Nevertheless, Hegel thought Schelling's absolute identity of Nature still left unanswered the question of the precise understanding of the relation between God and the world.

After criticizing the methodology of the predecessors, Hegel develops his own system and method. Taking dialectics from mathematics and logic, he endows it with an ontological significance—which becomes the secret of his revolution. Thus understood, dialectics becomes a compromise of subject and object. It brings together Fichte's ego ("Absolute I") and Schelling's creative energy of Nature. For dialectics is a path of negativity, a way of opposing the same and the other within unity. It brings together thinking and being in a mediation that recognizes distinctions and opposites to be reconciled. It synthesizes them and brings them into a unity.[39]

Undoubtedly, Hegel begins with Spinoza's substance, an independent and self-contained infinity. However, for him substance contains more than pure being that can be grasped intuitively. It cannot be isolated from individual beings. It is not an inert principle, but rather a cause, a ground and consequence, a thesis that passes out of necessity into an antithesis. It is in perpetual flux, its inner constancy is form-activity. This leads Andrew Reck to conclude that Hegel no longer needs the Greek-scholastic principle of analogy to explain the existence of independent beings in his system. For him, substantiality is a duality of constancy (sameness) and activity (differentiation). Substance is the Absolute, the Idea manifesting itself in the world. It expresses itself in accidents, which, unlike Spinoza's modes, are real and individual entities. Thus, we can see how dialectics changes the understanding of substance. Substantiality is now transcended in Subjectivity.[40]

The development of actuality is possible only through the way of negation. Hence, Spinoza's claim that "determination is negation" becomes a ground of Hegel's dialectics. Yet, unlike Spinoza, he gives negativity an affirmative character and reverses Spinoza's claim, saying that "negation is determination." Negativity is the way in which substance discovers its self-identity. Antithesis (difference, distinctness) is necessary on the way to synthesis: "Spinoza's substance becomes the dialectic Absolute Spirit, the infinite thought of all possible and actual objects."[41]

1.3.5. Hegel's Panentheism[42]

The analysis of Hegel's ontological dialectics leaves no doubts that he identifies the Absolute Spirit or the Absolute Idea with God. But what is the exact character of the relation of God to Nature in Hegel's system? He cannot be regarded as an orthodox classical theist for at least two reasons. His understanding of Absolute Substance unfolding itself in creation stands opposed to the classical view of divine attributes such as simplicity, infinity, and immutability. Moreover, if a dialectic manifestation of the Idea is necessary, God is not totally free in his act of creation—which contradicts one of the crucial assertions in classical creation theology. Hegel himself does not side with classical theism as he finds it—in reference to its opposing the finite and the infinite—trapped between the extremes of putting God altogether beyond nature and making him finite and therefore not God at all.

Knowing that he begins his philosophy with the notion of Spinoza's one substance, one may ask whether Hegel might be classified a pantheist. If this claim were true, Hegel would have to argue that either particular beings are not real—since in pantheism only God, the Idea, the Absolute Spirit, is real—or that God is merely the "inside" of the world. As far as the first assertion is considered, we know from Hegel's understanding of substance that he would reject this possibility. For he regards the emphasis on individuality of beings as one of the most important achievements of his philosophy. They do not vanish nor melt into the one Substance. Individual beings are real and external to God, although at the same time they are part of the dialectic process (namely, antithesis) and self-expression of his divine essence.[43]

If not the first, then maybe the second assertion might enable us to classify Hegel as a pantheist? Would he agree that God is "enclosed" within the world? Does Hegel's dialectic system preserve God's transcendence? The answer to this question is complex. On the one hand, unlike Spinoza, Hegel admitted and explained an element of contingency in Nature. Because individual beings are never perfect, mindless Nature cannot manifest the Absolute adequately and is only a stage on the way of the Idea's manifestation in the Spirit. Nature cannot exhaust or fully express God, which proves he is transcendent. Moreover, even Nature coming to self-consciousness in humans cannot entirely express the essence of the Spirit. This is another proof that the Subject (God) is transcendent. On the other hand, however, the question still remains whether all this depicts a real ontological difference between God and creation. If the answer were to be positive, we could defend Hegel's position against being classified as pantheistic.[44]

Trying to deal with this question, Robert Whittemore distinguishes between two meanings of "transcendence." If the term implies that God is not only above but also (ontologically) radically separated from the world—prior, independent, and unrelated to it (the idea of a God who is not related to the world in any possible way is foreign to the classical theism)—then Hegel fails in defending God's transcendence. However, Whittemore claims that "it makes sense to say of any whole that it *surpasses* or *excels* its parts; it *transcends* them even if it is not apart from them. To *encompass* something, to *absorb* it within one's being is surely to transcend it in a legitimate sense of the term; the absorber must *transcend* the absorbed."[45] Whittemore notes that according to this alternative definition of "transcendence," the word is not connotatively synonymous with "independence," and, applied to Hegel's philosophy of God, it enables us to state that he manages to safeguard God's otherness. However, the problem of this descriptive definition is that—in the case of God-world relationship—it loses (or does away with) the perspective of an ontological difference between God and creation. Although the relation of encompassing or absorbing suggests that both absorber and absorbed belong to the same ontological category, the definition of divine transcendence implies necessarily a radical ontological separation of God in his essence from all creation. This depicts a fundamental disparity in the very ground of God's being and that of creation, which eventually makes us regard God as totally independent from creation in his essence (which naturally does not mean he cannot be related to his creatures). Hence, I claim that any definition of God's transcendence that does not presuppose this ontological difference and independence on the side of the Creator should be classified as a relative or a limited transcendence.

Hegel's God (the Absolute, the Idea) appears to be transcendent precisely and only in the second, limited meaning of this term, as he is not entirely himself without creating the universe in which he becomes what he is "in very truth": "Of the Absolute it must be said that it is essentially a result, that only at the end is it what it is in very truth; and just in that consists its nature, which is to be actual, subject, or self-becoming, self-development."[46]

Those who share the opposite view—namely, that Hegel does maintain the ontological difference between the Spirit and the world—refer sometimes to the Christian doctrine of the Trinity, which Hegel uses to show the process of the Spirit manifesting itself in history. They claim that Hegel is aware of and tries to preserve the doctrine of both an ontological (transcendent) and economic (immanent) Trinity. Nevertheless, a careful investigation of his philosophical theism shows how far Hegel's kingdom of the Father, kingdom of the Son, and kingdom of the Spirit are from the orthodox concept of the Christian Trinity. Hegel denies the precedence of the Father. Christ's death and resurrection are not of a special significance for him, since everybody is redeemed by the act of awareness of being in God. The true divinity is the Spirit (Spirit is three-in-one and in fact it is *the* one, because it is three). But if that is the case, Hegel remains closer to gnostic than to orthodox Christian teaching.[47]

Consequently, most contemporary commentators claim that Hegel is neither a classical theist nor a pantheist, but a panentheist.[48] The explanation of this assertion and the strength of Hegel's position are hidden in his dialectics. God comprehending the world in himself is both dependent on and independent of it. He does not exist merely in himself but naturally posits himself in created beings. He is both transcendent and immanent, even if the former feature is interpreted in terms of the "limited" understanding of transcendence.

Thus, in Hegelian philosophy Spinoza's substance is finally replaced by Subject, and Spinoza's concept of God understood as supreme and independent Being is replaced by the notion of God who actualizes himself dialectically to comprehend himself in creation. Finally, Spinoza's failure in comprehending the concreteness and contingency of beings is overcome by Hegel's understanding of the role of each individual in the dialectical process of the unfolding reality. Hence, the essence of the shift in the philosophy of God from Spinoza to Hegel might be expressed by the following claim of the latter: "God is the one and absolute substance; but at the same time God is also a subject, and that is something more. Just as the human being has personality, there enters into God the character of subjectivity, personality, spirit, absolute spirit."[49]

As a result, although both Fichte and Schelling can be regarded as early precursors of panentheism—a term coined by Schelling and Krause—it was Hegel who explained and developed this position in a most coherent and effective way (even if he does not use the term "panentheism" himself). For this reason, he can be truly called the father of the contemporary versions of this theistic position.[50]

1.4. Dipolar God of Process Theism

The panentheistic philosophy of God originating from German idealism was adapted and modified by many influential thinkers of nineteenth-century academia in Germany, England, the United States, and France. What interests us most, however, is the twentieth-century versions of panentheism and their influence on the doctrine of the God-world relationship in the science-theology dialogue. Here, among panentheistic aspects of the thought of many philosophers and theologians, we find the process version of panentheism developed by Alfred North Whitehead (1861–1947) and Charles Hartshorne (1897–2000) and applied by John Cobb (b. 1925) and David Griffin (b. 1939). In what follows I will analyze the position of Whitehead and Hartshorne more carefully, because it seems to be favored by the participants of the dialogue between theology and science.[51]

1.4.1. Whitehead's Process View of Reality

If it is true that the whole history of the Western philosophical tradition "consists of a series of footnotes to Plato,"[52] Whitehead's own system goes back even further, to Heraclitus, for Whitehead, like Heraclitus, sees the universe in a constant process of change. Whitehead develops this general observation into the "philosophy of organism"—a new ontology proposed in the context of the scientific achievements and theories of the early twentieth century.

Whitehead sees the universe as an organic complex—rather than a dualism of "mind and mechanism"—constituted of basic events, "bits of experience," "actual occasions," "puffs of existence," that realize and actualize themselves (or "concresce") in one instant of time, to pass away or "perish" in the next instant. The enduring objects of our experience, such as atoms, molecules, material objects, living creatures, including human beings, are nothing more than organized "societies of actual occasions." Hence, it becomes clear that "process," rather than "substance" or "being," stands for the basic ontological category in Whitehead's system.

Despite his insistence on each "actual occasion" as a self-creating and self-actualizing individual, seeking the "satisfaction" or "enjoyment" of its own existence, Whitehead notes that actual occasions are related in space and time to other realities in their immediate environment. They "prehend" both past actualities and possible future actualizations, which is decisive for their dipolar nature. As they are neither mind nor matter taken separately, occasions of experience are implicitly both physical and mental.

Although many see this ontology as another version of panpsychism, David Griffin argues it should rather be classified as "panexperientialism," because the experience of actual occasions is not consciousness.[53]

However, the distinction he proposes does not seem satisfactory. If human minds and all other things in nature are complex organizations of actual occasions, then other beings' prehensions at the microlevel must be rudimentary forms of the cognition and perception taking place on the macrolevel in humans. Thus, the objects we perceive are subjects themselves, before they become objects for us. It is hard to defend this insistence on the reciprocity between object and subject in process philosophy from being classified as panpsychism. Moreover, as Deacon and Cashman note, "Critics argue that assuming that mind-like attributes are primitive merely postulates them into existence, and sidesteps the hard problem of accounting for their atypical properties. To this explanatory weakness is added the theory's inability to account for why there is no evidence of the unique properties of mind at levels of chemical or nuclear interactions, and why the distinctively teleological and intentional behaviors of mind, when they occur, do so only in animals with nervous systems."[54]

To complete this short presentation of Whitehead's philosophical system we must add that causation seems to be defined in his metaphysics as the appropriation of the past by the present, rather than the determination of the present by the past. Time is a by-product of actual occasions processing in sequence, rather than a basic metaphysical dimension, whereas space is a function of the proximity and availability of actual occasions to prehend one another. Whitehead also accepts the reality of universals defined as "eternal objects"—that is, possible kinds (ways) of existence that actual occasions can actualize. The question of the character and the origin of "eternal objects" brings us to the topic of the role of God in his "philosophy of organism."[55]

1.4.2. Whitehead's Panentheism

According to Cooper, Whitehead's concept of God changes and develops from *Science and the Modern World*, through *Religion in the Making*, to *Process and Reality*. In *Science and the Modern World*, Whitehead sees God simply as "the Principle of Concretion," which enables the actual world to relate to the realm of possibility. In fact, he identifies God with the realm of possibility but does not speak about God actualizing himself or actively relating possibilities to actualities (God is only a principle of such relation and not its originator or cause).[56]

In *Religion in the Making* God is described as an actual entity who makes possibility available to the creativity of other entities. He is now seen as an "actual but non-temporal entity." Nontemporal because as the Ground of possibility he includes in himself the realm of "eternal objects." Actual because ordering possibilities is an activity, and only actual entities can act. Moreover, it is already at this point of the development of Whitehead's thought that he names the two aspects of God's nature affecting each other: the eternal-ideal and the actual. He also states that God's determining possibilities do not violate the self-actualization of "actual occasions." His role is not to create entities but to offer them ideal possibilities and solutions.[57]

Whitehead's concept of God is fully developed and articulated in his *Process and Reality*, where he famously states that "God is not to be treated as an exception to all metaphysical principles.... He is their chief exemplification."[58] Faithful to this rule, Whitehead places God among other actual occasions. What distinguishes him is his nontemporal nature, in which God prehends all eternal objects and does not perish himself. Hence, God becomes the Ground of all possibilities in the world. Such is the primordial, conceptual nature of God, in which he has a subjective aim, namely, to constitute eternal objects through concrescence into relevant lures of feeling (propositions of future novelty for prehending actual occasions). Thus understood, God is the "unlimited conceptual realization of the absolute wealth of potentiality." However, all these characteristics of the primordial nature of God do not make him a transcendent Creator of the universe. According to Whitehead, "He is not *before* all creation, but *with* all creation." Moreover, in his primordial nature, God is "deficiently actual." Therefore, we "must ascribe to him neither fullness of feeling, nor consciousness."[59]

This lack of perfection in God is fulfilled in his consequent nature, which is "the physical prehension by God of the actualities of the evolving universe." This is the way in which the world becomes objectified in God, who is now "determined, incomplete, consequent, 'everlasting,' fully actual, and conscious." Thus, similar to actual entities, God has a dipolar nature. But that is not all. Both primordial and consequent natures of God are finally sublated to the superjective nature of God, which is "the character of the pragmatic value of his specific satisfaction qualifying the transcendent creativity in the various temporal instances."[60]

This concept of God, proposed by Whitehead, has at least four significant consequences:

1. As actual entities constructing reality have the power of self-actualization—they decide how to determine themselves based on what they prehend, and in so doing cease to be and give rise to other actual occasions—God ceases to be the Creator *ex nihilo* of the universe, which in the Whiteheadian system is simply everlasting: "God does not create the world, he saves it: or, more accurately, he is the poet of the world, with tender patience leading it by his vision of truth, beauty, and goodness."[61]
2. God "conceptually feels" which possibilities best fit each actual occasion, makes them available for prehension, and exerts a "pull," an attraction which can be seen as a type of final causation. "He is the lure for feeling, the eternal urge of desire," which strives to attract each actual occasion to appropriate his "initial aim" as its own "subjective aim."[62] But although he nurtures, attracts, and persuades, God never determines, while his own actuality is significantly shaped and limited by what creatures decide to be.
3. God seems to be, therefore, devoid of much of his transcendence. Because actual entities share with him the characteristic of self-causation, they can transcend all other actual entities, including God.[63] All he can do is (a) propose which eternal objects each actual entity should prehend in the future and (b) give to each of them an objective immortality in his consequent nature. Thus, the panentheistic metaphor of all being "in" God comes in the Whiteheadian system at the cost of a radical limitation, if not a complete redefinition of God's transcendence.
4. Consequently, because God is not seen as the "ultimate" source of meaning and being, the role of the Absolute is reserved in the philosophy of organism for "creativity," which becomes the most basic category of the universe, conforming everything, including God. It is "the ultimate principle of novelty, . . . another rendering of the Aristotelian 'matter,' and of the modern 'neutral stuff.' But it is divested of the notion of passive receptivity, either of 'form,' or of external relations." Although creativity is "conditioned by the objective immortality of the actual world"—that is, conditioned

by God, in whom all actual entities are objectified—at the end of the day, "in the philosophy of organism the ultimate is termed 'creativity'; and God is its primordial, non-temporal accident." God is merely "the outcome of creativity, the foundation of order, and the goad towards novelty."[64]

Even if Whitehead does not use the term, it goes without saying that his version of the God-world relation is panentheistic. "Each temporal occasion embodies God, and is embodied in God." This relation is intrinsic to the divine nature, rather than contingent or voluntary. "In God's nature, permanence is primordial and flux is derivative from the World." Moreover, both God and the world become parts of a greater, dynamic whole: "Neither God, nor the World, reaches static completion. Both are in the grip of the ultimate metaphysical ground, the creative advance into novelty."[65]

1.4.3. Hartshorne
As an assistant of Whitehead, influenced also by Charles Sanders Peirce, William James, Henri Bergson, and Martin Heidegger, Hartshorne introduces two important modifications to the process concept of God. He does not agree with Whitehead's principal rule treating God as a single actual entity. Rather, he proposes to reconceive him as a person, a monarchical society of occasions including the whole universe in himself: "The world consists of individuals, but the totality of individuals as a physical or spatial whole is God's body, the Soul of which is God."[66]

Hartshorne also develops Whitehead's dipolar view of divine attributes, in the context of a more explicit and elaborate criticism of classical theism, which he sees as "monopolar" in its definition of God's absolute, necessary, simple, eternal, immutable, and perfect nature. He classifies his theistic position as "panentheism" or "surrelativism." Unlike Whitehead, for whom the Absolute is creativity, Hartshorne sees God as "maximally absolute," but at the same time "maximally relative," that is, supremely relative, or "surrelative." His description of God's dipolar nature can be summarized thus:

1. *Divine necessity.* Although God's sheer existence is necessary, his actual existence entails the contingency and unpredictability of beings constitutive of the world—God's body.

2. *Divine omnipotence.* God does not do everything himself. He works through persuasion and enables creatures to exercise their causality: "His power is absolutely maximal, the greatest possible, but even the greatest possible power is still one power among others, is not the only power."[67] Because "only he who changes himself can control the changes in us by inspiring us with novel ideals for novel occasions . . . it is by molding himself that God molds us."[68]

3. *Divine omniscience.* God's knowledge is complete yet limited: "There is in God something absolute or non-relative, his cognitive adequacy. Nevertheless, in knowing any actual thing, God himself is related and relativized with respect to that thing."[69] Although he knows the past and the present completely through cognitive prehension, the future is for him merely possible.

Hartshorne openly identifies his own theistic position with panentheism and contributes significantly to its recent popularity. He localizes ontologically all beings (actual and possible, concrete and abstract) in God, noticing at the same time that "the mere essence of God contains no universe. We are truly 'outside' the divine essence, though inside God."[70] This is possible because as Creator God is both absolute and relative, being and becoming, potential and actual, abstract and concrete, necessary and contingent, eternal and temporal, immutable and changing, cause and effect, maximally excellent and yet always improving.[71]

2. ON THE MEANING OF "*EN*" IN PAN*EN*THEISM

Our investigation of the main stages of the philosophical development of panentheism can be enriched by an analysis of its applications in more recent theology. Here we find an important attempt at formulating more precisely the exact meaning and theological consequences of the acceptance of the panentheistic metaphor of the world being "in" God. In reference to the summaries offered by Brierley and John Culp we can list at least ten main aspects of this clarification:[72]

1. The world is "in" God primarily through participation in divine being. The world comes into being and continues in existence

through taking part in God's being.[73] Although this assertion is supported by classical theists as well, they nonetheless distinguish between God's uncreated *esse* and the created *esse* proper to contingent entities. I will say more about this distinction in section 1 of chapter 5.
2. If/because the world is "in" God, neither God's actions nor worldly occurrences are completely determined. This lack of complete determinism effects an unpredictable self-organizing relation of both God and the world based on prior actualizations of each. "The '*en*' designates an active indeterminacy, a commingling of unpredictable, and yet recapitulatory, self-organizing relations."[74]
3. If the world is "in" God, then it can be regarded as God's body. Proponents of this concept (Clayton, Griffin, Sallie McFague) make an analogy between the relation of God and cosmos and the relation of mind and body (God/world//mind/body), suggesting that this metaphor safeguards the distinction of the two sides of the relation, yet does not allow for their separation. In other words, just as the mind can control bodily actions without violating any physical laws of nature, so too God can act on the world in a way that is consistent with physical laws.
4. The language of people expressing themselves "in and through" their bodies can be applied as a metaphor of God acting "in and through" the world (a view shared by Peacocke and Norman Pittenger). The "through" is regarded as an important addition to the panentheistic "in," emphasizing both the immanence and transcendence of an agent.
5. The whole cosmos is the sacrament of God, who comes and acts "in" and "through" it. Particular sacraments of the church are specific intensifications of the general sacramental character of creation (a view shared by Peacocke, Paul Fiddes, McFague, Pittenger, and Kallistos Ware).
6. The language of dialectics and dipolarity is further developed and enriched, emphasizing embodiment as an intrinsic feature of divinity. God and cosmos are seen as "inextricably intertwined" (a view shared by Clayton, John Macquarrie, Leonardo Boff, and Pittenger). This entails a real reciprocity (bilateral relation) between God and his creatures—a doctrine that became a

source of contention between panentheists and classical theists. The reason for the controversy is the panentheistic reformulation of the classic divine attributes. Divine immutability means unchanging faithfulness, not unchanging being. Divine omnipotence and omniscience lose their absolute character due to the self-limitation of God's power and knowledge. Divine eternity does not exclude God's being dependent on time, and the concept of the self-emptying or "kenosis" of God leads panentheists to replace the attribute of divine impassibility with the notion of a vulnerable God who suffers in and with the creation.[75]

7. Accepting embodiment as an intrinsic feature of God leads many panentheists to assert that God is dependent on the world and needs it for the fulfillment of his love. To remain within the orthodox Christian tradition, those who follow the panentheism of Hegel and German idealism (Macquarrie, Fiddes) emphasize that, although not having a world is not an option for God, the cosmos represents, nonetheless, the preexistent free choice of God to create. They believe that Hegel's dialectics enables them to attribute to God both determinism and freedom, even if logically they are mutually exclusive. Followers of process panentheism (Hartshorne, Reese, Griffin) strive to save God's freedom by suggesting that even if divine embodiment is indispensable, God needs "a" world, but not necessarily "this particular" world.[76]

8. Proponents of panentheism argue that the tradition of classical Christian theism is characterized by a deep suspicion concerning the goodness of physical reality. They hold that the new version of theism they propose—in which the world is God's body—proves physical reality's intrinsic goodness. At the same time, however, they seem to be aware of the problem of evil present in the world, which the same panentheistic metaphor introduces into the very essence of God. I will discuss this dilemma in the next chapter.

9. Part of the panentheistic answer to the question of evil is the suggestion that divine love and embodiment imply that God suffers "in and with" creation (the long list of those who share this view contains Fiddes, Griffin, Hartshorne, Macquarrie, McFague,

Jürgen Moltmann, David Pailin, Peacocke, and Pittenger). For although creation is intrinsically good, it is subject to privation. Hence, contrary to the long-standing tradition of classical theism, panentheists claim that in his immanent nature God needs to be seen as "the fellow sufferer who understands."[77]

10. Followers of panentheism tend to see Jesus as different from other persons by degree rather than kind. Such "degree Christology" is a consequence of understanding the world as present "in" God. For if the world is truly present in God, God's work in Christ must be seen as simply a continuation of his work in the cosmos. It goes without saying that such an approach to Christology (supported by John Robinson, Macquarrie, Griffin, Peacocke, McFague, and Pittenger) departs from the classical teaching on the presence of both divine and human nature in the one person of Christ, which distinguishes him ontologically from all other human beings.

3. PANENTHEISM IN THE SCIENCE-THEOLOGY DIALOGUE

Panentheism has been welcomed by a number of thinkers engaged in the dialogue between theology and science, especially within the debate on the contemporary understanding of divine action. Our investigation of the thought of some major participants of this dialogue who embrace panentheism will prove that a tendency can be traced among them to favor the "dipolar" panentheism of Whitehead and Hartshorne. It will also reveal ways in which various theological clarifications of the meaning of the preposition "in" are used by these authors in their explanation of panentheism.

3.1. Ian Barbour

Focusing his research on the philosophical and theological implications of quantum mechanics, the theory of relativity, the big bang theory, evolutionary biology, and genetics, Ian Barbour—one of the first pioneers of the science-theology dialogue in the 1960s—criticizes classical theism and seems to follow the panentheistic proposition. Even if he does not use the very term "panentheism" in his writings, Barbour willingly embraces it in its Whiteheadian version, which he reinterprets in terms of Hartshorne's emphasis on divine "surrelativism." He begins with a reference to some basic concepts of the philosophy of organism, such as the primacy of time, the interconnection of events, the understanding of reality as an organic process, and the self-creation of every entity.[78] But what most influences Barbour's concept of the God-world relationship is the idea of God's relatedness and receptivity, elaborated by Hartshorne. In reference to Hartshorne's *Divine Relativity* Barbour says that "God is not merely influenced by the world; God is 'infinitely sensitive' and 'ideally responsive.' Divine love is supremely sympathetic participation in the world of process."[79]

Barbour also willingly embraces the process concept of a dipolar God, which makes him see God as a creative participant in the world, "the leader of a cosmic community," both social and ecological. At the same time, because he sees the world more as a community than as an organism, he prefers Whitehead's pluralistic ontology of actual occasions persuaded and nurtured by God to Hartshorne's mind-body analogy of a God-world relationship.[80]

3.2. Paul Davies

Paul Davies, an English astrophysicist, interested also in philosophical aspects of science, regarded himself a "deist" for most of his career. Nevertheless, he has recently embraced panentheism, claiming that it matches most closely his understanding of the God-world relation.[81] He contrasts it with both interventionist and noninterventionist views of divine action. The former looks unappealing to a scientist, because it assumes that "God acts as something like a physical force in the world" and "must violate the physical laws revealed by science."[82] Because it reduces God to an aspect of nature, it is questioned by theologians as well. The noninterventionist view of divine action assumes God can "act effectively in nature through specific events—making real difference in what actually come to pass—but in ways that do not interrupt these processes or violate the laws of nature."[83] This view includes at least three options: one locating divine action at the level of quantum mechanics, the second assuming the ontological openness of at least some chaotic systems, and the third based on EM and DC, which we will discuss below.

In contrast to these views Davies proposes his own modified (not deistic) version of uniformitarianism, which he classifies as panentheistic. It states that God created the universe with a very precise set of laws that combine uniformity and randomness, through which he continues the work of creation. According to Davies, these laws bring about new levels of complexity in nature. Most importantly, the EM of complexity is "lawlike, spontaneous, and natural, and not the result of divine tinkering or vitalistic supervision."[84] Davies claims that his point in appealing to panentheism is "to propose that in choosing these particular laws God also chose not to determine the universe in detail but instead to give a vital, cocreative role to nature itself."[85] God's creation of a self-creating universe is "teleology without teleology," a trend toward greater richness, the outcome of which, in all its details, is open and left to chance.

Although Davies does not clearly refer to any of the philosophical versions of panentheism, he seems implicitly to accept a dipolar concept of God. He sees the Creator as both immanent and transcendent, both temporal and eternal: "If God sustains the continually creative universe through time, then in this sense God possesses a temporal as well as an atemporal aspect."[86]

3.3. Arthur Peacocke

Arthur Peacocke, a British scientist and theologian, presents one of the most elaborate descriptions of panentheism among thinkers engaged in the science-theology dialogue. As a biochemist, he perceives the world as a complex and dynamic system of countless interactions inside and between different levels of organization of matter. He emphasizes the importance of the EM of complexity, the role of necessity and chance, and both bottom-up and top-down causation.[87] At the same time, however, Peacocke subscribes to a monistic ontological view, assuming that "all entities, all concrete particulars in the world, including human beings, are constituted of fundamental physical entities—quarks or whatever current physics postulates as the basic building constituents of the world (which, of course, includes energy as well as matter)."[88] Peacocke classifies his position as "emergentist monism," differentiating it from "nonreductive physicalism." He claims that those who embrace the latter ontology and speak about the "physical realization" of the mental hold a much less realistic view of the higher-level properties than he does. However, it seems to me that Peacocke's definition of "monism" is similar to the definition of physicalism offered by nonreductionist physicalists, whereas his notion of EM remains as vague, wanting, and unexplained as their notion of nonreductionism. For what exactly is the character and nature of emergent complexities if the whole reality is ultimately constituted of fundamental physical entities?

Peacocke develops his own version of panentheism in contrast to both classical theism—which he thinks places God entirely outside created reality and time and is unable to explain how God can act within the world—and those panentheists who claim that the world is literally part of God: "Since God cannot in principle have any spatial attributes, the 'in' (Gk. *en*) expresses an intimacy of relation and is clearly not meant in any locative sense, with the world being conceived as a 'part of God.' It refers, rather, to an ontological relation so that the world is conceived as within the Being of God but, nevertheless, with its own distinct ontology."[89]

With this ontological assertion in mind, Peacocke refers to Moltmann's use of the Kabbalistic doctrine of *zimsum*—that is, God's self-limitation and withdrawal from himself to himself to make possible creation, which itself is not of God's essence—and speaks about "the coinherent presence of God and the world." He finds the panentheistic "in" metaphor advantageous over what he calls the "separate but present to" terminology of divine immanence in classical theism, as well as the "world as part of God" language which identifies the world with God and merges too easily into pantheism. He thus specifies his own position: "God is best conceived of as the circumambient reality enclosing all existing entities, structures, and processes, and operating in and through all, while being 'more' than all. Hence, all that is not God has its existence within God's operation and Being. The infinity of God includes all other finite entities, structures, and processes; God's infinity comprehends and incorporates all."[90]

At this point Peacocke emphasizes that his panentheistic model is intended to be consistent with emergent monism and the causal closure of the world system. For him cosmic and natural processes "constitute a seamless web of interconnectedness and display emergence.... The processes are not themselves God, but the *action* of God as creator."[91] Consequently, says Peacocke, we need to acknowledge that God, acting from "inside" rather than from the "outside," "allowed his inherent omnipotence and omniscience to be modified, restricted and curtailed by the very open-endedness that he has bestowed on creation."[92] With respect to the "self-limitation" of divine omnipotence Peacocke claims that God so made the world that there are certain areas—such as human will—over which he has chosen not to have power. The "self-limitation" of divine omniscience "is meant to denote that God may also have so made the world that, at any given time, there are certain systems whose future states cannot be known even to him since they are in principle not knowable (for example, those in the 'Heisenberg' range)."[93] The "self-limitation" of God's omnipotence and omniscience speaks, according to Peacocke, of the "vulnerability" and "self-emptying" kenosis of God, who *"suffers in, with and under the creative process of the world."*[94] Finally, these considerations lead him to modify the classical attribute of divine eternity: *"God is not 'timeless'; God is temporal in the sense that Divine life is successive in its relation to us—God is temporally related to us."*[95]

Peacocke welcomes the analogy comparing the God-world relationship to the relation between mind and body. However, he does so with three qualifications, which show the limitations of the metaphor. First, he notes that unlike humans, who do not create their bodies, God does create the universe to which he is related. Second, unlike humans, who are not conscious of most processes that go on in their bodies' autonomous functions (e.g., breathing, digestion, heart beating), God "knows all that it is logically possible to know." Third, in using the analogy of human personal agency we are not implying that the "world is God's body" or that God is "a person" like us, but rather that he is "at least personal," "more than personal," "suprapersonal," or "transpersonal."[96]

Finally, following William Temple, Peacocke accepts one more theological qualification of panentheism, which sees the universe as a "sacramental" instrument of God's action: "For the panentheist, who sees God working in, with, and under natural processes, this unique result (to date) of the evolutionary processes corroborates that God is using that process as an instrument of God's purposes and as a symbol of the divine nature, that is, as the means of conveying insight into these purposes. But in the Christian tradition, this is precisely what its sacraments do. They are valued for what God is effecting instrumentally and for what God is conveying symbolically through them."[97]

Although the description of panentheism offered by Peacocke is fairly elaborate, he does not specify the philosophical background of his position. Knowing that his theology is regarded by many as remaining close to the thought of Whitehead and Hartshorne, Peacocke distances himself from process ontology and claims that panentheism as such is "not at all dependent on that particular metaphysical system."[98] Nevertheless, the lack of a more precise qualification of his definition of the panentheistic "in" metaphor as "an ontological relation" in which "the world is conceived as within the Being of God but, nevertheless, with its own distinct ontology" makes Peacocke's position vulnerable to criticism. What is at stake here—and in panentheism in general—is the problem of the defense of divine transcendence, which might not only be radically diminished but potentially lost with the introduction of the postulate of God's "self-limitation" of his divine attributes. An agent who knows considerably more than we but is not omniscient, who is considerably more powerful than we and yet not omnipotent, who guides creative processes of the world and yet suffers under its occurrences, seems to be a superknowledgeable and a superpowerful agent and yet not a truly divine one.

3.4. Philip Clayton

Philip Clayton, an American philosopher and theologian actively contributing to the science-theology dialogue, accepts and develops Peacocke's panentheistic theological cosmology. Taking a nonreductionist and nondualist view of the scientific description of the evolving and complex universe, he states that "emergence propels one to metaphysics, and metaphysical reflection in turn suggests a theological postulate above and beyond the logic of emergence."[99] In other words, Clayton believes that the emergent part-whole structure of the universe provides a model for the relationship between the universe and its ground, a relationship best explained in panentheism.

Similar conclusions can be drawn from the history of philosophical theology, which Clayton sees as a constant struggle between the concept of the "infinity-based ultimate unity"—which tends to separate all created, less-than-fully-perfect beings from the divine—and the concept of the "perfection-based irreducible pluralism," which tends to assume that because nothing can be "outside" the infinite, "all must be included within it." The former leads to deism, whereas the latter merges into pantheism. Again, Clayton sees panentheism as a middle-ground position avoiding both extremes.[100]

Clayton's training in philosophy leads him to an elaborate analysis and a more conscious choice of the philosophical base of his version of panentheism. Tracing it back to Neoplatonism, Eriugena, Eckhart, Nicolas of Cusa, and others who assert that "finite things are *in* God," he finds crucial for the recent panentheistic turn the philosophy of German idealism. Here we find him advocating in favor of Schelling and against Hegel, who remains—in Clayton's opinion—too close to the Aristotelian tradition, especially the idea of the first principle, which conditions all entities without being conditioned by them. He disapproves the necessity with which the Hegelian Absolute Spirit has to posit itself in creation—a requirement of the system of which the Absolute is an indispensable part. He thinks Hegel "fails to preserve the contingency and the mystery of the decision to create," and he contrasts Hegel's rational determinism with Schelling's personal notion of God's choosing freely to create and, thus, self-manifest himself, having history and destiny, growth and development in time. Influenced by Schelling, Clayton asserts that "God-as-personal emerges from the infinite ground; as a being involved in actual relations with the world, God undergoes real change and development through those relations."[101]

In his search for a notion of God more personal then Hegel's Absolute, Clayton sides not only with Shelling. He refers to the philosophy of organism, mainly for its emphasis on God's affectivity and the reciprocity of his relationship with actual occasions, which matches Clayton's strong emphasis on the personal character of God's nature. He implements the process concept of a dipolar God—which he finds present already in Schelling's idealism—and speaks about the dialectical unity-in-difference characterizing the ontology of "dipolar panentheism." He suggests that process theism's juxtaposition of the two poles in God (thesis and antithesis) needs a further development that will bring a better understanding of their reconciliation (synthesis).[102]

In addition, Clayton defines his version of panentheism as Trinitarian and kenotic. It is Trinitarian because it perceives the world as participating in God in a manner analogous to the way in which members of the Trinity participate in one another, except that the world is not and does not become God. It is kenotic because of God's free decision to limit his omnipotence in the process that results in the actuality of the world that is taken into God. At the same time, the freedom of God's decision to create enables Clayton to preserve the doctrine of *creatio ex nihilo*. For him the world does constitute the relational aspect of God but not the totality of God (God's essence as such).[103]

Clayton emphasizes the importance of panentheism in developing, articulating, and interpreting theological reflections inspired by the theory of EM. I will discuss this more in a separate section below. He accepts the idea of a God who both genuinely responds to and is affected by what his creatures do and who affects the world without supernatural interventions. He is also positive about the role of the mind/body metaphor in describing the God-world relationship.[104]

3.5. John Polkinghorne

John Polkinghorne—a theoretical physicist and one of the leading participants in the science-theology dialogue—is mostly known for his model of divine action developed in reference to the theory of chaos, which gives us a picture of the universe ruled by a structured randomness, an ordered disorder of occurrences at different levels of the complexity of matter.[105] Trying to save the principle of sufficient reason, Polkinghorne argues in favor of the reality of a top-down organizing causal principle, which—operating holistically through an "information input" and not through an input of energy—complements the energetic causality of natural causes. He ascribes such nonenergetic, pattern-forming influence to God.

What is most important for us, however, is Polkinghorne's later embracement of kenotic theology. As notes Ignacio Silva, Polkinghorne's assertion that "divine *kenosis* can ... be understood as having four dimensions—relating to the self-limitation of divine power, of divine eternity, of divine knowledge, and of divine participation in the causal nexus of creation"—clearly suggests the possibility of change in God in his relation to the world and places him, in a way, as a cause among causes.[106] The same approach to divine agency in creation finds its clear elucidation in Polkinghorne's *Faith, Science and Understanding* and *Science and the Trinity*, where he states that "acts of providence are to be understood in accordance with a recognition of the divine kenosis involved in creation, so that God is not supposed to be the agent of everything but, rather, a balance is struck between the actions of God and the actions of creatures. I have suggested, despite much theological argument to the contrary, that the Creator's self-limitation should be understood to extend even to God's condescending to act as a providential cause among causes."[107]

Consequently, in reference to the same kenotic theology—exemplified to the highest extent in the central event of the incarnation (in which "God submitted in the most drastic way to becoming a cause among causes")—Polkinghorne thinks that the intertwined agencies of the Creator and his creatures imply "a picture of undisentangleability [which] corresponds to God's loving choice to be a present cause among causes ... [where] kenotic providential causality is also exercised energetically as well as informationally."[108]

Does Polkinghorne's kenotic view of divine action bring him close to the panentheistic imagery of the God-world relationship? Although he appreciates the attempt of panentheism to find a proper way to express divine transcendence and immanence, Polkinghorne does not think it succeeds. Defending the true otherness of the world from God, he finds the language of panentheism "too-inclusive" and emphasizes that the Christian doctrine of creation is concerned with "the self-surrender of divine all-inclusiveness in the creating of a world genuinely other, to which God can be 'closer than breathing,' in the sense of continuously being aware of it and interacting with it, without being, even partially, identified with it."[109]

Critical of Neoplatonic overtones of panentheism—making the world an inevitable emanation of God's nature—Polkinghorne seeks to correct the overemphasis on transcendence in classical theism by referring to some other theological concepts. He reaches to the Eastern Orthodox distinction between God's essence ("God's being, ineffable to creatures") and his creative energies ("God's activity in creation"). He claims that "an appropriate understanding of the latter can provide a strong account of effective divine presence without endangering the distinction between creatures and their Creator." He suggests considering the energies as "immanently active divine operations *ad extra*."[110] The exact nature of the causation of God's creative energies, however, seems to remain obscure in Polkinghorne's writings.

The other theological point of reference in Polkinghorne's theory of divine action is Trinitarian theology, and especially "the celebrated theological aphorism called 'Rahner's Rule,' affirming the identity of the immanent Trinity (God in Godhead itself) with the economic Trinity (God known through creation and salvation)."[111] Polkinghorne believes that Rahner's theology is "a statement of theological realism" and confirms the fact that God's nature is truly made known through his revelatory acts.

When discussing God's relation to time, Polkinghorne rejects the classical concept of Boethius that God perceives all of history in *totum simul*, all at once. He sees it as promoting a concept of a God who knows the world from an atemporal viewpoint, and he claims that the world of becoming rather than static being—as we know it today—cannot be properly known this way. Interestingly, this assertion makes Polkinghorne embrace the basic ideas of the dipolar God of process theism, without fully adopting its metaphysics. Developing further his theology of divine self-limitation, he states: "God's acceptance of a temporal pole within divinity is another instance of the kenosis of the Creator in the act of creation, as the One who is eternally steadfast enters into a relationship with creatures that is conditioned by time. In turn this leads to a kenosis of divine knowledge.... In the kind of dipolar theism that I am seeking to espouse, God is understood to have *chosen* to possess only a current omniscience, temporally indexed."[112]

Finally, searching for an appropriate description of God-world relationship, Polkinghorne joins the camp of those who support the concept of "divine suffering in compassionate solidarity with the travail of creatures." Following Williams and Fiddes, he notes that "this divine possibility implies an openness to mutability, for 'suffering in the widest sense means the capacity to be acted upon, to be changed, to be moved, transformed by the action of, or in relation to, another.'"[113]

This short analysis of Polkinghorne's theology of divine action shows that he willingly embraces all the main theological conclusions flowing from panentheism. At the same time, his views on this theistic proposition remain fairly straightforward. Critical about its "too-inclusive" language, he is willing to accept the panentheistic image only as an eschatological fulfilment, and not a present reality.[114] He sees it as a *theosis*, that is, a true sharing in the life of God in the world to come: "God's final purpose is that creatures should enjoy fully the experience of the unveiled divine presence, and so share in the divine energies."[115] Polkinghorne's description of this state follows once again the process distinction of the eternal and temporal natures (poles) in God. He says about the beatific vision that "it will not be human participation in the ineffable life of the eternal divine pole. Rather, it will be an unending exploration of the reaches of the temporal pole of deity, made accessible to us in Christ."[116]

The study of Polkinghorne's attitude toward panentheism completes my general exposition of the thought of the main followers of this theistic position among scholars engaged in the science-theology dialogue. In the following section I will examine in more detail panentheistic emergentism.

4. PANENTHEISM AND EMERGENCE

We can name at least three reasons why emergentism has become a promising point of reference and inspiration for theologians in the science-theology dialogue. First, its intrinsically irreducible character challenges materialist eliminativism and stretches interpretational and explanatory boundaries of science, opening it to philosophy and metaphysics. Metaphysics, in turn, inspires a religious interpretation of the constantly emerging world.

Second, advocates of the reference to emergentism in theology of divine action emphasize that what EM says about the natural characteristics of inanimate beings, organisms, dynamic systems, and causal relations of their constituents reveals—through analogy—something of the truth about God and his relation to the world. At the same time, they underline the fact that their interpretation does not merge the scientific perspective with the theological, nor does it blur their boundaries and limits. It thus avoids the pitfall of the Intelligent Design movement, which, speaking of irreducible complexity, strives to see it as a scientific proof for the existence of an intelligent designer.

Third, supporting nonreductionist physicalism (or emergentist monism), proponents of emergentism in theology claim to remain free from both the philosophical mistake of introducing another version of Bergsonian élan vital in biology and the theological charge of supporting an interventionist account of divine action, according to which God interferes with the natural causation of creatures.

As we will see, the most popular theological interpretation of EM, offered by Peacocke, was developed within the framework of the panentheistic understanding of the God-world relation and divine action. But emergentism inspires also followers of religious and theistic naturalism, whose thought we shall analyze first.

4.1. Emergentism and Religious Naturalism

Even if the followers of ontological naturalism claim that "nature is all that is," some of them also speak of being "religious," without embracing any particular religion. They see world religions as cultural phenomena, grounded in social metanarratives, and they contrast them with a natural religious attitude expressed (1) in the interpretive sphere, acknowledging the importance of philosophical and existential questions about the source and meaning of everything; (2) in the spiritual response of awe, humility, reverence, assent, and transcendence to the phenomenon of existence; and (3) in the moral sphere, describing outward communal responses of care, compassion, fair-mindedness, responsibility, trust, and commitment.

Ursula Goodenough, in an article written with Deacon, says that, with regard to the interpretive sphere, "the emergence perspective offers us ways to think about creation, and creativity, that do not require a creator." The same emergent perspective, "while not ruling out purpose or plan, is coherent without invoking either." It also offers a fresh way to think about contingency in terms of dependency rather than being accidental or fortuitous and allows us to "see ourselves not as 'above' [the natural order] but rather as remarkably 'something else.'" "For all we know, and quite probably for all we will ever know, we are the only creatures of the universe who write psalms and sculpt marble and know how to work. We inhabit a virtual reality of symbols and ideas, and we are uniquely endowed with the capacity to teach as well as to imitate."[117]

Most importantly, irreducible complexity provokes a spiritual response of reverence, gratitude, and transcendence, which is related to EM in a special way, because it describes cases where "discontinuities (something elses) arise from, while remaining tethered to, their antecedents (nothing buts)." Goodenough and Deacon describe this transcendence as "horizontal" and add that "the emergentist perspective opens countless opportunities to encounter and celebrate the magical . . . a word often associated with supernatural miracles or the unexplainable . . . while remaining mindful of the fully natural basis of each encounter."[118] Thus, even if no miracles come from outside into the world and no covenant should be made with a transcendent God, a "covenant of silence" should be made with the Mystery of life, discovered within the perspective of EM, which enables us to discover "a way in which the universe is re-enchanted each time one takes in its continuous coming into being."[119]

In a similar vein, Jeremy Sherman—together with Deacon and in reference to his theory of emergent teleology arising spontaneously from a universe devoid of any such property—argues against the idea of the antecedent, transcendent, and disembodied *telos*. In the light of scientific EM, transcendent *telos* becomes "unnecessary to explain life, mind, and value," which are no longer "parasitic" on a "single divine source." Sherman and Deacon conclude that, although this view may be a source of discomfort, "it definitely shows that we belong here, that we are the legitimate offspring of this world, and that our experience of self-creation is what it seems."[120] They find the "open-ended *telos*" an essential feature of human experience and argue that "it could have arisen from material origins," which "bursts the myth" of an "ultimate transcendental *Telos*" on which the *telos* that we find in the world would depend.[121]

However, does this acceptance of a natural description of the EM of *telos* from purely material origins, delineated in physical terms, force an abandonment of theistic claims? Does it inevitably lead to naturalism devoid of religion? At first glance Sherman and Deacon seem to think so. They claim that "if *telos* can emerge from nonteleological beginnings, investing God with this property is redundant." At the same time, however, they soften this position, saying that what their investigation shows is only the fact that "teleological properties are not essential to conception of God or notions of Ultimacy."[122] Hence, in the concluding paragraphs of their article, rather than rejecting theism altogether, Sherman and Deacon invoke the apophatic traditions in Abrahamic religions, which deny "that any of God's traits can be described in ultimately idealized human terms." Accordingly, because they find *telos* as "a physical property defined in physical terms," they claim that it cannot be "the defining feature of a nonphysical God."[123] But this claim does not aspire to become a proof of God's nonexistence. Rather, it brings Sherman and Deacon close to theistic naturalism, which we will analyze next.

4.2. Emergentism and Theistic Naturalism

Those who wish to acknowledge the existence of God and yet—accepting the principle of the causal closure of physical reality—remain reluctant to speak of particular (special) divine action can be classified as theistic naturalists. Similar to Goodenough, Willem Drees combines his belief in the completeness of bottom-up scientific explanation with the sense of awe and mystery inspired by emergent properties, exercising top-down influence on their constituents. What makes him a theist, however, is his finding the source of all possibilities—realized in emerging reality—in God, a creative and sustaining "Ground of Being, who is also the ground of the natural order and its integrity."[124]

Another version of emergent theistic naturalism has been offered by Paul Davies. We have seen him arguing in favor of a God working through the laws of nature and bringing into existence new levels of complexity without "divine tinkering or vitalistic supervision." In his "modified uniformitarianism," no divine supervision of the details of evolution and the balanced proportion between order and necessity on the one hand and chaos and chance on the other is needed. Genuinely interested in emergent phenomena, Davies sees them as an outcome of teleology, but not a divine teleology. He proposes the image of God who "'initially' selects the laws, which then take care of the universe, both its coming-into-being at the big bang and its subsequent creative evolution."[125]

Going back to British idealism, we find one more version of theistic naturalism, which pays even more attention to evolution. It is Samuel Alexander's fully naturalized concept of providence and purpose rooted in "Space-Time." He states that "God is not the only infinite," because "we have, in the first place, the infinite Space-Time itself which is *a priori*" and generates "empirical infinites."[126] He readily applies EM to God without any qualifications and sees the world of nature as "tending to deity," which is a further step in the process of emergent evolution, gradually appearing, as the universe reaches a certain stage of complexity. Because deity is subject to the same laws that apply in the entire universe, God is never fully actualized, but always becoming. In Alexander's own words:

As actual, God does not possess the quality of deity but is the universe as tending to that quality. . . . Only in this sense of straining towards deity can there be an infinite actual God. . . . Thus there is no actual infinite being with the quality of deity; but there is an actual infinite, the whole universe, with a nisus to deity; and this is the God of the religious consciousness. . . . God as an actual existent is always becoming deity but never attains it. He is the ideal God in embryo. . . . As being the whole universe God is creative, but his distinctive character of deity is not creative but created.[127]

Although Alexander's version of evolving theistic naturalism did not gain much popularity at the time it was formulated, Gregersen notes that it reappears again and again in new forms. He gives the example of Harold Morowitz, a biophysicist who sees the laws of nature as statements of operation of the immanent God, whose transcendence is the result of nature's evolution and unfolding rather than the external source of the world.[128] What is more, Alexander's idea of God, who "as an actual existent is always becoming deity," returns as well, most notably in the Whiteheadian process version of panentheism.[129]

Going still further, we realize that Alexander may also be regarded as one of the precursors of the contemporary panentheistic revolution, for he says—reflecting on emerging deity—that "the body of God is the whole universe and there is no body outside his."[130] At the same time he tends to defend divine transcendence, saying that "empirical as deity is, the infinity of his distinctive character separates him from all finites. It is his deity which makes him continuous with the series of empirical characters of finites, but neither is his 'body' nor his 'mind' finite."[131] This attempt to express both divine immanence and transcendence recalls similar efforts of the panentheists I have analyzed above. We will now see how their ideas advance in the most developed proposition of theology inspired by EM, namely, the emergentist panentheism of Peacocke.

4.3. Emergentist Panentheism

What proves to be most important for Peacocke in his reflection on the emerging universe is the phenomenon of whole-part influence in natural systems. He states that, by analogy, "God could affect holistically the state of the world (the whole in this context),"[132] which, being "in" God, can be regarded—at least metaphorically—as his part. In reference to the image comparing the God-world relationship to the relation between mind and body, "God would then be regarded as exerting a continuous holistic constraint on the world-as-a-whole in a way akin to that whereby in our thinking we influence our bodies to implement our intentions."[133] This emergentist and panentheist concept of divine action thought of as a top-down (whole-part) influence becomes strategic for Peacocke's theology for two reasons.

First, because "the 'ontological gap(s)' between the world and God is/are located simply *everywhere* in space and time,"[134] Peacocke thinks that the idea of the holistic effect of God on the world helps to answer the troublesome question of the "causal joint" in the God-world relationship. In other words, "The ontological 'interface' at which God must be deemed to be influencing the world is, on this model, that which occurs between God and the totality of the world (= all-that-is)." What passes across this "interface," adds Peacocke, "may perhaps be conceived of as something like a flow of information—a pattern-forming influence," which does not "involve matter or energy (or forces)."[135]

Second, "such a unitive, holistic effect of God on the world could occur without abrogating any of the laws (regularities) which apply to the levels of the world's constituents." In another words, "If God interacts with the world-system as a totality, then God, by affecting its overall state, could be envisaged as being able to exercise influence upon events in the myriad sublevels of existence of which it is made without abrogating the laws and regularities that specifically apply to them. Moreover, God would be doing this without intervening in the supposed gaps provided by the in-principle inherent unpredictabilities."[136]

Trying to be even more specific in expressing the same intuition, Peacocke adds, "The proposal implies that patterns of events at the physical, biological, human, and even social levels could be influenced by divine intention without abrogating natural regularities at any of these levels."[137] Therefore, contrary to the proponents of religious and theistic naturalism, Peacocke defends the possibility of "special divine action," understood as God's causing "particular events and patterns of events," which express his particular intentions, "as distinct from the divine holding in existence all-that-is."[138] However, because he limits such special divine acts to the "personal level," Peacocke suggests a gradation and "an increasing intensity and precision of location" of divine top-down influences "from the lowest physical levels up to the personal level, where they could be at their most intense and most focused." Consequently, even if this model of divine action remains "clearly too impersonal to do justice to the *personal* character of many (but not all) of the profoundest human experiences of God," Peacocke hopes it has "a degree of plausibility."[139]

Peacocke's panentheistic interpretation of emergentism shows not only that the patterns of EM are grounded in the divine order and that God is always responsive to the evolutionary process but also that the emerging complex systems, structures, and organisms are located within God's being. Peacocke's concept of divine action based on the analogy to the whole-part influence in natural systems becomes an important voice in the debate on the nature of God's causation in the world in the age of science. This concept has been analyzed more carefully by Gregersen, who emphasizes—in the context of the distinction between weak and strong EM—that "Peacocke is not only seeing whole-part causation as an epistemological analogy for divine action, but is suggesting that God actually works in and through a world of emergent processes."[140] Clayton sees Peacocke's emergentist revision of traditional theism as being fairly close to the process theology concept of God. He finds it assuming that "God is 'co-creator' with finite agents, luring them without coercion and without pre-determining the outcome. Having established the parameters for the emerging universe that our sciences are now discovering, God allows the open-ended process of evolution to construct more and more complex organisms and interacting systems, leading to (but not necessarily culminating with) rational, moral, aesthetic beings such as ourselves."[141] Clayton concludes, "A dipolar theology allows one to conceive God both as the ongoing Ground of the process of emergence *and* as involved in and responsive to the entities that emerge within that process."[142]

Although it gained considerable acknowledgment and support, Peacocke's careful development of emergentist panentheism and Clayton's praise of its potential to "formulate a single ontological vision rather than sharply separating the becoming of the world from the timelessness and aseity of the divine being"[143] were criticized by James Haag. He thinks that emergent monism comes, ironically, with "a healthy dose of theological dualism," and he claims, "While Peacocke's is a theology that takes seriously the emergentist world-view, the inclusion of an ontological distinction between God and the world will make it difficult, if not impossible, to find ultimate reconciliation between emergence and theology."[144] This remark moves our inquiry toward one more theological model of divine action that strives to be even more compatible with EM in avoiding the alleged dualism of the position shared by Peacocke and Clayton. It was offered by Gordon Kaufman.

4.4. Emergence and Serendipitous Creativity

Contrary to Peacocke and Clayton, who strive to emphasize the personal aspect of God's divine action in emergentist panentheism, Kaufman radically criticizes traditional, personal metaphors, preferring to view God as existing before and apart from evolutionary processes. He speaks about the fault of the anthropomorphic and "reifying talk of God as Creator," understood as an "agentlike being, one who 'decides' to do things, who 'designs' projects and then brings into being new, extraordinarily complex realities."[145] He sides with Clayton in proposing a process model of God, but at the same time goes beyond the ontological split between the Creator and creation still present in emergentist panentheism. He suggests that we should think of God not as the Creator, but as "the religious name for the profound *mystery of creativity*—the mystery of the EM, in and through evolutionary and other originative processes, of novelty in the world."[146]

To avoid the accusation of reintroducing a form of vitalism (a new élan vital) Kaufman adds that creativity is not a kind of force working within or beyond evolutionary processes, for "to regard creativity as a kind of 'force' is to suggest that we have a sort of (vague) knowledge of an existing something-or-other when in fact we do not."[147] He sees it rather as a process that leads to genuine novelty in the universe, from big bang to the present day. He adds that although this "serendipitous creativity" is inherently mysterious, it is not mystery itself. It "differs from 'mystery' in that it directs attention to the coming into being of the new, whereas 'mystery' (when used in a theological context) refers to fundamental limits of all human knowledge and carries no such further meaning."[148]

Kaufman distinguishes three modalities of creativity, which are both "serially and dialectically interconnected": (1) the beginning of the universe in which we find ourselves, which is commonly called the big bang; (2) the creativity manifest in evolutionary processes, that is, the ongoing EM of increasingly complex and novel realities; and (3) the creation of extraordinarily complex cultures by human beings (human "symbolic" creativity).[149]

Similar to the religious naturalism of Goodenough, Kaufman's "serendipitous creativity" does not imply or presuppose a creator. What differentiates his thought from flat religious naturalism, however, is Kaufman's not eliminating God/Creator altogether, but identifying him with the "mystery of creativity."[150] This same fact differentiates his concept of God from the theistic naturalism of Alexander, who identifies God with nature "tending to deity." Answering the challenge of falling into pantheism, he emphasizes that God as creativity should not be confused with any of the realities of the created order: "God is *creativity*, not one of the creatures."[151] To the soteriological and eschatological question, Kaufman replies that true faith can no longer live with a conviction that in the end everything is going to be okay: "True faith in God is, rather, acknowledging and accepting the ultimate mystery of all these things in our lives and, precisely in the face of that mystery, going out like Abraham (as Hebrews 11:8 in the New Testament puts it) not really knowing where we are going but nevertheless moving forward creatively and with confidence."[152]

This short analysis of questions concerning Kaufman's idea of "serendipitous creativity" ends our investigation of theology inspired by the theory of EM. At the same time, it opens a way to a more elaborate critical evaluation of this theology, which will be pursued in the next chapter.

CHAPTER 4

Problems of Panentheism and Theology Inspired by Emergentism

We have discussed in the previous chapter the proposal for a panentheistic revision and reformulation of the image of God and his relationship with the world. We have also analyzed various theological responses inspired by the theory of EM. The time has now come to ask whether these propositions offer a consistent and plausible form of theism, which can be suggested as a legitimate alternative to and transformation of classical theism. I think that such a thesis, though supported by many, is overly optimistic.

In what follows I will first address some major theological problems of panentheism. This general approach will then be developed and further specified in my criticism of the shortcomings of the theology inspired by EM, and of emergentist panentheism in particular.

1. CRITICAL EVALUATION OF PANENTHEISM

The analysis of philosophical and theological panentheism offered in the previous chapter might seem to imply that panentheism is a coherent and plausible theistic proposal in the age of science. The truth is, however, that panentheism raises numerous doubts and questions which seriously challenge its explanatory value. They include, among others, (1) reservations concerning the interpretation of the panentheistic "in" metaphor, (2) questions with regard to a proper expression of both divine immanence and transcendence, (3) the question of theodicy, and (4) a concern about the relevance of the soul (mind)/body metaphor as applied in the panentheistic imagery. I will now discuss each of these issues in some detail.

1.1. The Vagueness of the "In" Metaphor

First of all, we must note that panentheism is not one particular view but a large family of views on the God-world relationship. The resulting ambiguity can be traced back to a lack of clarity regarding the exact meaning of the preposition "in" ("en") in the very term pan-*en*-theism. Despite many attempts at clarification, the assertion that the world is somehow "in" God remains vague and leads to seemingly insurmountable problems. Moreover, since the major advocates of panentheism tend to speak of the preposition "in" as a metaphor, it is important to ask whether it is really employed as merely a metaphor, even by them.

Those who, like Brierley, truly think about the panentheistic "in" metaphorically acknowledge that an attempt at closer explication of the metaphor in question leads unavoidably to a continuum of further metaphors, which—coming out of different ways of understanding the word "in"—accommodate a whole variety of theistic positions, including classical theism, eschatological panentheism, process theism, and pantheism.[1] The list of metaphors suggested by Brierley contains the following propositions:

1. God is separate from the cosmos.
2. The cosmos will be in God.
3. God is present to the cosmos.
4. God contains cosmos.
5. God is affected by the cosmos (e.g., God suffers).
6. God acts in and through the cosmos.
7. The cosmos is a sacrament, or sacramental.
8. God penetrates the cosmos.
9. God is the ground of the cosmos.
10. The cosmos is God's body.
11. God includes the cosmos, as a whole includes a part.
12. God and the cosmos are inextricably intertwined.

13. God is dependent on the cosmos.
14. God is dipolar.
15. God is totally dependent on, or coterminous with, the cosmos.[2]

In his article summarizing the volume *In Whom We Live and Move and Have Our Being*, Clayton speaks about "a sort of family-resemblance relationship between the various usages of the word 'in' by panentheists." In reference to Thomas Jay Oord he lists a number of meanings of "in" that seem to be entailed by the contributors to the volume. The list ranges from literal to purely metaphorical interpretations of the panentheistic preposition:

The world is "in" God because
1. that is its literal location;
2. God energizes the world;
3. God experiences or "prehends" the world (process theology);
4. God ensouls the world;
5. God plays with the world (Indic Vedantic traditions);
6. God "enfields" the world (J. Bracken);
7. God gives space to the world (J. Moltmann, drawing on the *zimsum* tradition; A. Peacocke and many of the authors in this text);
8. God encompasses or contains the world (substantive or locative notion);
9. God binds up the world by giving the divine self to the world;
10. God provides the ground for emergences in, or the emergence of, the world (A. Peacocke, P. Davies, H. Morowitz, P. Clayton);
11. God befriends the world (C. Deane-Drummond);
12. all things are contained "in Christ" (from the Pauline *en Christo*);
13. God graces the world (all of the above).[3]

This plurality of assertions does not clarify the precise meaning of the panentheist position and the preposition "in" used in its definition. Moreover, a closer analysis of some of the propositions listed by Brierley and Clayton suggests that their followers tend to treat the panentheistic "in" as something more than just a metaphor. Points 4, 5, 10, 11, and 12 on Brierley's list and points 1, 7, and 8 of Clayton's list seem to interpret it in terms of physical position or location, which may lead to a false "mereologization" and "spatialization" of the ontological (and theological) discourse.[4] Similarly, points 4, 8, 10, and 11 on Brierley's list and points 1, 4, and 8 on Clayton's list may suggest that God and the world are the same substance or at least the same type of substance, which puts the transcendence of God into question.[5] Following these interpretations, Lataster claims, "In fact, there are panentheistic scenarios in which the universe is of the substance of God (this is possibly a common attribute of all pantheisms and panentheisms in general), which is entirely compatible with the 'in' in panentheism being used as a spatial preposition."[6]

But is the lack of clarity and precision in specifying the meaning of the preposition "in" in pan*en*theism such a big problem? If we, nonetheless, choose (as many panentheists claim to choose) to treat it metaphorically, should not its openness to various interpretations be treated as its strength and virtue, rather than its flaw and weakness? Moreover, claims Göcke,

> Taken at face value, the interpretative burden is not due to this very preposition 'in,' and is therefore not a special problem only of panentheism, but is due to the fact that none of the prepositions, used in a philosophical or scientific context, has a clear-cut meaning. That is to say, a similar request could be brought against every attempt at determining the relation of God to the world: if it is claimed that God *is* the world, then the objection would target the identity and ask what it then means that God *is* the world. If the world is thought as *outside* God, then the objection would be the following: what does that mean, that the world is *outside* God?[7]

Göcke is right that our predication of God and his attributes is always analogical and that, hence, any spatial metaphor will always beg the question of a proper interpretation. However, the difficulty with the term "pan*en*theism" is that it introduces the preposition "in," which is intuitively referred to spatial relations, into the name, and in fact into the definition of a particular theistic position—as the term pan-*en*-theism already defines and expresses its core argument (presupposition). This may, and in fact does, easily turn the attention of many from the philosophical-theological toward spatial-empirical understanding of God-world relationship. Hence, I claim that the panentheistic "in" does need clarification and precise definition. Moreover, the assertion that creation is "outside" God—often attributed to the proponents of classical theism—does not seem to faithfully represent their position. Nor is it a way in which they would commonly describe their understanding of the God-world relation. They prefer to carefully suggest that the substance of any contingent (created) entity is not ontologically equal to the substance of God. It does not imply that creatures are independent from God or that God is not present in the world (is "outside" of it).

Going back to those who promote panentheism, it turns out they themselves do find it necessary and important to further clarify and explain the meaning of the preposition "in" within the term defining their theistic position. For instance, we have seen Peacocke emphasizing strongly that panentheism is not intended "in any locative sense." When saying this he seems to stand in agreement with an important assertion of Krause, who states:

> Following present linguistic usage, I use "in" here ... of finite essences and essentialities, and mean by it that the higher whole is this finite thing, as its part, in such a way that this finite thing, as a part, is the same, according to its essentialities, as the higher whole it is part of. However, as a part, it is limited in so far as its limits are the limits of the higher whole, as a higher whole, but do not exhaust the limits of this higher whole as such.... Of course, all the words in our ordinary language [*Volkssprahe*] which designate relations between things are first derived from space, as "in," "beside," "on," "below," "beside," "out." Or, rather, in the ordinary, pre-scientific consciousness they are mostly understood only from space. But all these words must be understood in an abstract way, and taken in a way that transcends their use in relation to sense, when they are used in connection with philosophy. It is, therefore, not permitted to distort these words of the philosopher, as if he were speaking of spatial relations, if he also uses these words, to denote the relation of the finite to the infinite.[8]

However, despite distancing himself from the spatial interpretation of the "in" metaphor, Peacocke—in delineating its "ontological" meaning—seems to return, one way or the other, to construing it in spatial terms.

This becomes evident when he speaks about the world conceived "within the Being of God," whose infinity "includes all other finite entities, structures, and processes ... comprehends and incorporates all," or when he defines God as "the circumambient reality enclosing all existing entities."[9] In what sense (way)—we want to ask—does God include, comprehend, enclose, and incorporate all beings? If Peacocke wanted to interpret these terms metaphysically, he should have provided a more precise definition of each one of them within the context of his theology. Otherwise, the use and connotations of these terms may put God's transcendence into question, especially in reference to Peacocke's ontology of "emergentist monism."

Clayton's assertion that "Emergence ... represents a powerful answer to misgivings about the preposition 'in'" is far too optimistic and does not solve the problem either. Because Clayton sees the emergent self-inclusion relation "⊂" as meaning "belonging to" or "being a member of"—that is, as a relation of "logical inclusion rather than (primarily) one of location"—its application to the God-world relationship suggests that this relation is merely logical as well.[10] It is true that on another occasion Clayton strives to define the panentheistic "in" in ontological terms, saying that "the infinite God is ontologically *as close to finite things as can possibly be thought without dissolving the distinction of Creator and created altogether.*"[11] Nevertheless, without a more precise description of the character of this ontological closeness between God and creatures and of the corresponding ontological difference between them, Clayton's interpretation of the preposition "in" remains deficient. In a more recent attempt at dealing with the same problem of clarifying the meaning of the core panentheistic metaphor Clayton comes dangerously close to pantheism, which reveals an inherent ambiguity of the panentheistic theological research program even more. He states, "Even though you are a devout panentheist, you still admit that things *appear* to be separate and discrete; they appear to exist as individual substances. But ultimately, you believe, they are grounded in the One Ultimate Reality. For you what we call 'objects' are, ultimately speaking, modes or manifestations or expressions of that One. Hence their appearance as separate, independent existents is not ultimately real."[12]

The highly novel and technical language of process theism is of no help in this regard either. We have learned from Whitehead that through

God's prehension of each creature, the world becomes objectified in him. Hartshorne further develops the language of inclusivity and states: "A supreme person [God] must be inclusive of all reality. We find that persons contain relations of knowledge and love to other persons and things, and since relations contain their terms, persons must contain other persons and things.... In God, terms of his knowledge would be absolutely manifest and clear and not at all 'outside' the knowledge of the knower."[13] I side with Owen Thomas in his criticism of this assertion. Knowing someone or something and having them in mind does not mean other people or things are ontologically in us. What is then the meaning of Hartshorne's idea of containment in relation to persons and things? Is the relation of God prehending actual occasions and their societies, described in analogy to human knowledge or love, strong enough? Many panentheists, following Peacocke and Clayton, appear to want to define their position ontologically, and not merely logically or epistemologically.[14]

Another attempt at clarifying the meaning of the panentheistic "en" comes from Joseph Bracken. In reference to the thought of Colin Gunton, he proposes the concept of Trinitarian field theory, which perceives the world as an extended yet finite field activity within the all-encompassing field of activity constituted by the three divine Persons, remaining in an ongoing interrelatedness with one another and with all creation.[15] John Culp notes that Bracken's theory draws on a systems approach in natural sciences, Whitehead's concept of complex entities as societies of actual occasions, and the Christian Trinitarian doctrine.[16] Bracken sees reality as an all-encompassing society in which subsocieties operate in their own ways yet remain related to God understood as the regnant subsociety, enriched by the information coming from creatures.

Despite its originality, Bracken's proposal does not escape, in my opinion, the above-mentioned difficulties of mereologization and spatialization of the theological discourse, as well as the problem of specifying the ontological nature and the character of the relation between God and his creatures. This becomes apparent in reference to his asserting that "the three divine persons and all their creatures are together constituent parts or members of an expanded divine life-system" and that "the ongoing community of the divine persons with one another within the divine life-system is the topmost process/system within the hierarchically ordered set of processes/systems constituting the overall

God-world relationship."[17] Consequently, it seems to me that, so far, the question about the exact meaning of the core ontological assertion of panentheism remains unanswered.

1.2. Divine Immanence

Nevertheless, panentheism does not concentrate only on the fact that everything is "in" God. It looks at this assertion from the other side and emphasizes God's immanence in created things as well. The panentheist "in" can be thus used in two different ways, says Clayton: "The world is in God, and God is in the world."[18] John Cobb, however, rejects this notion of mutual immanence and maintains that "the two ideas are quite distinct, and it is possible, even common in the Christian tradition to affirm that God is in all things without affirming that all things are in God."[19] He claims that what panentheism really wants to emphasize is God's immanent presence in the universe. But if this is the case, what is its superiority over classical theism, which panentheists so vehemently criticize and reject?

Gregersen rightly notes that "classical theism, even in the form of substance theism, entails a very strong doctrine of divine immanence."[20] To illustrate this point, he notes that Aquinas says, at some point, that "God is in all things; not, indeed, as part of their essence, nor as an accident, but as an agent is present to that upon which it works. . . . Hence it must be that God is in all things, and innermostly."[21] To this we might add that Aquinas further specifies the nature of God's presence, saying he is in everything

1. by his power, inasmuch as all things are subject to his power (in opposition to those who claim that visible and corporeal things are subject to the power of a contradictory principle);
2. by his presence, as all things are bare and open to his eyes (in opposition to those who deny God's presence in inferior bodies);
3. by his essence, as the cause of being of all things (in opposition to those who assume that there are creatures mediating being from God to the lower creatures).[22]

Gregersen emphasizes the importance of Aquinas's teaching, which states that without becoming a creature ("part of their substance") and without being an emergent property of the world (an "accident"), God creates the world "as if from within." Moreover, Gregersen shows that Aquinas actually uses panentheistic imagery with a clear understanding of its metaphorical status. Here, Gregersen once again quotes Aquinas: "That in which bodily things exist contains them, but immaterial things contain that in which they exist, as the soul contains the body. So God also contains things by existing in them. However, one does use the bodily metaphor and talk of *everything being in God* inasmuch as he contains them."[23]

If neither the notion of radical divine immanence nor the panentheist imagery of the world being "in" God is absent in classical theism, one must ask what actually differentiates contemporary panentheism from the classical theism? Gregersen accurately names the difference, noting that, for Aquinas, the natures and activities of creatures cannot affect nor have a real feedback effect on God. Although the world utterly depends on God for its existence, it cannot affect the being and knowledge of its Creator. Consequently, even if the relations in the heart of the Trinity are real, and the world's relation to God is real, "there is nothing God can 'learn' in relation to the creatures, no 'challenges' to be met, no free acts to 'wait for'"[24]—as God's relation to the world is "in idea" (*secundum rationem*) only. It is, *per analogiam*, as a relation of a knower to the thing known, which means it is nonmutual. The knower is really related to and truly dependent (for knowledge) upon the knowable thing, but the thing is not in any way affected by the fact he or she knows it. Similarly, since God does not depend on creatures in any possible way, he is not really related to them, but only "in idea." Let us quote Aquinas himself on this crucial issue: "Since therefore God is outside the whole order of creation, and all creatures are ordered to Him, and not conversely, it is manifest that creatures are really related to God Himself; whereas in God there is no real relation to creatures, but a relation only in idea [*secundum rationem tantum*], inasmuch as creatures are referred to Him."[25]

This statement of Aquinas has become a source of a controversy, leading to a severe criticism and rejection of his theology by most contemporary theologians. They accuse Aquinas of overemphasizing divine transcendence. Among them we find those who are willing to go as far as William Lane Craig and David Tracy, who call Aquinas's teaching "absurdity" and claim that it jeopardizes Christianity's "most fundamental religious affirmation."[26] Following this same criticism, says Gregersen, the proponents of "all versions of panentheism . . . more or less . . . share . . . the claim that there exists a real two-way interaction between God and world, so that (1) the world is somehow 'contained in God' and (2) there will be some 'return' of the world into the life of God."[27] He contends this is precisely the claim that differentiates Aquinas's theism from panentheism:

> What constitutes the common aspiration of the [different] versions of panentheism? I suggest that they all *share the intuition of a living two-way relation between God and world, within the inclusive reality of God.* . . . The real demarcation line between panentheism and classical philosophical theism is neither the immanence of God nor the use of the metaphor of the world's being "in" God. The real difference . . . is that [according to classical theism] the nature and activities of the creatures do not have a real feedback on God. There is, in other words, no return from the world into God.[28]

Klaus Müller contends that "panentheism . . . implies something like the denial of the lack of consequences of the world and the finite for God as such. The world-transcending self-identity of God does not rule out a determination of God through the universe."[29] Similar is the view of Göcke, who, interpreting Krause, contends,

> While *Orwesen* cannot change, and eternally is what it is, . . . what *Orwesen* is eternally is what it is in virtue of its intrinsic and temporal constitution that, of course, is influenced by the free decisions of free creatures. Free choices, in other words, determine what the Absolute as such eternally is, and because everything is in the Absolute, another way to spell this out in the dialectic of temporal freedom and eternal determination is to say that the Absolute freely determines what it yet always already will have been. The course of the world, as a consequence, is not determined and not knowable by God as such.[30]

For an example of a theologian working in the field of science-theology dialogue who shares this understanding of panentheism, we may quote

Clayton, who proposes the concept of the "*inter*dependence" of God and the world and asserts that "the world depends on God because God is its necessary and eternal source; without God's creative act it would neither have come into existence nor exist at this moment. And God depends on the world because the nature of God's actual experience depends on interactions with finite creatures like ourselves."[31]

But is the fierce rejection of God's absolute immutability in order to defend his involvement in created reality justified in light of what we have just seen in Aquinas's description of the divine immanence of the transcendent God? It seems to me that the controversy arises from the apparent misunderstanding of Aquinas's reflection on the nature of the God-world relationship, which may occur when it is read outside the context of his entire philosophical and theological system. Those who approach Aquinas's philosophical theology in this way may mistakenly interpret his thought as being in accord with a particular and distorted interpretation of classical theism that places God entirely outside created reality and time, views him as uninterested in what happens to his creatures, and is therefore unable to explain God's action in the world.

A fitting answer to the panentheistic criticism of classical theism in its teaching on God's immanence can be found in the thought of some prominent contemporary interpreters of Aquinas's philosophical theology. Among them we find Michael Dodds, who, in his major work dedicated to the question of the relevance of the Aristotelian-Thomistic notion of causation and divine action in the context of contemporary science, notes that Aquinas's statement describing the God-world relationship as real on the side of creatures and in idea on the side of the Creator "is in fact an affirmation of God's transcendence and intimate involvement in creation" and "implies not that God is more distant from creatures than they are from one another, but infinitely closer."[32] He continues, "Real relations of mutual dependence arise between creatures that are never more than 'beside' one another. They cannot capture the closeness of divine presence that arises from the action of God who is never simply beside but most deeply within the creature. To predicate such a relation of God would be to reduce God to the level of one creature existing beside another. In effect, the notion of 'real relation' is simply too remote to express the intimacy of God's presence."[33] On another occasion Dodds explains once again:

As the source of created being, God is most intimately present to each thing. His presence is far more intimate than what is possible between the terms of a mutually real relation among creatures. In a mutually real relation of cause and effect, the created agent is related to the effect through the medium of the motion that it produces and that is distinct from its essence. God, however, as the source of *esse* in creatures, is present to each one by his very essence [cf. *ST* I.8.3 ad 1]. And because *esse* is the innermost actuality of the creature, God is most intimately present to each thing [cf. I.8.1 co.].[34]

Another contemporary specialist in Aquinas's philosophy and theology, Brian Shanley, adds: "Aquinas clearly thinks that God is 'related' to the world in the sense that he creates, loves, knows, wills, governs, and redeems the world. The denial that God is 'really related' to the world does not undermine any of these claims. It simply denies that God's causal activity, and any relational terms thereby ascribed to him, implies any alteration in his being. When God acts so as to bring creatures into relationship with him, all of the 'happening' is located in creation rather than in God."[35]

In the light of these explanations it becomes clear that Aquinas does not compromise the truth about God's involvement in his creation when stating that the relation between God and the world is neither real in both of its terms (which would imply that "action" is predicated univocally of God and creatures) nor merely of idea in both terms (since God's action entails a real ontological change in creatures) but is real on the side of creation and of idea only as we understand it on the side of God. Rather, in making this distinction, his position gives right to God's immanence without being at risk of losing the perspective of the Creator's radical transcendence, which seems to be put in jeopardy in contemporary panentheism.[36]

1.3. Divine Receptivity and Transcendence

Panentheists who criticize classical theists on their understanding of the God-world relationship argue in favor of replacing the teaching of classical theism with a new theology that would put more emphasis on divine immanence. Nonetheless, their understanding of God's immanence is rather peculiar. We have seen them arguing that it entails a real reciprocity, a bilateral relation between God and creatures that affects and changes both sides of the relation. It makes God somehow dependent on the world—as if he needs it for the fulfillment of his love—and requires a reinterpretation of the classic divine attributes such as immutability, omnipotence, omniscience, eternity, and impassibility. The new definition of "impassibility" becomes one of the crucial arguments of the panentheistic revolution. God is now seen as suffering "in and with" creation, as being "the fellow sufferer who understands."

This way of argumentation, present already in the panentheism of German idealism, finds considerable development and support in the process theism of Whitehead, and especially in Hartshorne's contrast between the "monopolar" God of classical theism and the "dipolar," "surrelative" God of process thought: a God who is necessary but contingent, omnipotent but limited in his power, omniscient but limited in his knowledge, eternal but dependent on time. Among theologians building on these philosophical ideas we have seen Barbour speaking about God as "infinitely sensitive" and "ideally responsive," participating sympathetically in the world of process. Peacocke elaborates on God's self-limitation, which renders him vulnerable and self-emptied, whereas Clayton emphasizes that "God depends on the world because the nature of God's actual experience depends on interactions with finite creatures like ourselves."[37] A kenotic view of God becomes important for Polkinghorne's view of divine action as well.

The redefinition of divine immanence suggested here raises some serious questions concerning divine transcendence, allegedly protected by the second part of the core definition of panentheism, which states that God is more than the created world. But is God's transcendence really guaranteed in panentheism? If we define divine transcendence in terms of (1) radical ontological difference and independence, (2) unchangeability, (3) absolute actuality with no potency, (4) timelessness, (5) absolute omnipotence, and (6) absolute omniscience, then panentheism seems to put it into question. I have said before (section 1.3.5 of chapter 3) that any definition of God's transcendence that does not presuppose a radical ontological difference and independence on the side of God should be classified as a relative or a limited transcendence. The question remains whether a God who is merely relatively transcendent can be still regarded a God of Christian theism? Similar is the case of the other five aspects of the definition of divine transcendence listed here. When redefined in terms of the panentheistic dipolar image of God, they seem to be compromised, which puts divine transcendence into question.

1.3.1. Divine Transcendence in Philosophical Panentheism
Philosophically speaking, the problem goes back to Hegel and German idealism. For even if Hegel states that Nature cannot fully express God, the real ontological independence and difference between God and creation—required by the definition of divine transcendence—seems to get lost in his philosophical theism. Hence, I have classified his understanding of God's transcendence as limited (or relative). This can be proved by the example of Hegel's reflection on the act of creation. Although he has a notion of God before the origin of the world, he insists that God could not exist without it, as he needs dialectically to posit himself in contingent beings. This presupposes the possibility, if not necessity, of change in God.

The question of divine transcendence becomes even more problematic in process theism. We have seen that the Whiteheadian God is almost devoid of transcendence. As an actual occasion among other actual occasions that can transcend him, "with" creation and not "before" it, God is not a creator of the universe, which in Whitehead's system is everlasting. As a gentle persuader, God can only provide the stimulus for actual occasions prior to their act of prehension and absorb worldly events after their occurrence, which means he cannot be present simultaneously with actual entities in their act of concrescence. As a primordial, nontemporal occasion and outcome of creativity—which plays the role of the Absolute in Whitehead—God seems to lose all divine transcendence in favor of immanence. Even if in his primordial nature God, unlike any other actual occasion, prehends all eternal objects, he is deficiently actual and needs to prehend and objectify the world (not necessarily the present one but some world) to become fully conscious.

A number of advocates of process theism present arguments defending divine transcendence in the Whiteheadian notion of God. William Christian argues that according to Whitehead, in spite of everything being in process, God (1) is free, (2) has his own subjective immediacy, and (3) is perfect in scope, quality, and intensity of his experience.

Burton Cooper attributes to the God of process theism transcendental qualities of (1) ontological priority in providing definition, (2) universal relation to all actual occasions, (3) grounding all novelty, (4) grounding and preserving all values, and (5) integrity in seeking to increase value in the world and in loving it. Palmyre Oomen defines the power and almightiness of Whitehead's God in terms of (1) his originating all actual occasions by presenting to each one of them its initial aim (in reference to eternal ideas that may help them to avoid evil in their concrescence), (2) his preserving all that can be preserved, and (3) the everlasting character of his nature, which does not allow any of the actual occasions to overcome (transcend) him forever (even if Whitehead's universe is everlasting, actual occasions are not).[38] However, even if these attempts succeed in establishing an ontological difference between God and the world, the notion of God's transcendence so defined can be regarded—in the light of what I have reported about Whiteheadian theism—as, at most, relative and not absolute.

Culp claims that "transcendence may be either horizontal, between like entities, or vertical, involving different entities." He thinks "Whitehead's understanding of God's transcendence is horizontal and limited because God only influences events before or after the decisions of the events. God cannot be present simultaneously with the event."[39] He asserts, further, that Clayton (among others) answers this objection by identifying top-down causation as indicating God's vertical transcendence, which Alexander Jensen finds inadequate when taking into account the difficulty of specifying the nature of the "causal joint" between God and the world (see section 2.4 below). Trying to answer the same criticism, Clayton notes that, in fact, few process panentheists accept a full equality between finite actual occasions and the divine actual occasion or occasions. He thinks God, being the chief exemplification of creativity, indicates a difference between himself and actual occasions, which preserves his vertical transcendence. However, his claim might be still challenged by the fact of Whitehead's attribution of metaphysical ultimacy to creativity.[40]

Hartshorne strives to defend God's transcendence when he recognizes and names some significant differences between divine and human embodiment: (1) Unlike humans, God foresees all possibility and prehends all realized actuality. (2) He is not mortal and does not forget the past. (3) Whereas human relations to other beings are external, God's relations are internal to himself, because he is "incarnate" in the entire cosmos. (4) Human relatedness is always selective, constrained, mediated, and focused on particulars, whereas God sees, is immediately aware of, and is responsive to all individuals in his entire body.[41] But does this list of differences between God and humans truly safeguard divine transcendence? We cannot forget about the philosophical consequences of Hartshorne's insistence on the "surrelativity" of God and the dipolar character of his divine nature. These consequences introduce contingency and unpredictability into the actual existence of God, limiting both his supreme rule over the world—since God is both omnipotent and limited in his power—and his knowledge of its future states, since at least some of those states are for God merely possible. This suggests the possibility of change in God, which reduces and puts into question his transcendence. Accordingly, we must say that in Hartshorne's theistic proposition—similar to transcendence's status in German idealism and the Whiteheadian system—divine transcendence is merely relative.

As we have seen (section 1.1 above), Bracken strives to defend divine transcendence within his field understanding of panentheism by giving priority to God as the regnant subsociety. This, however, would qualify, once again, as a relative and not absolute transcendence. One might argue that this is precisely what Whiteheadian (and all process) panentheists want to defend. If this is true, then I would like to emphasize that in my opinion transcendence, when referred to God, must be absolute. Otherwise God ceases to be God and becomes like one of the creatures (contingent entities), a supercreature maybe, but not the transcendent source of everything.

1.3.2. Divine Transcendence in Theological Panentheism
Theologians participating in the science-theology dialogue who support panentheism emphasize unanimously that it safeguards divine transcendence. Ian Barbour, who willingly accepts Whiteheadian ontology, realizes it puts God's transcendence to a test. Nevertheless, he claims that God's transcendence "is still strongly represented," even if Whitehead sees God as merely one factor among others and rejects *creatio ex nihilo*.[42] He says that "Whitehead attributes to God the all-decisive role in the creation of each new occasion, namely provision of its initial

aim."[43] But we cannot forget that the notion of "creativity" as the ultimate category in Whitehead's ontology and his idea of God as an "outcome of creativity"—primordial and nontemporal, but still merely an "outcome"—show that God cannot have a decisive role in the becoming of the world in the Whiteheadian system. Moreover, we must remember that in Whitehead's cosmology every "actual occasion"—as it determines itself—is perfectly able to reject or ignore God's proposal of its initial aim. In what sense, one may therefore ask, is God's provision of an initial aim "all-decisive"? This analysis renders Barbour's defense of divine transcendence in panentheism questionable.[44]

Davies affirms God's transcendence in two ways. First, he notes that most scientists, including atheists, agree that "the universe as it exists is not necessary—that it could have been otherwise"[45]—which emphasizes God's freedom in creating it. Second, paying much attention to the laws of nature, Davies states: "God acts to create all that is, including space, time, and the laws of nature, and thus these laws are in this sense eternal, too. Indeed, one of the purposes in choosing these laws is that they permit the universe—including space and time—to originate spontaneously 'from nothing' in a lawlike manner, without the need for a further, special divine act. Thus the eternal selector God is, in this function at least, outside of time altogether."[46]

The problem in Davies's view of the laws of nature is that he tends to attribute to them an ontological and causal character, as when he claims that the laws of nature "permit the universe ... to originate" or that the laws of physics "produce order" or that they "can themselves be regarded as expressions of God" and his divine agency.[47] Given that, philosophically speaking, laws of nature have rather a descriptive, not prescriptive character, their status in Davies's thought is doubtful. This weakens his panentheism all the more, as he seems to replace the core panentheistic thesis of the world's being "in" God with the statement that "God can be thought of as logically prior to the universe and responsible for the set of laws that allow self-organizing complexity to emerge in the universe. That is, the rational order of the physical universe is grounded in God."[48] "Logical priority" of God and grounding "the rational order of the universe" in God do not seem to entail his ontological divine transcendence.

Moreover, even if—as Davies claims—the function of God as the "selector" of the "eternal" laws of nature places him "outside of time altogether," Davies's defense of God's eternity is challenged by his acceptance of the dipolar, temporal-eternal view of the divine nature proposed by process theism. Unfortunately, Davies does not elaborate more on his ontological commitments concerning God's eternity and temporality, which leaves the question of the plausibility of his defense of divine transcendence open.

Divine transcendence is also endangered in Peacocke's panentheism. For although he argues that the "fundamental 'otherness' of God in his own inscrutable, unsurpassable and ultimately incomprehensible Being is essential to what we mean by God," who cannot be thought as "any ordinary 'cause' in the physical nexus of the universe itself,"[49] we have seen that Peacocke's interpretation of the panentheistic "in" metaphor as "an ontological relation" is ambiguous and vague. Because he does not clarify his own ontological position, his postulate of the world conceived as "within the Being of God but, nevertheless, with its own distinct ontology" remains unclear. Without a precise philosophical foundation, it is impossible to prove that this statement truly protects divine transcendence. Moreover, Peacocke's careful description of the "self-limitation" of the divine omnipotence, omniscience, impassibility, and eternity brings him close to the dipolar view of the God of process theism. That he distances himself from this philosophical tradition does not help clarify his version of panentheism. Nor does it prove that he manages to safeguard God's transcendence.

Considering the problem of divine transcendence in Clayton's panentheism, we have seen him embracing the philosophical theism of both German idealism and process thought. His theological position faces thus the same challenges as the philosophical systems on which he founds it. These we have already discussed above. What is more, as philosopher, Clayton must be aware of the differences between the two schools of philosophical theism he follows. These differences are expressed in, among other things, process theism's rejection of *creatio ex nihilo* and the world's origin in time, as well as in its reducing God's transcendence to a minimum—which contrasts with a more transcendent view of God in German idealism. Clayton tends to embrace the single idea of the dipolar God without accepting the whole ontology grounding process theism, which would otherwise prevent his arguing that "God is not dependent on the world for existence but preceded the world and created it."[50] Thus, whether his position is ultimately consistent and coherent and whether it safeguards divine transcendence remain questionable.

Finally, in reference to Polkinghorne and his eschatological panentheism, one may argue that it defends God's transcendence more than other versions of panentheism. At the same time, however, his position remains exceptional and departs from the view of other contemporary proponents of panentheism, who want to see the world being "in" God at present, and not only in the final consummation of the whole of reality at the end of time. What is more, we have seen that Polkinghorne's insistence on the temporal character of God's knowledge of the world and his suffering in solidarity with creation bring him close to process theism. This definitely challenges Polkinghorne's notion of divine transcendence, similar to his assertion—based on the kenotic theology he embraces—that God is a cause among causes, working both informationally and energetically (through a nonenergetic information input, as well as through causal changes requiring an exchange of energy). Such agency introduces the possibility of change in God, which compromises his divine transcendence.

Summing up, my analysis shows that panentheists have some serious difficulties in saving the reality of divine transcendence. This fact seems to seriously undermine the validity and credibility of their position.

1.4. The Problem of Evil

Another question which arises in panentheism is that of evil. If all aspects of the world are "in" God, then he must contain all forms of natural and conscious human evil. According to the "expressivist panentheism" of German (Hegelian) idealism, the world is an alienation of the Spirit, which is reconciled with the Eternal Being through self-consciousness. Thus, because the world is a negation, an antithesis, it may contain evil, which is overcome in synthesis. However, this does not offer a satisfactory theodicy because—as Gregory Vlastos shows—Hegelian dialectics is homogenous; that is, all three stages of the dialectical triad "are ontologically *homogenous*. Thesis, antithesis, and synthesis are all of the nature of Idea."[51] But if this is the case, then the evil present in antithesis enters the very essence of God's nature.

Peacocke argues that because all evil is "internal to God's own self," God "can thereby transform it into what is whole and healthy," as he "heals and transforms from within."⁵² But again, this does not explain how evil can become a part of God's nature and be reconciled with its ultimate goodness. Since neither Peacocke nor any other proponent of the dipolar character of God's nature mentions limitation of God's goodness along with the limitation of other divine attributes, that goodness should remain unchanged with the creation of the universe. Peacocke's acknowledgment of evil being "internal to God's own self" seems to contradict this line of argument.

The answer of process panentheism seems the most satisfactory, at least at first. The persuasive God of Whitehead cannot force actual occasions to avoid evil, which makes it real but not intended by God. Griffin adds that we can still say in process theism that all evil (physical and moral) is in God, so long as we remember it does not affect God's primordial eternal essence, but is present in his experience of the world, that is, in his consequent nature. He argues that although present in God, evil does not enter his intentions. However, process theodicy faces a major challenge, as the whole picture is devoid of an eschatological hope. One cannot be sure that God's love will eventually and ultimately conquer evil. As Gregersen points out, in process theism "there seems [to be] no redemption possible for the tragically un[ful]filled aspirations of life, nor for the problem of the horrendous evils of wickedness.... A soteriological deficit is obvious."⁵³ Hence, the question of evil in the panentheistic view of reality and the God-world relationship remains challenging and unanswered.

1.5. Soul (Mind)/Body Metaphor

As part of our critical evaluation of panentheism we need to discuss in more detail the analogy made by its proponents between the God-world relationship and the relation of soul and body (or mind and body).⁵⁴ We have seen (section 1.2 of this chapter) that the metaphor in question was used already by Aquinas. In his description of divine immanence, he states that it is possible to say that "all things are in God; inasmuch as they are contained by Him," just as "the soul contains the body."⁵⁵ Applying this analogy in contemporary panentheism, Clayton states:

> The world is in some sense analogous to the body of God; God is analogous to the mind which indwells the body, though God is also more than the natural world taken as a whole. Call it the panentheistic analogy (PA). The power of this analogy lies in the fact that mental causation, as every human agent knows it, is more than physical causation and yet still a part of the natural world. Apparently, no natural law is broken when you form the (mental) intention to raise your hand and then you cause that particular physical object in the world, your hand, to rise. The PA therefore offers the possibility of conceiving divine actions that express divine intentions and agency without breaking natural law.[56]

It is worth noting that Clayton reverses the relation of containment. Whereas in Aquinas's version of the analogy "the soul [as higher and immaterial principle] contains the body," according to Clayton's panentheistic analogy "the mind indwells the body." The latter suggests that the immaterial principle is somehow contained within the material (bodily) principle, which might put into question the immaterial principle's status of being a higher principle (a ground for the defense of divine transcendence when the analogy is applied to the God-world relationship).

Most importantly, the analogy in question—when interpreted in terms of the classical hylomorphic philosophy of Aristotle and Aquinas—brings to our attention the category of formal causation. The human soul, as the principle of actuality, forms primary matter in an organic human body, thus grounding dispositions for all its efficient causal actions and reactions. This metaphysical context allows us to define precisely the type of causation exercised by the soul and to relate it to the physical (efficient) causation of an organism (folk description attributes it to human body rather than to the organism as a whole). The difference between these two types of causation and the metaphysical dependence of efficient on formal causes provide the ground for an interpretation of the analogy of a soul/body//God/world relationship in which God's transcendence and immanence are properly expressed and defended.

It seems to me that Clayton's interpretation of the same analogy lacks precision. He does distinguish two types of causation when he states that "every human agent knows" that mental causation "is more than physical causation." At the same time, however, he does not explain what the category "more" actually means in this context, that is, whether the distinction between mental and physical types of causation is merely quantitative or, rather, qualitative. This issue seems to be crucial for the proper expression of divine transcendence when the analogy mind-body is applied to the God-world relationship. Similarly, Clayton's description of "divine intentions and agency without breaking natural law" needs a more precise and positive qualification explaining the actual nature of God's causation, as well as the way in which his agency differs from the natural agency of the creatures. Otherwise Clayton's description of the panentheistic analogy might be in danger of falling into religious or theistic naturalism, described in sections 4.1 and 4.2 of the previous chapter.

Other contemporary panentheists remain less optimistic about PA than Clayton. I have already mentioned three qualifications restraining its use given by Peacocke, who notes that

1. unlike God, who creates the world, the human mind does not create the body;
2. unlike God, who is aware of everything, the human mind is unaware of many processes and operations of the body;
3. the world is not literally the body of God, who is also more than personal ("suprapersonal," or "transpersonal").[57]

What is more, the adequacy of the panentheistic analogy seems to be further challenged by contemporary neuroscience and philosophy of mind, which strive to explain the phenomenon of mind as a supervenient or emergent phenomenon of the brain.[58] Thus, in addition to the concerns listed by Peacocke, we can name another three limitations of the soul (mind)/body//God/world analogy:

1. While the soul in antiquity was seen as a life-supporting part of the human person, mind is regarded today—in light of the most recent hypotheses in philosophy of mind—as either a "supervenient" reality based on the "subvenient" causal basis of the human brain or an emergent phenomenon, which, if applied in panentheism, may suggest that God is a supervenient or emergent phenomenon with respect to the universe. God would then strongly depend on the universe. The metaphor, taken to its

logical conclusion, would imply not that God creates the world but that the world creates God.[59]

2. Many philosophers of mind and of science treat the mind/body metaphor as an anomaly that is unintelligible, which may diminish its explanatory value as the analogy of the God-world relationship in theology entering into a dialogue with natural science.

3. Because the body is essential to being a human, the question arises whether the world is essential to or a part of God and whether it is divine.[60]

Mindful of these difficulties, Gregersen concludes, "Attractive as the soul-body metaphor may have been in the past, it no longer commends itself as an adequate contemporary model for the God-world relationship."[61] This observation confirms my contention that the analogy in question does not bring more clarity to the meaning of the central ontological doctrine of panentheism.

2. SHORTCOMINGS OF THEOLOGICAL INTERPRETATIONS OF EM

In the previous chapter I showed that emergentism has inspired several theological proposals. Although they contribute greatly to the science-theology dialogue, they raise some important reservations and questions that cannot be left without comment. They will be the subject of my investigation in the remaining part of the present chapter.

2.1. Problems of Religious Naturalism

The agenda of religious naturalism, departing from any institutional religion, concentrates on a phenomenological description of natural religious attitudes and spiritual experiences, leading to a moral response. However, the question remains about the character and source of the phenomena in question. In her article coauthored with Deacon, Ursula Goodenough follows his concept of emergent teleology, which inspires her to suggest the possibility of creation and creativity without a creator. She seems to suggest that personal discovery of creation and creativity thus understood is grounded in a natural religious attitude and inspires spiritual experiences.[62] These attitudes and experiences open the possibility of human transcendent experience without divine transcendence, which triggers, in turn, a conscious moral response in relation to the world on our part. However, her position lacks a more precise definition and clarification of some of the terms she uses.

Concerning the naturalist understanding of "transcendence," if we assume Goodenough wishes to argue in favor of its reality without any reference to divine transcendence, the nature of human transcendental experience in her writings seems to remain undefined. Is it related to the discovery of Aristotelian transcendental attributes of being, such as beauty, goodness, or truth? Or does it refer to Kantian epistemology and Kant's idea of transcendental knowledge, that is, knowledge of how it is possible for our mind to "constitute" objects of our experience? Goodenough does not specify her position on this issue.

With reference to the concept of creation, if it is not understood as coming into being out of nothing (*ex nihilo*) and as the work of a divine Creator, it seems to be limited to a merely natural transformation of already existing entities. Even if we welcome Deacon's dynamical depth model of EM and his idea of causal absences, we must not forget that they also need to "operate" on already existing systems (see chapter 2, section 4.3). Their causality does not point toward nor presuppose *creatio ex nihilo*. Thus, Goodenough's use of the terms "creation" and "creativity" is meaningful and sustainable only in reference to the position arguing in favor of an everlasting world, whose constituents can be naturally rearranged in a way that inspires us to describe it in terms of creation and creativity and leads us to a spiritual response of reverence and gratitude. If it does introduce an ontological category of creation, its character is merely horizontal and not vertical. Whether religious naturalists are willing to accept such ontology remains unclear.

What is more, the same concept of creation proposed by Goodenough seems to either eliminate any question about the ultimate (and transcendent) origin and source of creation or to answer it in reference to immanent mundane reality (the system of the universe with emergent teleology intrinsic to it is all that is). Such an attitude is certainly naturalistic, but is it still religious? Is seems to me that the mere use of terms such as "creation," "creative," "spiritual," or "transcendent"—deliberately devoid of reference to any theological tradition—does not make one's position religious. Goodenough does not appear to be clear about what being "religious" without participating in a religious tradition actually means. In fact, her reflection on the phenomenon of EM in the evolving universe, distancing itself from any notion of purpose, plan, or divine transcendence and favoring a purely naturalistic approach, leads—from the theological point of view—to agnosticism, if not atheism. For what is most crucial in all religions is not the cultural aspect or the social metanarrative they offer, but the acknowledgment of and reference to the ontologically real transcendence or transcendent being, which many of them understand and recognize as a personal God.

Concerning the other version of religious naturalism, presented by Sherman and Deacon and based once again on the latter's idea of emergent teleology, I refer the reader to the analysis and answer to Deacon's concept of dynamical depth offered in chapter 2. There I discussed Deacon's insistence on the spontaneous character of orthograde and contragrade changes across the scale of all three levels of EM. I argued that the spontaneity of all processes bringing about new emergent phenomena needs to be rooted in the intrinsic character (nature or essence) of entities that enter these processes. Their intrinsic nature determines the propensities, dispositions, and natural teleology that characterize them. Hence, I have concluded that formal causation cannot be reduced to morphological shape or the geometry of phase space and that teleology is present at all levels of dynamical depth. Consequently, both formal causation and teleology cannot be reduced to physical properties.

If this is the case, however, the purely naturalistic explanation of finality that dismisses any notion of a transcendent *telos* might be put into question. If new and higher levels of teleology depend on the nature of entities and processes that "produce" them, the philosophical question of why these entities and processes have certain features and organize themselves in a certain way remains valid and can inspire a theological reflection. Obviously, one can refrain from pursuing this sort of reflection, remaining at the level of a purely natural description of the EM of various levels of novel teleological phenomena. Such an attitude is typical and commonly accepted in natural sciences. At the same time, teleology is also one of the central categories analyzed in philosophy of nature. This fact keeps reminding scientists that, strictly speaking, intrinsic goal-directedness is a metaphysical rather than a purely physical phenomenon (it is a metaphysical principle that finds its expression in the physical dispositions and behavior of a given natural entity). But to say that the natural description of teleology—be it physical or philosophical—makes theological reflection redundant and unnecessary crosses the boundaries and the methodology of both natural science and the philosophy of nature. Both, in addition to asking "how?" questions, pose "why?" questions, and neither is able to give an ultimate answer to the latter.

Thus, the fact that physical and philosophical investigation is able to name some basic features of the growing complexity of teleological phenomena that belong to the natural order does not preclude the reality of other aspects of teleology that may belong to the supernatural order, explored by theology. It seems to me that Sherman and Deacon may misunderstand apophatic theology when claiming that the physical aspects of *telos* prevent us from referring teleology to a "nonphysical God." We cannot forget that the way of negation (*via negativa*) in theological predication is usually contrasted—within the Western tradition of Christian theology—with positive assertions about God (*via positiva*). Following these two complementary paths our theological discourse enters the third way, which characterizes divine attributes as supereminent (*via eminentiae*).[63] Hence, although it is true that natural teleology cannot be attributed to God as a "defining feature" of his essence, which is totally unlike any *telos* observed in nature, nothing prevents God as the pure act from being a transcendent source and supereminent fulfillment of all teleological phenomena in the realm of mundane reality.

2.2. Problems of Theistic Naturalism

Theistic naturalism has been often criticized for its deistic inclinations. We have seen that the self-sufficiency of naturalistic accounts in Drees's interpretation of reality makes him reluctant to speak of particular instances of divine action. This attitude brings him close to the deistic assumption that God created the world *ex nihilo* at the beginning of time and then left it to its own rights and laws of operation. Drees does speak of God as a creative and sustaining "Ground of Being," which suggests that he, in fact, goes beyond deism. Nevertheless, his strong commitment to naturalistic explanations offered by science seems to prevent his exploring the broader view of causation, which reaches beyond what is given in scientific data, opening the way to new perspectives in our understanding of divine action.

Despite Davies's claim that his modified panentheistic uniformitarianism enabled him to leave behind deism, which he embraced for most of his academic career, one may doubt whether he succeeded in developing a truly nondeistic view of divine action. For although he argues in favor of "God's continuing role of creating the universe afresh at each moment," he adds that God does this "without in any way bringing about particular events which nature 'on its own' would not have produced."[64] Davies's metaphor of modeling the universe as a chess game tends to be deistic as well. He sees God as setting the rules—that is, laws of nature—which ensure a trend toward greater richness, diversity, and complexity of emergent phenomena. But the final outcome of "the game" depends on creatures' playing it and on the specific character of some rules, which serve to "constrain, canalize, and encourage certain patterns of behavior, even if they do not fix the goal in advance" (e.g., nonlinear or inherently statistical laws that make room for chance).[65] I claim that without the distinction between the primary causation of God and the secondary and instrumental causation of creatures, which I will discuss in the final chapter, the metaphor offered by Davies has a strongly deistic flavor. It seems to limit the role of God to merely establishing the rules—that is, the laws of nature—without specifying what God's continuing action of creating the universe afresh at each moment actually means.

2.3. The Trap of Emergentist Pantheism

The most problematic, and thus the least favored, version of theistic naturalism is Alexander's naturalized concept of an emerging God. His view of the world of nature as "tending to deity," which is always becoming and subject to the same laws that apply in the entire universe, shows strong affinity toward pantheism. This may be the reason his version of theism did not gain much popularity after it was formulated at the beginning of the twentieth century.

Nevertheless, we have seen that Alexander's idea of God's transcendence as a result of nature's evolution "reemerges" in the thought of

Morowitz, and his concept of an "always becoming deity" goes hand in hand with the Whiteheadian process version of panentheism. Moreover, we have pointed toward Alexander's definition of the whole universe as the body of God, whose distinctive character of infinity separates him from all finites. Although this attempt at expressing both divine transcendence and immanence makes Alexander one of the pioneers of contemporary panentheism, his pantheistic inclinations show that the same tendency may be present in the latest versions of panentheism as well. The danger becomes all the more evident in the context of the difficulties with formulating a precise definition and understanding of the preposition "in" in pan-*en*-theism, which I have discussed above.

Kaufman's definition of God as "the religious name for the profound *mystery of creativity*—the mystery of the EM, in and through evolutionary and other originative processes, of novelty in the world,"[66] may be in danger of falling into pantheism as well. For although he emphasizes that creativity "is not a quasi-scientific *explanation* of why and how new realities come into being,"[67] its status as a philosophical and theological term remains unclear. Kaufman describes it as "the mystery of new realities continuously being created, the mystery of complex things emerging from things less complex, the mystery of the coming into being of the universe and ourselves in that universe, the mystery of an open and unknown future into which we are all moving."[68]

Since, in Kaufman's position, creativity identified with God is omnipresent and pervasive through all levels of complexity of the universe, this position can be understood as pantheistic. His claim that creativity operating in the world is mysterious does not suffice to defend its transcendent character.

2.4. Problems of Emergentist Panentheism

In light of the challenges facing religious and theistic naturalism, as well as Kaufman's concept of God as "serendipitous creativity," the proposition of emergentist panentheism developed by Peacocke and supported by Clayton seems to be much more plausible, well considered, and theologically nuanced. However, it too faces some important challenges. Peacocke's concept of the "ontological interface" between God and the totality of the world (all-that-is), as well as the idea of the "flow of information" that passes through this interface, without any transfer of matter and energy and without the involvement of any physical forces, raises some serious scientific, philosophical, and theological questions.

From the scientific point of view, the concept of the world-as-a-whole does not make sense within the context of contemporary cosmology, which states that the universe does not have a boundary. Referring to the basic models in big bang theory, Robert Russell shows that "the topology of the universe is a three-dimensional hypersurface expanding in time. The hypersurface is either "spherical" (closed/finite in size, eventually recontracts), 'flat' (open/infinite in size, expands forever) or 'hyperbolic' (open/infinite in size, expands forever)."[69] Consequently, the expanding universe does not have a boundary on which God could exercise a top-down (whole-part) divine action "from the outside." Moreover, only with difficulty may we conceive scientifically a flow of information that occurs without a transfer of matter and energy and without the involvement of any physical forces. Peacocke seems aware of this problem when he acknowledges that at all levels observable by us in nature any input of information requires some input of matter/energy, even if only minimal. He thus acknowledges that his theory of divine action faces still the "causal joint conundrum," which is now analyzed at its ultimate level, involving "the very nature of the divine being in relation to that of matter/energy."[70] Nevertheless, although he acknowledges the problem, Peacocke leaves it unresolved.

From the philosophical point of view, Peacocke's "causal joint conundrum" raises an ontological question concerning the ultimate nature of God and the universe. His idea of the world-as-a-whole remaining "in" God and his concept of the whole-part influence of God on the world both lead inevitably to a spatial interpretation in which God and the world need to be ontologically the same *type* of being/substance (not necessarily the same *particular* being/substance). We saw Peacocke arguing that the world is within the "Being of God," but "with its own distinct ontology." The terminology used in this statement seems to be grounded in Aristotelian-Thomistic metaphysics, but Peacocke appears to misunderstand its meaning. Thomism speaks about the participation of the creation in God's being, but it avoids the spatial concept of remaining "within" the being of God. The distinctiveness of the world's ontology means that, unlike God, who "is" being himself, all created animate and inanimate entities "have" being, that is, "participate" in God's being. Spatial categories used by Peacocke are irrelevant here, which proves once again that the ultimate ontological status of God and creation in his writings remains unclear.

Philosophical analysis of causality as it is described by Peacocke raises other important questions as well. He pays much attention to the top-down (whole-part) causation in natural systems, which is an equivalent of DC in EM theory. However, similar to many other thinkers, he does not specify, either scientifically or philosophically, the ultimate nature of this causation. At the same time, Peacocke's claim that the God-world relationship does not involve an exchange of "matter or energy (or forces)" suggests that in the case of natural systems top-down causation does involve such an exchange, which makes us interpret it as a physical, efficient causation. Chapter 1 demonstrated that such interpretation of DC is highly problematic.

Another problem of Peacocke's understanding of causation in complex systems and across the "interface" between God and the totality of the world is that he apparently concentrates his analysis on the classical, mereological version of EM, based on part-whole relations and DC. We have seen that this type of emergentism faces insurmountable difficulties. I have also distinguished it from the "increased constraints" emergentism proposed by Deacon, which emphasizes more the dynamic aspects of reality and the role of specific forms of absences, increasing systemic constraints. It is important to note that, unlike Deacon, Peacocke does not pay attention to the revival of the Aristotelian plural notion of causation in contemporary philosophy of nature. Although he acknowledges the importance of the systems approach in biology, he seems to be unaware of the complexity of its philosophical presuppositions.

Finally, from the theological point of view, Peacocke's analysis of and struggle with the "causal joint conundrum" proves he does not escape the tendency of treating God as a univocal cause that belongs to the same order of causation as natural causes. What becomes obvious in this context is Peacocke's concern that God's action should not violate any laws and regularities applying to the various levels of the complexity. But one cause can violate another only if they are of the same sort. Only univocal causes need to "leave room" lest they interfere with one another. Moreover, Peacocke's conviction that divine being and divine attributes must be redefined in the context of contemporary science flows from the same univocal understanding of God's nature. I will say more about the problems of such an attitude in the next chapter.[71]

Another challenge to Peacocke's theology comes from his concept of God's exercising divine action on the world-as-a-whole, which seems similar to the theological naturalism of Reese and Davies and shows the same deistic tendencies. Analyzed from the human perspective, Peacocke's notion of divine action remains indistinguishable from a natural operation of the processes in the universe. Consequently, his defense of "special divine action" is also unconvincing, as it cannot be distinguished from "the divine holding in existence all-that-is." The measure of "an increasing intensity and precision of location" of divine top-down influences in Peacocke's view of special divine action remains unspecified, which renders the whole argument questionable.

Finally, Peacocke's concept of the "flow of information—a pattern-forming influence," passing through the "ontological interface" between God and the totality of the world, seems to refer to a flow that is directed one way. Aside from the problem of the undefined causal nature of this "pattern-forming influence," it seems to flow from God, who is able to pass information without an exchange of "matter or energy (or forces)" to his creatures. Since no creature can act in the same way (which Peacocke himself acknowledges), one cannot see how information might flow in the other direction. This fact challenges Peacocke's panentheistic commitment, for—as we have seen above—reciprocal causal influence between God and the world is one of its most important presuppositions. What is more, one would expect that this particular weakness of Peacocke's panentheism should prevent Clayton from easily accepting and affirming it, because he wants to link it to the dipolar notion of the God of process theology, who is by definition influenced by the world. Yet, contrary to this expectation, Clayton does not seem to see this problem, which significantly weakens the plausibility and coherence of his position.

This note closes our analysis of the major problems of panentheism and the difficulties concerning theological concepts inspired by EM theory, with special attention paid to the challenges of Peacocke's proposal of emergentist panentheism. We saw that his concept of divine action, defined in terms of the DC-based version of ontological EM and the panentheistic metaphor of the world being "in" God, who still surpasses it, raises important scientific, philosophical, and theological questions. This investigation, together with the critical analysis of Deacon's "increased constraints" version of emergentism and my attempt to reinterpret it in terms of both the classical and the new Aristotelianism—offered in the second chapter—inspire me to offer a constructive proposal of a new theological interpretation of the EM theory. I will develop it in the last chapter of this project, to which I shall now turn.

CHAPTER 5

Emergence and the Thomistic View of Divine Action

Among the philosophical and theological challenges facing the emergentist panentheism of Peacocke, described in the previous chapter, we must emphasize once again the three main obstacles that make his position difficult to accept. First, despite his attempts to defend divine transcendence, Peacocke's struggle with the "causal joint conundrum"—resulting from his description of the "ontological interface" between God and the totality of the world—shows his tendency to see God as a univocal cause that belongs to the same order of causation with natural causes. This univocal way of thinking about the causation of God and creatures explains Peacocke's strong emphasis on the completeness of natural causes and his ultimate wariness about the liability of God's interference with them—a view shared by Davies, Kaufman, and theistic naturalists such as Drees.

Second, the emergentist panentheism proposed by Peacocke suffers from a lack of precision in metaphysical terminology used to define the ontological relation between God and the universe. His suggestion to see the world-as-a-whole remaining within the "Being of God," whose infinity "includes all other finite entities . . .[,] comprehends and incorporates all," and "encloses" all existing entities, invokes inevitably the spatial interpretation of the panentheistic preposition "in"—the interpretation that Peacocke wants to avoid. Thus, even if he emphasizes that the world has "its own distinct ontology," Peacocke's project—due to its lack of ontological precision—runs into the risk of classifying God and his creation as the same type of being/substance (even if not the same particular being/substance).

Finally, the third major problem with Peacocke's position—shared by other advocates of theology inspired by EM theory—is his commitment to the DC-based version of ontological EM, which was challenged by Deacon's "increased constraints" emergentism. Based on the mereological description of part-whole relations and arguing in favor of irreducible DC of emergents, the classical type of emergentism faces the problem of reconciling its nonreductionist agenda with the physicalism to which it is dedicated. Although it has been suggested that this tension can be resolved if DC is reinterpreted in terms of Aristotelian formal causation, we were able to see in section 4.2 of chapter 1 that such a project has not been further pursued among classical emergentists.

Interestingly, Deacon's version of EM, with its strong emphasis on the importance of both formal and final causation, seems to follow the advice given by the classical emergentists. However, unlike the classical emergentism, the dynamical depth model of EM characterizes emergents in terms of the dynamic rise of constraints in natural systems, and not in reference to the notion of irreducible DC. At the same time, although it rejects the eliminativist approach to formal and final causation, the "increased constraints"–based model of EM is, nonetheless, open to their reductionist analysis. Reinterpreting them in the context of the "machinery" of the dynamical depth model of EM, Deacon sees formal causation as the function of the geometry of phase space and sees final causation as an emergent outcome of nature, that is, a result of mere mechanical, physico-dynamic processes. In the course of my investigation I have discussed some major metaphysical difficulties challenging Deacon's project. At the same time, however, I have proved that it bears a potential for bringing a true recovery of the Aristotelian view of causation and its reconciliation with contemporary science.

Although the DC-based model of ontological EM is able to recognize the role of formal cause only at higher emergent levels of complexity, the emphasis on the dynamical aspects of growing complexity in Deacon's dynamical depth version of EM—when reinterpreted in terms of both the classical and the new Aristotelianism of dispositional metaphysics— helps us realize that formal causation and teleology are rooted in the very nature of all inanimate and animate entities, at all levels of growing complexity of matter. Moreover, unlike DC-based emergentism, Deacon's "increased constraints" emergentism provides us with a theoretical model of the way in which new and higher types of teleology emerge, typical of higher dynamic systems and organisms, which are characterized by higher types of formal causation. It thus bridges the gap between lower and higher levels of ontological complexity, which remains problematic in the mereological type of EM.

Hence, I want to contend that Deacon's dynamical version of emergentism—reinterpreted in light of classical Aristotelianism and contemporary dispositional metaphysics—opens a way to a new theological interpretation of EM, which enables us to avoid the problems of Peacocke's emergentist panentheism. This new interpretation, which I wish to propose, is related to Thomas Aquinas's classic understanding of God as the source of being (*esse*), exercising his divine will through all four types of causation, listed by Aristotle and commented on by Aquinas himself. My interpretation will concentrate especially on Aquinas's concept of God as the final end (*telos*) and the creator of all substantial forms, acting through his divine ideas as exemplar causes of inanimate and animate beings. It will also refer to Aquinas's crucial distinction between the primary and principal causality of God, on the one hand, and the secondary and instrumental causation of creatures, on the other, as well as his distinction of four aspects of God's agency through secondary and instrumental causes. In what follows I will delineate the main points of my constructive proposal, beginning with some general remarks on Aquinas's understanding of God's nature and action.

1. AQUINAS ON THE NATURE OF DIVINE ACTION

Because according to the rule *agere sequitur esse* (action follows being) divine action must be a consequence of who God is, we must begin with a short reflection concerning the character of divine nature. Here, most fundamentally, we learn that—unlike created things, both animate and inanimate—God is *ipsum esse subsistens*, that is, self-subsistent, the shear act of "to be" itself. For Aquinas, the material and formal causes described by Aristotle explain the essence (*essentia*) of every material being. In addition to essence (which explains *what* a thing is) each thing also requires a distinct principle to explain the fact *that* it is. Aquinas names this principle *esse* (the "act of being"). What is at stake here for him is the fact that all creatures receive their *esse* from God. In other words, we may say that, as such, they depend on God in their existence (*esse*) and therefore "participate" in his divine being. However, we must immediately qualify our assertion, emphasizing that creaturely *esse* is, nonetheless, a created *esse* that cannot be predicated univocally with divine *esse*. For unlike God, a self-subsistent being (*ipsum esse subsistens*) whose essence (*essentia*) is identical with his existence (*esse*), creatures have essences that are not identical to their acts of existence. Put differently, the act of existence of each contingent entity is created, while at the same time it participates in the existence of God, the first and ultimate source of "all being" (*esse*).[1] As Aquinas explains, "Subsisting being must be one. . . . Therefore, all beings apart from God are not their own being, but are beings by participation."[2] Moreover, unlike creatures, God as *ipsum esse subsistens* does not have any potentiality. He is, therefore, a pure actuality that cannot change.[3]

This philosophical reflection on essence and existence becomes a point of departure for Aquinas's theology of creation and divine action. Aquinas insists that on God's part there is only one act, the act of God's being (*ipsum esse subsistens*), though from our human perspective we rightly distinguish the act of creation (*creatio ex nihilo*) and the act by which the world, once created, is kept in existence (divine *conservatio*, which in more recent theological reflection is often called *creatio continua*). Aquinas emphasizes that the act of *creatio ex nihilo* is not a change in created being (a modification of being) but is rather the production of all created being. The act of divine *conservatio* can be described as the sustaining of creatures in being.[4] Aquinas attributes both aspects of creation to the providence of God, which he sees as including not only the eternal plan of God ("the reason of order," *ratio ordinis*) but also the execution of that plan, which he calls "governance."[5] Moreover, because God is fully actual, we must acknowledge that his action is identical with his being: "God's power is His essence. Therefore, His action is His being. However, His being is His substance. Therefore, God's action is His substance."[6] If this is the case, however, it becomes clear that God cannot be treated as a univocal cause, acting along with the causality of creatures. Because "nothing agrees with Him either in species or in genus,"[7] God cannot be classified with creatures within the same ontological category. On the contrary, although immanently present in his creation, in the very essence of his divine nature, God must be a transcendent agent, the "cause hidden from every man."[8]

This conclusion touches on the thorny issue of God's infinite and innermost transcendence and immanence, often a cause of division between the followers of classical and some lines of more contemporary theology. In the previous chapter (section 1.2), I discussed Aquinas's understanding of divine immanence. Here, I add that his emphasis on the immutable and unchanging character of God's nature, radically different from that of all creatures, leaves no doubt that in Aquinas's theology God radically transcends the world he creates. Thus, we can see how Aquinas manages to find a way to affirm both divine immanence and divine transcendence, a way which seems to be lost in both historical and contemporary versions of panentheism—whether philosophical or theological.

Moreover, this conclusion, supported by a proper understanding of Aquinas's description of the God-world relation—real on the side of creation and of idea only with reference to God (see once again chapter 4, section 1.2)—proves that his theological system provides a ground for a consistent and coherent concept of divine action, which I find still valid in the age of science. One of that system's main points, strategic for this project, is the analysis of the causality of God in reference to Aristotle's four causes as well as his notion of chance, as interpreted by Aquinas. I will discuss these issues in the following section.

2. DIVINE ACTION IN RELATION TO THE FOUR CAUSES

I introduced in the previous section a key metaphysical distinction between substance (essentially considered as the composition of matter and form) and the act of existence. Aquinas saw that the essence (*essentia*) of each material thing—that is, primary matter actualized by a particular substantial form—is still, in its own order, in potency to the act of existence (*existentia*).[9] The nature of this act is not easy to grasp, because it is not a "thing" or an "essence," but that which actualizes essence— the principle that explains why essence ("*what* it is") in fact is ("that by which it *is*").[10] I have also emphasized the importance of Aquinas's characterization of God as the first source of *esse* for his theology of creation and divine action.

However, the causality of God is not limited, for Aquinas, to the Creator's enabling creatures to participate in his divine *esse*. God's action is also manifested in all four modes of causation described by Aristotle, as well as in events attributed to chance.

2.1. Divine Action and Material Cause

Beginning with material cause, we must note that because primary matter is pure potentiality, it would be erroneous to assert that God (total actuality) is the ultimate primary matter of each being.[11] At the same time, however, it remains clear for Aquinas that primary matter comes from God and retains a likeness to him. Aquinas thus must acknowledge that "also primary matter is created by the universal cause of being."[12] Thomas repeatedly emphasizes that God's action is expressed in creating and providing primary matter, understood as a source and principle of potentiality, a necessary principle for the possibility of the occurrence of all changes in nature. To this he also adds that because primary matter cannot exist without form, even in God the idea of primary matter is not distinct from the divine idea of the composite of matter and form.[13]

2.2. Divine Action and Formal Cause

Because formal cause reduces primary matter from potentiality to act, we may appropriately refer to God as the ultimate source of formal causation. Hence, states Aquinas, "form is something divine and very good and desirable." The reason we can say it is divine is that "every form is a certain participation in the likeness of the divine being, which is pure act. For each thing, insofar as it is in act, has form."[14] In other words, through their substantial form, creatures possess, in part, the actuality that is infinite in the Creator. Consequently, God can be said to act in the world as the creator of all forms and the source of all actuality.

At this point Aquinas goes beyond Aristotle's philosophy and his theory of intrinsic formal causation, introducing the Platonic idea of external exemplar forms (causes), which he sees not as subsisting entities, but as ideas in the mind of God: "In the divine mind there are exemplar forms of all creatures, which are called ideas, as there are forms of artifacts in the mind of an artisan."[15] I will now analyze the causality of divine ideas in more detail, because this will prove important in my theological interpretation of EM.[16]

2.2.1. Ideas as Exemplars
We must begin with an important distinction. Although we tend to think about ideas in cognitive terms, they can be also regarded as ontological principles. Following Augustine, Aquinas observes that according to the proper meaning of the word (in medieval Latin), an "idea" is called a "form" (Greek εἶδος, *eidos*), and thus includes knowledge of that which can be formed. It belongs, therefore, to the practical intellect, the end of which is action, that is, the production of things. Moreover, an idea belongs to practical knowledge either actually (when it is realized) or virtually (when it can be realized).[17]

The simple definition of an exemplar as "that in the likeness of which something is made"[18] requires further specification. Exemplarism is, in fact, an analogous notion. It can refer to (1) natural exemplars, that is, cases of natural agents causing effects sharing the same species as themselves (e.g., man generating man); (2) external exemplars, that is, models external to the agent's mind and to his nature (e.g., models imitated in art); and (3) intellectual exemplars, that is, ideas as forms in the mind of an agent (e.g., ideas in the mind of an artisan leading him to an artistic creation). Aquinas states that only the last kind of exemplarism defines "exemplar" according to the primary meaning of the term and that that sense provides the best analogy to the theory of divine ideas as exemplar causes: "The divine knowledge which God has of things can be compared to the knowledge of an artist, since He is the cause of all things as art [or artisan] is the cause of all works of art."[19]

Since an exemplar is that in imitation of which something is made ("the principle of the making of things"—*principium factionis rerum*),[20] and God does not make everything he knows, "only those notions [*rationes*] understood by God in imitation of which he wills to produce a thing in existence [*esse*] can be called 'exemplars.'"[21] In other words, Aquinas explains that "as an exemplar, . . . [an idea] has respect to everything made by God in any period of time."[22] Hence, only those ideas in the mind of God that are actually productive are exemplars.[23]

2.2.2. Ideas in the Mind of God

Defending the reality of divine ideas, Thomas presents three arguments, which we can classify as follows:

1. The "argument from natural teleology," which states that the natural order of things toward an end requires an intelligent agent directing the whole process by means of exemplar ideas.[24]
2. The "argument from divine similitude," which states that because every agent causes things similar to itself, the effects of the causation of God as an immaterial agent must preexist in him in an immaterial way, that is, as exemplar ideas.[25]
3. The "argument from divine self-knowledge," which assumes that as God has a perfect knowledge of himself, his power and its effects, this knowledge must consist of the ideas in the divine intellect.[26]

As one can see, all three arguments deduce the existence of divine ideas from the existence of God. As Gregory Doolan notes, the argument that is most general is the "self-knowledge" argument, which shows that God possesses ideas of anything that it is in his power to create, whether he creates it or not. The other two arguments are more specific as they refer to God's ideas of only those things he actually does create at some point. In other words, they concentrate on divine ideas that are operative and act as exemplar causes.

Aquinas appears to favor most the argument from teleology, which he finds ultimately inexplicable apart from a reference to the divine intellect. By means of this argument Aquinas both responds to Aristotle's rejection of exemplarism and provides an alternative to the exemplarism of Plato, whose *via abstractionis* made him assume that anything that can be separated in thought can exist as separated in reality. Aquinas shows that Plato's erroneous conclusion that ideas subsist independently of an intellect comes from the fact that his universal abstraction does not acknowledge the integral role of efficient and final causation in exemplarism. Aquinas, by contrast—in his argument from natural teleology—begins with the efficient causation of natural agents, which helps him discover the phenomenon of natural finality, which opens, in turn, the way to the discovery of the first intelligence and the involvement of exemplar ideas.[27]

2.2.3. Identification of God's Exemplar Ideas

Distinguishing between the divine essence as it is in itself and the divine essence as it is known by God as imitable, Aquinas affirms—from his earliest works—the multiplicity of the divine ideas, including ideas of genera, species, individuals, accidents, and even primary matter and pure possibles.[28] At the same time, however, he restricts the term "exemplar" to refer to the practical ideas of individual things that God makes (creates) at some point. These are ideas of particular substances or separable accidents, that is, accidents brought into being by an operation other than the one by which the subject was produced (e.g., a man's becoming a grammarian). In other words, Aquinas sees exemplars as necessary in the production of all concrete things so they can have a determinate form. He argues the necessity of "notions" (*rationes*) of all things in the divine wisdom, "which types we have called ideas—i.e. exemplar forms existing in the divine mind."[29] Thus, while Platonic Ideas are of universals, for Aquinas divine exemplars are of individually existing things.

Aquinas is careful to note that the plurality of divine exemplars is identical with the one divine essence inasmuch as that essence is known by God as imitable. For even if there can be no real multiplicity of ideas in the divine essence, there can be a logical multiplicity of them as objects of God's understanding, which accounts for the diversity of creatures. They imitate God's essence in different ways, as each living and nonliving entity has its own being (*esse*), distinct from that of every other entity. Hence, the multiplicity that Thomas attributes to ideas, although logical, is ultimately rooted in an ontological reality: "The one divine essence thus acts as the idea for everything that God makes, but it does so differently for different things according to the relationship that each thing bears to it. Since there is a diversity of such relationships, there must also necessarily be a diversity of ideas. If we consider the divine essence as it is in itself, then it is true that there is only one idea imitated by all things; but if we consider the various relationships that creatures have to that essence, there are many ideas."[30] Moreover, this reasoning enables us to distinguish between the two types of divine exemplarism: the exemplarism of divine ideas, which involves a perfect likeness between a finite being and its representative divine idea, and the exemplarism of the divine nature, which involves the degrees of imitation (greater or lesser), because none of the finite creatures can imitate the fullness of God's essence.

2.2.4. Causality of Exemplar Ideas

The most general characteristic of an exemplar idea in Aquinas's thought is that it is what he calls a form (*forma exemplaris*).[31] But in what way is an exemplar a form? It is a form not so much as it informs the mind of the knower, but as a principle of production—an operative form "in regard to which" (*ad quam*) a thing is formed.[32] It should not be identified with the substantial form (the intrinsic part of the very nature of the composite thing), because it is extrinsic and exists apart from the thing itself. Whereas substantial form forms a thing by way of inherence, an exemplar idea does so by way of imitation. Its action is expressed in a thing's being assimilated to its exemplar and the agent that possesses it and in receiving a determinate form from the exemplar (even if the thing in question is assimilated to its exemplar by means of its own intrinsic substantial form).[33]

Inasmuch as exemplars are, properly speaking, productive ideas, their causality is also necessarily related to efficient and final modes of causation. Because the efficient causation of an agent (e.g., an artisan) is a realization of exemplar ideas present in the agent's mind, we can state that exemplars play a role in efficient causation.[34] Moreover, an assimilation to the exemplar occurs because of an agent's intention, because a form in the intellect cannot be productive of anything except through the mediation of the will. The action of the will of an agent, in turn, is related to the goals intrinsic to his nature. Thus, what becomes crucial for a proper understanding of exemplar causation is finality or teleology, that is, the intentionality of an agent who predetermines the end for himself.[35]

Aquinas's reference to all three causes (formal, efficient, and final) in describing exemplar causation comes as no surprise in the context of his emphasis on the interrelatedness of causes (see chapter 2, section 1.5.2). At the same time, however, we must remember that at the end of the day, an exemplar is first and foremost an external formal cause for a particular entity. It can be referred to an efficient or a final cause only "in a certain way" (*quodam modo*), that is, with qualifications.

2.2.5 Causality of Divine Exemplars

Contrary to the Neoplatonist concept of necessary emanation, Aquinas emphasizes that God does not act by the necessity of his nature. Although he knows all things by the necessity of his nature—as the divine perfection requires that all creatures are understood in the Creator—this necessity does not entail a necessary creation. Quite the contrary, divine exemplars are ideas that are "determined by the divine will to an act."[36]

Because an exemplar is an extrinsic formal cause, divine ideas are not parts of creatures. If they were, God's essence would be part of them as well, because divine ideas are ontologically identical with God's essence. Thus, even if God, through exemplar ideas in his divine mind, is the formal cause of creatures, he is so only in an extrinsic manner. At the same time, however, divine exemplars as extrinsic forms are also causes of intrinsic forms in created things. What is more, God is the cause of the material principle of things as well and knows them not only according to their universal natures but also with regard to their individuality. Thus, Aquinas explains that the intelligible species of the divine intellect is "the principle of all the principles which enter into the composition of things, whether principles of the species or principles of the individual."[37] This fact reveals the limits of the analogy between divine exemplarism and human art. Whereas God is the cause of all the principles that enter into his work, a human artisan is not, because he knows and produces his work by means of its form alone (its matter has been prepared by nature).[38]

The exemplar forms in the divine intellect are productive of whole things, with respect to both their form and their matter. Therefore, the divine exemplars are related to individual creatures not only as to their specific natures but also as to their singularity. In other words, God's ideas of individuals correspond to particular creatures regarding their individual, specific, and generic nature. Doolan suggests that each divine exemplar, thus understood, might be described as a *forma totius* (the form of the whole), since it corresponds to all of the principles in a creature and is distinguished from a *forma partis* (the form of the part), which is found together with primary matter, entering with it into the constitution of the essence of a given created entity (actualized by a proportionate act of existence).[39] But if such is the nature of the causation of divine exemplars, does not it exclude the causality of natural agents? This question brings me to the consideration of divine action through efficient causality of creatures.

2.3. Divine Action and Efficient Cause

Speaking of Aquinas's understanding of God's action through the efficient causality of created beings, we need to put it in the context of the medieval debate between occasionalism and divine concurrentism.

2.3.1. Medieval Occasionalism
The Middle Ages gave rise to an intriguing opinion on causality, which remained in radical opposition to the developments proposed in the Western centers of studies in Oxford and Paris. Medieval occasionalism, related to *kalām*—a type of philosophizing in Islam, developed by Al-Ash'arī (d. 935) and Al-Ghazālī (1058–1111)—denied any kind of causation in creatures, on the assumption that things do not have natures that can guarantee enough constancy in their behavior. Questioning Aristotelian philosophy, which, in their opinion, reduced supernatural intervention to the notion of the "Unmoved Mover"—understood as the origin of all movement in nature—medieval occasionalists attributed causation in nature to the one true agent, God. In other words, they thought what we observe as cause-effect dependencies in nature are merely "occasions" for God to act.

The ideas of Al-Ash'arī and Al-Ghazālī went hand in hand with the occasionalist metaphysics of the Muslim Mutakallims, which originated in the middle of the ninth century. Seeking to prove that God was the sole power and active agent at work in the universe, they embraced a form of atomism, claiming that the existence of indivisible magnitudes in space entailed the existence of indivisibles of time. Mutakallims argued that God re-creates the world from one moment to the next. What we perceive as a continuous pageant of changes in accordance with natural laws of cause and effect is, in fact, merely the result of God's way of re-creating the world, with numerous changes, and yet in accordance with strict patterns and rules. Consequently, when a natural entity is seen to act, it does not operate by its own power; it is rather God who acts through it. There is no other meaning to the notion of cause and effect.[40]

2.3.2. Divine Concurrentism
The occasionalist view of causation was radically opposed by Averroës (1126–1298) and Moses Maimonides (1135–1204), both of whom

insisted on a self-evident character of the fact that things do have essences and attributes determining their functions and actions, which can be described using Aristotle's four causes. Additionally, Averroës objected to the suggestion that activity is legitimately attributable only to an agent having will and consciousness. He insisted that the distinction between natural and voluntary activity must be maintained, since natural agents always act in a uniform way (fire cannot fail to heat), while voluntary agents can act in different ways at different times and in different circumstances.

Aquinas sided with Averroës and Maimonides, arguing that occasionalism deprives natural things of the actions that belong and are proper to them. With regard to God as the creator of substantial forms, Aquinas reminds us that the likeness between the agent—that is, the efficient cause—and its effect observed in nature makes it unreasonable to pass over the natural generators of substantial forms and to claim that God obviates the causality of natural agents. He refers to the position of occasionalists who claim that natural agents cannot generate new substances and thus make new forms exist where they were not present before (creating *ex nihilo* and giving something they do not themselves have). He claims that this error of occasionalists arises from their misunderstanding and ignorance regarding the nature of form. They do not realize that what is produced *per se* is a composite thing and that the form is not, properly speaking, made, but is that "by which" something is made. Hence, what is made comes to be through acquiring form that is not subsistent but educed from the potentiality of primary matter, in causal processes that are proper for natural agents.[41]

Referring once more to divine exemplars, we can say they "con-create" form and matter, whereas natural agents "generate" things by causing forms to be educed from the potentiality of primary matter. Consequently, Aquinas states that natural agents are the cause of the coming-to-be (*causa fiendi*) of a thing, whereas they cannot be the ultimate cause of its being (*causa essendi*).[42] But the question remains still with regard to the nature of the efficient causation of creatures in the coming-to-be of inanimate and animate entities. Does efficient causation interfere or concur with God's action? In the previous sections we have seen God as the source of being, primary matter, and intrinsic substantial forms. We have also discussed the causation of divine exemplars, which Thomas defines as extrinsic forms, that is, divine ideas in the mind of God. The time has come to ask whether God acts through the efficient causation of natural agents as well and, if he does so, how we can explain the apparent problem of the double agency of God and creatures in causing one and the same effect.

2.3.3. Primary and Secondary Causation

Aquinas sees God as the first source of all efficient causation. He states that "all agents act in virtue of God himself: and therefore He is the cause of action in every creature."[43] At the same time, he introduces important metaphysical distinctions which help him specify the nature of efficient divine action in relation to the efficient causality of creatures. Following Aristotle (see chapter 2, section 1.4), Aquinas distinguishes first between God as the "primary cause" and creatures as "secondary causes," emphasizing that "God's immediate provision over everything does not exclude the action of secondary causes; which are the executors of His order."[44] Since God as the Creator has gifted every creature with its proper causality, according to its nature, his influence cannot interfere with this causality, but must rather be its source. Consequently, Although we can say that a particular natural effect comes to be both through the agency of God and through the agency of a created entity, we must remember "that the same effect is not attributed to a natural cause and to divine power in such a way that it is partly done by God, and partly by the natural agent; rather, it is wholly done by both, according to a different way, just as the same effect is wholly attributed to the instrument and also wholly to the principal agent."[45]

This is possible once we acknowledge that, metaphysically speaking, the transcendent God does not belong to the same order of causes as creatures. Even if "all created things, so far as they are beings, are like God as the first and universal principle of all being"[46] that is immanently present in their operations, the causation of the Creator transcends that of the creatures infinitely. The influence of the first cause is therefore not only more intense, so that we can state with Aquinas that "God is more especially the cause of every action than are the secondary agent causes."[47] In addition, we must realize that God's agency belongs, in its essence, to an entirely different ontological order, which makes it transcend infinitely the causation of creatures.[48]

2.3.4. Principal and Instrumental Causation

The passage from Aquinas's *Summa contra gentiles* quoted above, in which he attributes causal effects observed in nature to the agency of both God and creatures, introduces a further distinction in the realm of secondary causes, which remains crucial for his theory of efficient causation. Whereas some secondary causes act according to their natural dispositions, others produce effects beyond their capacities. Aquinas classifies the latter as instrumental causes and emphasizes their dependence on principal causes for their operation. He notes that the nature of an instrument is to move something while being moved itself by a principal agent. A saw working upon a bench has two operations, one belonging to its own form (to divide), and another "which belongs to it in so far as it is moved by the principal agent and which rises above the ability of its own form" (to make a straight cut agreeing with the pattern).[49]

To give another example, a flute played by the flautist produces an effect that exceeds its capacity. Hence, it is an instrumental cause of the sound, which depends on the principal agency of the flautist. However, when we consider the same flautist playing her flute in an orchestra, we realize that her action does not exceed her capacities. To the contrary, it is proportionate to her skills. But at the same time, guided by the gesture of the conductor, she contributes to a greater effect of the sound of the symphony. Hence, the flautist can be classified as the secondary cause of the symphony, a cause that contributes to it only under the influence of the primary cause (a conductor), while the flute she plays is the instrumental cause of the same symphony, contributing to it under the principal causation of the flautist.

Aquinas applies the distinction between principal and instrumental causation in his theology of divine action. Dodds notes this is a fact of great importance. Since all actions of efficient causality involve a bestowal of being, whether substantially or accidentally, distinguishing between principal and instrumental causes, we describe God as the source of absolute being, bringing the world into existence *ex nihilo* in the act of the initial creation, and keeping it in existence afterward. At the same time, we ascribe to creatures an instrumental causality in the instantiation of new particular beings, through substantial or accidental change.[50]

Consequently, we can further specify and nuance Aquinas's statement that natural agents cannot be the ultimate cause of being of a thing (*causa essendi*). Together with John Wippel I hold that, although Aquinas insists that creatures cannot produce the act of being (*esse*) *ex nihilo*, one can nevertheless say that they are—in a sense—causes of *esse*. Since being is always given by God in proportion to form educed from primary matter, if the latter happens through the operation of the instrumental causation of secondary causes, we can state that they are—analogously speaking—instrumental causes of *esse*. Hence, we find Aquinas saying: "Being is the proper product of the primary agent, that is, of God; and all things that give being do so because they act by God's power."[51] In other words, they do so as instrumental causes, acting under the principal causation of God.

2.3.5. Four Aspects of God's Agency through Secondary and Instrumental Causes

Aquinas's further explication of how exactly God acts in the world through secondary and instrumental causes can be found in his *Q. de pot.* 3.7. Summarizing the crucial part of the response (the corpus of the article), Ignacio Silva suggests distinguishing four aspects of efficient divine action in the world, in reference to Aquinas's list of four ways of being the cause of action of something else:[52]

1. To be a cause of something else means, first of all, to give it power to act, since every action, as a manifestation of a certain power, is ascribed to the giver of that power as effect to cause. "In this way"—says Aquinas—"God causes all the actions of nature, because he gave natural things the forces whereby they are able to act."[53]
2. To be a cause of something else means, secondly, to uphold (preserve) a natural power in its existence. Thus, "a remedy that preserves the sight is said to make a man see."[54] In this sense we may say that "God not only gives existence to things and their causal powers when they first begin to exist, but also causes existence in them as long as they exist, by preserving or sustaining them in existence. If the divine causality were to cease, all operation would come to an end."[55]
3. To be a cause of something else means, thirdly, to apply the power (a thing) to act. "A thing is said to cause another's action by

moving it to act"—says Aquinas—"as a man causes the knife's cutting by the very fact that he applies the sharpness of the knife to cutting by moving it to cut."[56] Hence, God can be thought to cause the action of every natural thing by moving and applying its power to act.

4. To be a cause of something else means, finally, to act as a principal agent working through the thing in question as an instrument. "In this way"—adds Aquinas—"again we must say that God causes every action of natural things."[57] What Thomas has in mind here is the fact of which I have spoken before, that is, that each efficient action includes causing ("giving") being (*esse*) in one way or the other. Since it is an effect (a perfection) that belongs to God alone to produce by his own power, created efficient agents must be considered as instrumental causes of (*esse*).[58]

Having listed all four aspects of efficient divine action in the universe, Silva suggests classifying the first two of them as "founding" and the other two as "dynamic."[59] He also claims that both opponents and proponents of Aquinas's doctrine tend to forget about the latter.[60] Among the members of the first group he mentions Thomas F. Tracy, who—paying attention only to the founding aspects of efficient divine action in Aquinas's description—comes to the conclusion that within his system "God acts exclusively at the absolute ontological ground of all events, and never acts directly to affect the course of history." Hence, when we ask whether "God responds to the dramas of human history, to the struggles of the Church, to the cries of the oppressed, to the restless human heart"—our answer must be in the negative.[61] In response to this objection, Silva notes that although Tracy properly explains the "static" (founding) aspects of God's action in creating and sustaining all entities and their powers, he leaves out the key dynamic aspects of the same divine action, which allow us to answer his question about God's providence and guidance of the universe through particular occurrences (special divine action) in the affirmative.

Silva contends that contemporary proponents of Aquinas's view of divine action show the same tendency to concentrate more on the first two founding aspects of God's agency in the world. I do not think, however, that this is always the case. As Silva himself notices, Rudi te Velde does offer a commentary (similar to his own) on four aspects of efficient divine action in the universe in Aquinas's *Q. de pot.* 3.7 co.[62] Although it is true that physicist and theologian William Stoeger concentrates more on the fact that "God operate[s] on a secondary cause . . . by bringing it into existence and conserving it in existence, . . . [i.e.,] continuing to endow it with the nature or properties it has" (which emphasizes the two founding aspects of efficient divine action), we must not forget that he also mentions God's applying created causes in particular causal occurrences—as he states that "God directly causes or constrains some created beings to act as secondary causes"—and speaks of "the nexus between God and the secondary causal instruments" (which alludes to dynamic aspects of efficient divine action).[63]

William Carroll, who concentrates more on the topic of creation and evolution, does seem to overlook the importance of the dynamical aspects of efficient divine action. Emphasizing the founding aspects, he states, "God does not only give being to things when they first begin to exist, He also causes being in them so long as they exist. He not only causes the operative powers to exist in things when these things come into being, He always causes these powers in things."[64] At the same time, he does not seem to discuss the fact that, nor underline the importance of the way in which, God applies these natural powers to act as instrumental causes in particular circumstances.[65]

The case of Dodds's detailed and careful description and analysis of Aquinas's position on divine action in the context of contemporary science is nuanced. Even if he does not apply directly the fourfold typology of founding and dynamic aspects of efficient divine action based on *Q. de pot.* 3.7 co., he does discuss them indirectly in his argumentation. He definitely acknowledges the importance of the first two founding aspects of God's agency in the world when he states that "every creature has a characteristic activity in virtue of its substantial form. . . . Aquinas sees such action as a manifestation of God's power and goodness, since God not only causes each thing to exist but also endows each with its proper causality," which he naturally sustains in his creatures as long as they exist.[66] Dodds does not forget about the dynamic aspects of efficient divine action either. His inquiry into the way in which God as primary cause acts through secondary causes alludes to the fact of God's applying natural causal powers in concrete situations. He also discusses—this time in more detail—God's principal agency working through instrumental causation of creatures in the bestowal of being (he gives a very helpful example of dogs producing puppies).[67]

2.3.6. Response to Contemporary Criticism of Aquinas's View on Efficient Divine Action in Creation

The participants of the dialogue between science and theology tend to identify Aquinas's view on efficient divine action with his entire teaching on God's agency in the world. They do not seem to pay attention to Aquinas's emphasis on God's agency through material and formal causes, as well as final causes and occurrences attributed to chance, which I will discuss in the next two sections. Nor do they consider the significance of his argument concerning God as the ultimate source of *esse*, bestowing it on creatures through the instrumental causation of other created entities. Consequently, even within their examination of Aquinas's view on efficient divine action, they seem to limit their interest to the distinction between primary and secondary action, with virtually no references to the other fundamental distinction between principal and instrumental agency of God and creatures. They give their interpretation of Aquinas's position the name "double agency" and bring against it several charges:

1. This notion of divine action is not able to do justice to both God's and creatures' causal agency. It forces us, at the end of the day, to admit that it is either God or nature that produces the effect. Hence, notes Nancey Murphy, it "inevitably slides back into occasionalism or else assigns God the role of a mere 'rubber stamp' approval of natural processes," which makes it fall into deism.[68]
2. Double agency theory raises the question of intelligibility. On the one hand, when interpreted metaphysically, it does not seem to say much about how exactly God's primary action works in/through created entities. Hence, contends Polkinghorne, "this leaves the idea looking like mere fideistic assertion, compatible with any known facts about the way things happen and so lacking any interpretative force in relation to the way things happen."[69] In fact, adds Peacocke, instead of answering the causal joint conundrum, it leaves an "ontological gap" at the causal joint between the agency of God and that of creatures.[70] On the other hand,

when interpreted from the scientific point of view, double agency brings back the conflict, as it seems to see God as "one of the efficient causes affecting (and effecting) every event." Hence, adds Clayton (summarizing Tracy's view), "it envisions a type of continuous divine intervention in the world that is no weaker than the classical accounts of miracles."[71]

3. An additional charge brought by Polkinghorne states that double agency "makes God directly responsible for all that happens, thereby intensifying the problems of theodicy."[72]

4. Finally, Murphy claims that the same double agency model of God's agency in the universe "leaves no room for any sort of special divine acts."[73]

In answer to these charges it is important to note, at first, that all of the authors mentioned here take as an object of their criticism the account of double agency offered by the twentieth-century English, Oxford-based theologian Austin Farrer. Polkinghorne is the only one who mentions—yet with no reference to primary or secondary sources—that double agency "is a theological tradition at least as old as Thomas Aquinas."[74] On the one hand, Farrer's account does follow, to some extent, the view of divine action offered by Aquinas when he speaks of

> two agencies of different level taking effect in the same finite action, the finite agency which lives in it, the infinite agency which founds it.... On the theistic hypothesis everything that is done in this world by intelligent creatures is done with two meanings: the meaning of the creature in acting, the meaning of the Creator in founding or supporting that action.... Where the creature concerned is non-intelligent there are not two meanings, for only the Creator has a meaning or intention. But there are still two doings, it is the act of the Creator that the creature should either act or be there to act.[75]

Moreover, one could argue that Farrer is still in line with Aquinas's thought when he states that "the causal joint (could there be said to be one) between God's action and ours is of no concern in the activity of religion.... Both the divine and the human actions remain real and therefore free in the union between them; not knowing the modality of the divine action we cannot pose the problem of their mutual relation."[76]

On the other hand, however, the last sentence from the second quotation makes Farrer's position vulnerable to the critical charges raised by Polkinghorne, Clayton, Peacocke, and Murphy. Following the line of skepticism about our ability to describe more precisely the way in which God's agency concurs with creaturely causal activity, Farrer concludes that this aspect of our theological reflection remains a mystery: "What sense is there in demanding an exact account of an action which, by hypothesis, is outside our knowledge? . . . But as soon as we try to conceive [divine agency] in action, we degrade it to the creaturely level and place it in the field of interacting causalities. The result can only be monstrosity and confusion."[77]

It becomes obvious that this kind of approach to divine action falls foul of the charges of not being able to find a middle ground between occasionalism and deism and of supporting modern fideistic tendencies in theology. Nonetheless, we must not forget that Farrer's position is not equal to the one offered by Aquinas. We have seen already that the point of departure of Aquinas's notion of divine action in the universe is God's agency in bestowing *esse* and being a source (the Creator) of primary matter (principle of potentiality) and each substantial form (principle of actuality), which is a realization of a divine exemplary idea and provides for a given natural entity's identity and dispositions. Hence, all efficient actions of an entity in question (living or nonliving), which are manifestations of its dispositions, are at the same time expressions of the primary agency of God, working through the activities of his creature acting as a secondary cause. If we want to apply the category of "causal joint"—the relevance of which in this discussion remains still highly questionable—it must be said that it is of a metaphysical nature rather than a physical one. It is through bringing all created entities into existence, keeping them in existence, and being a metaphysical source of their identity (essence) and all the efficient actions that realize their natural dispositions that God is present and acts in the universe.

Moreover, as emphasized by Silva, God's agency—applying created causes in particular situations—becomes also a principal type of agency with regard to the special kind of secondary causation of creatures, that is, their instrumental participation in God's bestowal of *esse* on new entities that come into existence in mundane reality. This fact becomes another important expression of divine action in the world.

These two significant assertions enable us to answer the main charges against Aquinas's views of efficient divine action and of divine action in the universe in general:

1. First of all, they show that Aqinas's proposition avoids the fallacy of both occasionalism and deism. It does not see God being in charge of everything in a way that would put into question the causal autonomy of creatures. Neither does it perceive God as leaving the universe entirely to its own causal operations after having created it.
2. Secondly, a precise and careful metaphysical investigation—in reference to some crucial ontological categories of the Aristotelian-Thomistic philosophical system—enables Aquinas to give a quite comprehensive answer to the question about the way in which God's agency concurs with the causality of his creatures. Although this answer enables him to escape the charge of fideism, it does not eliminate the notion of ineffability, because it acknowledges that God is ultimately beyond human reason and that our predication of him must be analogical.
3. Thirdly, in reference to Polkinghorne's charge concerning theodicy, one must not forget that for Aquinas (as for Augustine and other church fathers) evil is not an actuality in itself, but the privation or lack of an actuality that should be present. Aquinas applies this truth to both natural (physical) and moral evils (sins). Consequently, in his account evil occurrences and actions are not causes *per se* but merely (if they can be regarded causes at all) *per accidens*.[78]
4. Finally, answering Murphy's objection about the possibility of special divine action within the double agency account of efficient divine action, Silva's fourfold way of understanding Aquinas's position shows that God's primary and principal agency in each creaturely secondary and instrumental action is objective—that is, general—but it does not preclude it being special, that is, applied in a particular situation. Moreover, Aquinas acknowledges also the reality of unique cases of special divine action that we usually classify as miraculous: "Since the order of nature is given to things by God; if he does anything outside [*praeter*] of this order, it is not against [*contra*] nature."[79]

Consequently, notes Dodds,

> in addition to causing the natural actions of secondary causes, God can also "act independently of the course of nature in the production of particular effects" [*Q. de pot.* 6.1 co.]. Through such action, God may produce effects that are beyond the power of nature [ibid.]. He may also restrain secondary causes from producing their normal effects or produce those same effects directly by his causal power alone [see *SCG* III.96.14; III.99.2 and 7; *ST* I.105.1 ad 3; I.105.2 co.; I.109.1 co.]. In addition, he may use secondary causes, such as the saints as his instruments in working miracles [*Q de pot.* 6.4 co.].[80]

Summing up the section on the efficient mode of divine agency in the world, I want to emphasize once again that the two metaphysical distinctions offered by Aquinas (between primary and secondary and between principal and instrumental types of causation) enable him to argue in favor of divine concurrentism, attributing the source of efficient causation fully to both God and natural agents. His reasoning avoids the charge of overdetermination, because it is based on the ontological distinction between the causation of God and that of his creatures. Moreover, it becomes an important alternative to theologies that view God acting in the world as a cause among causes and therefore tend either to limit divine action to the realm of indeterminism or to introduce a kind of divine self-limitation that would prevent God from interfering with the causality of creatures. The use of secondary and instrumental causes "does not bespeak any divine limitation, but (if anything) a divine exuberance in willing to share 'the abundance of his goodness' with creatures."[81]

2.4. Divine Action and Final Cause

The last of the four causes, which both Aristotle and Aquinas regard as the "cause of causes," is the final cause. According to Aquinas, similar to other modes of causation, all forms of natural teleology find their ultimate source and fulfilment in God. Commenting on the normative aspect of teleology, he notes that "the end of all things [by which he seems to understand the totality of the universe] is some extrinsic good," which is "outside [extrinsic to] the universe."[82] It is possible to say that the same end is also desired by each creature individually, as they are looking for the fulfillment of their particular natures. In other words, ἐντελέχεια (*entelecheia*), an ultimate actualization of form in the final state of development (actualization) of an entity, bears some likeness to God and his goodness. It brings Aquinas to the conclusion that "everything is . . . called good from the divine goodness, as from the first exemplary effective and final principle of all goodness."[83] Consequently, we must acknowledge that

> All things, by desiring their own perfection, desire God Himself, inasmuch as the perfections of all things are so many similitudes of the divine being. . . . And so of those things which desire God, some know Him as He is Himself, and this is proper to the rational creature; others know some participation of His goodness, and this belongs also to sensible knowledge; others have a natural desire without knowledge, as being directed to their ends by a higher intelligence.[84]

> All things desire God as their end, when they desire some good thing, whether this desire be intellectual or sensible, or natural, i.e. without knowledge; because nothing is good and desirable except forasmuch as it participates in the likeness to God.[85]

Most importantly, as Dodds notes, this influence of God as the first final cause is much more profound than "the force that moves the atoms in Newtonian science."[86] It does not involve any force, or the physical pushing-and-pulling (change of mass and energy) characteristic of efficient causality. It is simply a communication of the perfection and goodness of God, who—unlike the God of process-based panentheism—does not act to attain any fulfillment in his divine action:

> Some things . . . are both agent and patient at the same time: these are imperfect agents, and to these it belongs to intend, even while acting, the acquisition of something. But it does not belong to the First Agent, Who is agent only, to act for the acquisition of some end; He intends only to communicate His perfection, which is His goodness; while every creature intends to acquire its own perfection, which is

the likeness of the divine perfection and goodness. Therefore the divine goodness is the end of all things.[87]

2.5. Divine Action and Chance

Having studied Aquinas's account of divine action in reference to all four Aristotelian causes, we must now consider the question of God's causality and chance. Following Aristotle, Aquinas affirms a kind of mitigated determinism in nature. He distinguishes between events that happen always, for the most part, and those occurrences which normally require a cause—unconscious or conscious—and yet such cause cannot be found. He classifies the latter, respectively, as chance and fortune.[88] He thus acknowledges that although natural causes are necessary, insofar as they are determined to bring about particular effects, their necessity is suppositional, which becomes the source of contingency in nature. On several occasions Aquinas lists three main reasons for the reality of contingency:

> (1) First, because of the conjunction of two causes one of which does not come under the causality of the other, as when robbers attack me without my intending this; for this meeting is caused by a twofold motive power, namely, mine and that of the robbers. (2) Second, because of some defect in the agent, who is so weak that he cannot attain the goal at which he aims, for example, when someone falls on the road because of fatigue. (3) Third, because of the indisposition of the matter, which does not receive the form intended by the agent but another kind of form. This is what occurs, for example, in the case of the deformed parts of animals.[89]

It is important to note that contingency is good and wanted by God. As Creator and primary cause, he agrees that in the natural course of events "an effect does not follow from a first cause unless the second cause has already been placed." Consequently, adds Aquinas, "the necessity of a first cause does not introduce necessity into its effect unless the second cause is also necessary."[90] But it can be contingent. Hence, we might say that God agrees for the secondary (or instrumental) causes to determine, in a way, the primary (or principal) cause toward particular effects. At the same time, secondary and instrumental causes are never above the primary and principal causation of God, who freely decides that some occurrences in nature will come about as contingent and not necessary.

At the same time, we must not forget that—unlike a number of contemporary theologians discussing divine action in reference to natural sciences (including Polkinghorne, Russell, Murphy, and Tracy)—Aquinas would disagree that the universe needs to include a certain level of indeterminacy for special divine providence to have a space for maneuver.[91] Rather, he would say that God wants indeterminacy and contingency because a universe that includes all modes of being and acting is simply more perfect than a universe that lacks any of them: "It would be against the perfection of the universe if no corruptible thing existed, and no power could fail [in producing its effect]."[92]

This is true on the assumption that God's agency in the created world refers to all entities and events at all levels of the complexity of the universe. Hence, even if, concerning causality of chance and fortune, Aquinas agrees that they are not causes *per se*, he agrees with Aristotle that, as *per accidens* types of causality, they must be related to proper material, formal, efficient, and final causes, relevant to beings and processes in which they occur.[93] As such, chance events should be classified as contingent events, which Aquinas sees as remaining under God's providence:

> God, Who is the governor of the universe, intends some of His effects to be established by way of necessity, and others contingently. On this basis, He adapts different causes to them; for one group of effects there are necessary causes, but for another, contingent causes. So, it falls under the order of divine providence not only that this effect is to be, but also that this effect is to be contingently, while another is to be necessarily. Because of this, some of the things that are subject to providence are necessary, whereas others are contingent and not at all necessary.[94]

This explanation leads Aquinas to conclude that, in terms of God's action, nothing really happens by chance: "Things are said to be fortuitous as regards some particular cause from the order of which they escape. But as to the order of Divine providence, 'nothing in the world happens by chance,' as Augustine declares."[95] Dodds rightly observes that this assertion may be misinterpreted as suggesting radical determinism of all events in nature, which would de facto make chance and contingency imaginary and unreal. Such may be an interpretation of Aquinas's example of the fortuitous meeting of two servants which was actually foreseen by their master: "So far then as an effect escapes the order of a particular cause, it is said to be casual or fortuitous in respect to that cause; but if we regard the universal cause, outside whose range no effect can happen, it is said to be foreseen. Thus, for instance, the meeting of two servants, although to them it appears a chance circumstance, has been fully foreseen by their master, who has purposely sent them to meet at the one place, in such a way that the one knows not about the other."[96]

Answering this objection, Dodds states we should not forget that chance is a category which applies to secondary causality and which, as such, cannot be attributed to God. Chance events retain their character as random and "uncaused" in the order of natural causation, while they are intended, along with other kinds of contingency, necessity, and freedom, in the order of God's action. Therefore, "God's transcendent primary causality can act through chance itself as a sort of secondary cause just as it does through other secondary causes. As God is not the immediate cause of effects that occur through other secondary causes, so God is not the direct cause of events that happen by chance. As God does not distort the causality of other secondary causes, neither does God alter the causality of chance. . . . We might say that God's causality acts precisely through the 'noncausality' of chance."[97]

3. A CLASSICAL ACCOUNT OF EMERGENCE AND DIVINE ACTION

If it is true that God is the first and ultimate source of existence (*esse*), the Creator of primary matter and of all substantial forms, the first and principal efficient cause, working through secondary and instrumental causation of his creatures (including the "noncausality" of chance), the source of all teleology and the end of all things—then the new emergent levels of complexity, observable in nature and redefined in terms of the

Aristotelian fourfold notion of causation, should be regarded as expressions of divine action as well.

This fundamental observation can be applied, first, to the classical account of ontological EM. Since the coherence and plausibility of the irreducible character of EM and DC are defensible with reference to the retrieved classical division of four causes, the dialogue between natural sciences and philosophy can be further extended to include the theological perspective of divine action in and through emergent entities/organisms and dynamical processes they enter. Following Aquinas's line of reasoning concerning the nature of divine action, we may say that God communicates his goodness and perfection through all four types of causation, at all levels of complexity. The classical theory of EM and DC highlights some particular stages of one and the same process that is an expression of the abundance of God's actuality, which he wants to share with and bestow upon his creation.

3.1. Essence and Existence of Emergents as a Communication of Divine Actuality and Perfection

Most importantly, although God works/acts in the universe through the secondary causation of natural agents, what he communicates is his very being (*esse*), his ultimate actuality and perfection. In other words, our reflection on divine action in reference to the hylomorphic composition of matter and form helps us realize that even if substantial forms proper for emergents are educed from the potentiality of primary matter through the efficient causation of secondary causes, their actuality is ultimately rooted in the pure actuality of God. Consequently, substantial forms of emergents can be regarded as an expression of the abundance of new ways in which created entities can participate in the ultimate perfection and actuality of God's essence.[98] Similarly, the act of existence (*esse*) of concrete emergent nonliving entities and living organisms, even if caused instrumentally by creatures, is never actually produced (created) by them but is ultimately rooted in the actuality of God, whose existence (*esse*) is identical with his essence (*essentia*). Hence, the fact of coming into being and the continual existence of new emergent entities become new expressions of God's free will to allow his creatures to participate (based on the analogy of attribution) in his divine *esse* (his divine life).[99]

Moreover, as we consider divine exemplars of all emergent entities, we arrive at the same conclusion. Even if exemplary ideas in the mind of God are multiplied according to their relations to things, concerning their reality they are nothing other than the one divine essence, inasmuch as its likeness can be shared in many different ways by different things.[100] The fact of the continual EM of new entities and phenomena in nature shows the abundant fecundity of exemplary ideas in the divine mind, that is, the infinite richness of God's essence.

3.2. Efficiency and Teleology of Emergents as an Expression of Divine Agency in the Universe

As I said in my new philosophical interpretation of the classical view of EM (see chapter 2, section 3.2), efficient causality becomes an indispensable explanatory aspect of DC, because new types of activity and reactivity proper to an emergent entity are manifestations of kind-specific powers (dispositions), inseparably related to the entity's very nature (essence). In the context of theological reflection on divine action in EM, it becomes apparent that these new types of activity and reactivity are also new ways in which God, as the primary and principal cause, works in the created universe through the secondary and instrumental causality of his creatures. For God not only is an ultimate ground and sustainer of the new powers of emergents but also applies them in particular causal interactions and uses them as instruments in bestowing *esse* on other contingent entities that come into being.

Concerning theological interpretation of new kinds of teleology proper for emergents (see chapter 2, section 3.3), I want to recall my assertion that the natural finality of all created entities becomes another way of communicating the ultimate perfection and goodness of God. Complementary to the view of the Creator perceived as the ground and point of departure for all created entities and their perfections (the way of *exitus*), natural teleology enables us to see God also as the ultimate end and goal of all creatures (the way of *reditus*). Hence, whatever perfection is achieved as the fulfillment of a rational or a nonrational creature's natural tendency to achieve an ultimate actualization of its form (i.e., its ἐντελέχεια, *entelecheia*), that perfection bears similarity to and brings this creature closer to the ultimate perfection of God's being, which is "desired" by all created entities. Referring this rule to emergent entities and processes which they enter—with new expressions of natural teleology, proper to their natures (essences)—we might perceive them as new ways in which God directs the whole of creation toward himself. It is important to note that his divine action through teleology of emergents does not abrogate their contingency and freedom (in the case of conscious emergent creatures).

3.3. Advantages of the New Theological Interpretation of the Classical Account of Emergence

In the light of this analysis, I hold that all important aspects of DC—highlighted in my reinterpretation in terms of Aristotle's fourfold typology of causes, which confirms its irreducible character (see chapter 2, section 3)—become an expression and an indication of God's action in nature. I hold that this new interpretation of the theological implications of the classical account of ontological EM protects and properly expresses the intimacy of God's immanent presence in his creation, as well as the transcendence of his nature and divine ordering of things, in both their stability and their changeability.

Thus, this interpretation becomes an important and viable alternative to the emergent panentheism of Peacocke, discussed earlier. Most importantly, unlike his model of divine action, my own view of divine agency through changes and phenomena described in natural sciences and in philosophy of science as emergent does not limit God's agency to a general influence on the world-as-a-whole (in analogy to DC having an influence on lower levels of organization of matter, where God's influence "upon events in the myriad of sublevels of existence" would have to be intermediated). Quite contrary, it perceives God's originating and sustaining action in and through all particular cases of emergent transitions and their outcomes, in all aspects of essence and existence of emergent entities and processes they enter, which I was able to describe in reference to the reach index of metaphysical categories offered by the Aristotelian-Thomistic system of philosophy and theology. Moreover, my model avoids the problem of the univocal predication of God and his agency, as well as the scientific, ontological, and theological difficulties of the causal joint conundrum.

At the same time, however, I want to acknowledge, once again, that the mereological foundation of the DC-based account of EM becomes its serious limitation, because that foundation seems to recognize and localize formal and final aspects of DC only at higher, emergent levels of complexity. Consequently, it associates formal and final causes with wholes, and not with their parts, which leaves unanswered the question about the ultimate nature and causation of parts, building up wholes. In this context, my reinterpretation of Deacon's dynamical depth model of EM in terms of both the classical and the new Aristotelianism of dispositional metaphysics may prove to be even more adequate and fruitful in the investigation of the theological implications of emergentism.

4. THE DYNAMICAL DEPTH ACCOUNT OF EMERGENCE AND DIVINE ACTION

As has already been said, the secret of Deacon's "increased constraints" emergentism lies in his emphasizing the dynamic interactions and dependencies of constituents of complex structures and organisms rather than simply analyzing their whole-part relations. We have seen him paying a great deal of attention to the EM of teleology at the higher levels of complexity of matter, in reference to his concept of the "causality of constructive absences." However, although I found Deacon's argumentation convincing in its basic assumptions, my critical evaluation of his project in terms of classical Aristotelian and contemporary dispositional metaphysics inspired an important supplement and further metaphysical refinement of his theory delineated in chapter 2.

I have argued that the spontaneity of natural orthograde and contragrade processes, which underlie each stage of the growing complexity in Deacon's dynamical depth model of EM, needs to be grounded in potencies and tendencies characteristic of all natural beings, from elementary particles, molecules, and simple and more complex inorganic and organic substances and structures to living and conscious organisms. I have also said that it is precisely their nature that makes homeodynamic, morphodynamic, and teleodynamic processes possible. Consequently, I have emphasized that teleology is an irreducible qualitative characteristic of all levels of dynamical depth, inextricably related to dispositions of concrete beings that are constitutive for the processes taking place at each of those levels. Dispositional features of these entities are a function of their unity and identity, provided by their substantial form, and irreducible to the accumulation of quantitative aspects of dynamic systems.

These conclusions led me to suggest that although what is not present does matter—which makes Deacon's emphasis on the reality and informative character of constraints valid and important—constraints are nonetheless effects of causal interactions rather than causes themselves (they belong to the category of quality rather than to the category of action). What is more, the true causal dependencies taking place in dynamical systems require a description that goes beyond efficient causation and is irreducible to it. I have argued that such description is possible within the new Aristotelianism of dispositional metaphysics, which opens the way back to the classical notion of formal and final causation and shows that the account of dynamic and process aspects of living and nonliving entities is lacking without the notion of their substantial unity.

This reflection becomes an inspiration for the constructive proposal of a new theological interpretation of EM, which I will now offer.

4.1. Dynamical Depth as Communication of God's Perfection

The emphasis on the importance of formal and final causation, which characterizes all inanimate and animate entities participating in orthograde and contragrade changes at each level of dynamical depth, enables us to consider these types of causation as an expression of divine action. In line with my analysis of divine action in the context of the classical account of EM, I claim—together with Aquinas—that through formal causation and teleology God communicates his divine goodness and perfection at all levels of complexity.

On the one hand, similar to the DC-based view of EM, the "increased constraints" emergentism proposed by Deacon emphasizes some particular stages of one and the same process that is an expression of the abundance of God's actuality, which he wants to share with and bestow upon his creation. On the other hand, formal and teleological aspects of all entities/processes providing for orthograde and contragrade changes in the rising scale of dynamical depth reveal even more thoroughly and with a greater emphasis the depth of the immanent presence and action of God in originating novelty through secondary and instrumental causation of contingent entities. At the same time, while through this activity of secondary and instrumental causes God communicates his very being (*esse*), he remains always its ultimate, unchangeable, and eternal origin and source. This fact protects divine transcendence, providing for a proper expression of the complementarity of these two crucial attributes of divine essence.

Again, as in my theological interpretation of the DC-based account of EM, I want to emphasize that the growing complexity and perfection of emergents at each and every level of dynamical depth is grounded in their participation in the pure and ultimate actuality of God. The same divine actuality is the source of their efficient dispositions (powers) to educe new forms from the potentiality of primary matter and to instrumentally participate in God's bestowal of *esse* on new contingent entities that come into being. Hence, the important steps on the ladder of complexity specified by Deacon (homeodynamics, morphodynamics, and teleodynamics) become indicators of primary matter being properly disposed to receive substantial forms proper for the particular exemplars of both more static and thoroughly dynamic entities. These same entities find their ultimate blueprint in divine exemplars in the mind of God.

However, if the whole process of dynamical depth, and the various stages of the growing complexity of natural systems in particular, can be described as a communication of divine goodness and of the ultimate perfection of God's actuality and being, what is the principle of ontological plurality and diversity (multiplicity) that enables us to distinguish and derive many entities from the one being of God?[101] Interestingly, to answer this question Aquinas points toward the importance of nonbeing (*non-esse*) as having an essential role in answering the query concerning the meaning of multiplicity in the unity of divine exemplar ideas. His teaching on this subject becomes even more intriguing in the context of Deacon's theory of the causation of absences.[102]

4.2. Dialectics between Being and Nonbeing

When he speaks about the distinction between the many and the one and about the derivation of the many from the one, Aquinas notes first—in his commentary on Boethius's *De Trinitate*—that something is said to be plural (or many) by reason of its being divisible or divided.[103] In the case of things that are posterior and composite, the cause of division is the diversity of things that are simple and prior. But what about the latter? Things that are simple and prior must be divided of themselves. When he refers to the Parmenidean paradox, Aquinas notes that being (*ens*) cannot be divided from being insofar as it is being, for nothing is divided from being except nonbeing. In other words, "from this-being, this-being is not divided, unless in this-being there is included negation of the same being." Hence, "just as first being, inasmuch as it is undivided, is immediately recognized as one, so after division of being and non-being there is immediate recognition of the plurality of first simple beings."[104] This plurality in turn accounts for the diversity of created things that are secondary and composed.

Consequently, Thomas argues that the fundamental opposition (or division) between being and nonbeing is necessary to distinguish the first created effect from God, the uncreated cause. We may say that such an effect "is" or "enjoys being," presumably because it exists, and yet "it is not" or "is nonbeing" in some other way, as it falls short in perfection of its divine cause. Other effects created immediately after the first one will also combine being and nonbeing in a similar fashion. Moreover, each of them may be described as nonbeing in another sense, insofar as it cannot be identified with other immediately created effects.

From what was said in section 1 of this chapter we know that whatever proceeds from God, who is pure being (*actus purus*), falls short of God's simplicity. Unlike God, in whom *essentia* and *esse* are identical, contingent entities and substances are composed of essence (*quod est*) and existence. Now, in light of Aquinas's emphasis on the role of nonbeing in the derivation of many from the one, we may add that the very possibility of distinguishing between any creature's essence and the divine essence (which is identical with God's existence) rests upon and presupposes the division and the opposition between being and nonbeing. Moreover, the possibility of distinguishing between essence and existence within a complete being itself rests upon this most fundamental opposition as well.[105]

Therefore, from the standpoint of creatures, we must acknowledge the importance of nonbeing in the distinction and derivation of many creatures from the one being of God. This assertion, however, raises a crucial question asked by Wippel: "If God is to understand anything as other than himself, will not Thomas hold that God too must know it in terms of being and nonbeing?"[106] In other words, assuming that every actually existing creature which corresponds to a concrete divine exemplar is truly distinct both from the divine essence and from other creatures, we must ask whether God can view himself as imitable to a limited degree, by a given creature, without combining being and nonbeing in his very act of understanding.

This question is important because if the answer is positive, one might argue that—since God's act of understanding is identical to his act of creation—nonbeing (absence) becomes an integral aspect of divine creative power. This idea might be readily welcomed by those who support Deacon's argument concerning causation of absences. However, I think that the opposite is true, that is, nonbeing does not actually enter God's act of understanding. The defense of my position takes the form of a fairly complex and quite speculative argument, which might be difficult to follow for those less familiar with the Aristotelian-Thomistic tradition of philosophy and theology. Nevertheless, I hope that it will not discourage the reader but rather help him/her understand my explanation of the substantial difference between Aquinas and Deacon on the topic of nonbeing, which I will discuss in the following subsection. But first, the argument against the thesis of nonbeing entering God's act of understanding.

4.3. Does Nonbeing Enter God's Act of Understanding?

Among Thomists discussing the topic of divine ideas we find Vincent Branick, Wippel, and Doolan emphasizing that God must know things different from himself in terms of both being (insofar as they imitate the divine perfection to a given degree) and nonbeing (insofar as they fall short of the fullness of divine perfection). This causes these scholars to conclude that nonbeing actually does enter God's act of understanding. The early presentation of the argument by Branick in 1968 is the most bold and straightforward: "God, by knowing what he is, *Ipsum Esse*, knows what he is not and what is other than him, nonbeing in its most absolute sense. Absolute non-being is what is most directly other than absolute being. In this way the dialectic negating principle enters into the mind of God."[107]

Although Wippel, commenting on this issue in 1985, replaces absolute nonbeing with relative nonbeing, the ultimate shape of his argument in favor of the opposition between being and nonbeing in the mind of God is reminiscent of the one offered by Branick: "It seems that it is this most radical opposition or division between being and nonbeing which is required for Thomas to account for God's knowledge of creaturely essences. This in turn is required for there to be creaturely essences which can receive and divide *esse*. And this, finally, is necessary if there are to be created beings or a many."[108]

More recently (2008), the same argument finds considerable support in Doolan's investigation of divine exemplar ideas, in which he states:

> Just as creatures fall short of the infinite perfection of the divine essence, so too do the divine ideas. One idea is distinguished from another according to the degree of perfection that it represents, each one falling short of the complete perfection of the act of being that is the divine essence. In this respect, non-being in a relative sense must in some way enter even into the mind of God as a negating principle; for, in knowing things other than himself, God knows how they are *not* himself. In short, it would seem that God can only consider himself as imitable inasmuch as he combines in his act of understanding both being (*esse*) and non-being (*non-esse*).[109]

I find the argument presented by Branick, Wippel, and Doolan doubtful for at least two reasons. First, as Wippel acknowledges, the reasoning introducing the dialectic between being and nonbeing into the mind of God goes beyond Aquinas's *ipsissima verba*. A closer analysis of the passages from Aquinas used by all three scholars in support of their thesis shows that none of those passages actually speaks about nonbeing entering the mind of God. In the most suggestive argument quoted by Branick, Wippel, and Doolan we find Aquinas stating: "Created things, however, do not perfectly imitate the divine essence. Consequently, His essence as the idea of things is not understood by the divine intellect unqualifiedly, but with the proportion to the divine essence had by the creature to be produced, that is, according as the creature falls short of, or imitates, the divine essence."[110]

The exegesis of the argument leaves no doubts. Although Aquinas admits that ideas of things are understood by the divine intellect "with the proportion to the divine essence had by the creature to be produced," he does not define this proportion in terms of nonbeing. I am sure that there must be reasons why Aquinas—who in his commentary on Boethius's *De Trinitate* speaks openly about the importance of the idea of nonbeing for our human way of understanding the multiplicity of entities—does not introduce the same category of nonbeing when explaining God's knowledge. We must not forget that ideas in God (including exemplar ideas) are not "that by which" God understands (as ideas [expressed species] in the human mind are that by which we understand), but rather "that which is understood," that is, "what" God understands. In other words, unlike humans, who understand "by" ideas, God does not understand "by" anything; he simply "is" his act of understanding. Thus, although it is not contrary to divine simplicity to claim that God understands many things, it would be contrary to divine simplicity to hold that such understanding requires many ideas "by which" God understands. For divine ideas are simply God's understanding of many things, that is, God's understanding of the many ways his being/essence/goodness can be participated in.

Consequently, this way of reasoning does not introduce any plurality in God's knowledge as such, since he knows all things by the one act of knowing his essence, which is identical with his being. In this context, the argument presented by Branick, Wippel, and Doolan—who introduce nonbeing in God's act of understanding—may be in danger of imposing human categories of knowing on God, who does not understand by putting ideas together, even the ideas of being and nonbeing. For even if every creature participates in the perfection of God's being/essence/goodness only to some limited extent, and this limitation is essential to creaturehood, we must be careful in discussing how this limitation is related to divine knowledge. In that respect, it seems better to describe the limited creature as a mode of being rather than nonbeing. It is only from our point of view that the distinction and derivation of many creatures from the one being of God implies a reference to nonbeing. This may be the reason why Aquinas refrains from introducing nonbeing into God's act of understanding.

The other critical and problematic aspect of this conversation concerns the distinction between the intellectual and ontological levels of analysis. Both Doolan and Branick stress the fact that introducing non-being in God's mind does not mean that it enters the very essence of divine nature:

> Positing of non-being that allows for the production of a multiplicity of creatures can *only* occur by intellection. Since God is being itself (*Ipsum Esse*), non-being cannot enter into his essence except through intellection, since the only ontological status non-being has is an *ens rationis*.[111]

> For Thomas this diremptive state of the "positing of nothing" by *Ipsum Esse Subsistens* occurs only in intellection. The non-being in no way affects the reality of the original infinite being. Non-being can only be an *ens rationis*. It does have some meaning, namely, the absence of being and real truth.[112]

Doolan's statement defending the ontological immutability of God's essence remains highly ambivalent as he claims that "non-being" in fact does enter God's essence "through intellection." Although we saw him saying that nonbeing in this case must be understood in a "relative sense," he does not explain what exactly he has in mind. The whole idea becomes even more suspicious in terms of Doolan's thesis quoted earlier in this section, in which he tends to reify divine ideas, saying that "just as creatures fall short of the infinite perfection of the divine essence, so too do the divine ideas." To "fall short" of the divine essence expresses an ontological condition. If God's own ideas "fall short" of the divine essence, they cease to be God's ideas, since God's ideas "simply are" God. Branick's position suffers from the same criticism. If God's knowledge and being are identical, it is hard to understand how "non-being" as *ens rationis* does not affect the infinite essence of God.

The reflection of Wippel, as presented in the opening of this section, poses similar questions. It is true that his argument (that the "most radical opposition or division between being and nonbeing ... is required for Thomas to account for God's knowledge of creaturely essences") seems to refer to God's intellect and not to God's being. At the same time, however, we find Wippel saying that the suggested opposition between being and nonbeing in the divine intellect "is required for there to be creaturely essences which can receive and divide esse." This suggests—according to the rule of the unity of God's essence and intellection—that the whole argument has important ontological consequences. It seems to introduce into God's essence a dialectic of being and nonbeing—an idea somewhat closer to Hegel but foreign to Aquinas.

Summarizing my critical investigation of the proposition supported by Branick, Wippel, and Doolan, I prefer to follow the original thought of Aquinas, who does not introduce the idea of nonbeing into divine intellect. I find this fact crucial for my project, because it confirms once again that nonbeing (absence) has an important, and yet merely descriptive role in our human distinction and derivation of the many from the one. It is informative and tells us something about the nature of living and nonliving entities. It describes the limited degree of their participation in the perfection of God's being, in terms of the substantial form and natural teleology proper to every creature. At the same time, nonbeing as such cannot be causal and cannot enter the realm of God's exemplar ideas, which are causal by definition.

The realization of this fact becomes crucial for my comparison of the meaning of nonbeing in the thought of Aquinas and Deacon, which I pursue in the following section.

4.4. Aquinas and Deacon on the Nature and Role of Nonbeing

What certainly becomes intriguing, in the light of my investigation so far, is that both Aquinas and Deacon recognize the importance of nonbeing in defining the nature of living and nonliving entities. Nevertheless, the question remains whether they agree as far as the character and nature of nonbeing is concerned.

With regard to Aquinas's understanding of nonbeing, we must acknowledge and emphasize that he does not treat it as a causal principle. In reference to his distinction between *per* se and *per accidens* modes of causation (see chapter 2, section 1.6, and chapter 5, section 2.5) Thomas applies the same differentiation when discussing the origin of change and of the coming to be of things in nature. In his commentary on Aristotle's *Physics*, we find him saying:

Nothing comes to be from non-being simply and *per se*, but only *per accidens*. For that which is, i.e., being, is not from privation *per se*. And this is so because privation does not enter into the essence of the thing made. Rather a thing comes to be *per se* from that which is in the thing after it has already been made.... Thus a thing comes to be *per se* from being in potency; but a thing comes to be *per accidens* from being in act or from non-being.... He [Aristotle] says this because matter, which is being in potency, is that from which a thing comes to be *per se*. For matter enters into the substance of the thing which is made. But from privation [non-being] or from the preceding form, a thing comes to be *per accidens* insofar as the matter, from which the thing comes to be *per se*, happened to be under such a form or under such a privation. Thus a statue comes to be *per se* from bronze; but the statue comes to be *per accidens* both from that which does not have such a shape and from that which has another shape.[113]

It becomes clear that Aquinas sees the *per se* source of change, novelty, and coming to be of things in the principle of potency, and not in nonbeing. Because it is defined as negation or nothingness—according to the principle saying that nothing comes from nothing—nonbeing can be thought of as a source of a new being only *per accidens*. Similar to the principle of actuality in an actualized being (i.e., its substantial form), which by itself does not have any potency, nonbeing as sheer nothingness does not have potency either. Yet it is only what is in potency that can be regarded as a *per se* source of change and coming to be of new things in Aquinas. Now, the principle of potency is described in his metaphysics in three complementary ways. First, with respect to the hylomorphic composition of matter and form, primary matter is a source of potency, which is actualized by substantial form. Thus, if an actualized entity changes into something else, it happens due to the potentiality of primary matter and the activity of an efficient cause, educing a new substantial form from it. According to the other distinction, the proximate (secondary) matter is in potency to receive a new accidental form, as in the example of a statue coming to be from bronze. Finally, the essence of a thing (i.e., the composition of primary matter and substantial form) is still in potency with regard to the act of existence (*esse*), which is a participation in God's infinite *esse*.

Most importantly, neither primary nor secondary (proximate) matter nor essence, when described as distinct from, respectively, substantial form, accidental form, and the act of being (*esse*), can be regarded as nonbeing, except in some analogical sense of this term. In other words, potency, as a *per se* source of novelty and change, is not "nothing" in Aquinas's metaphysics.[114] At the same time, nonbeing—as sheer nothingness—is not in potency to become anything, and thus it cannot be regarded as a *per se* principle or cause of change. If we still find nonbeing characterized as a source of change in some sense, we must remember that nonbeing can be called a principle of change only *per accidens*. Consequently, although nonbeing remains an important category in the human intellectual distinction and derivation of the many from the one, it does not have in Aquinas's metaphysics an ontological and causal character. Nor does it enter into the realm of exemplar ideas in the mind of God. For he always sees the many limited degrees of imitation of his perfection by creatures as participations in his infinite being, and thus does not need to define/understand these degrees of imitation in reference to the principle of nonbeing.[115]

Concerning Deacon's understanding of nonbeing, we saw him defining it in terms of "constraints," limiting possibilities of action in dynamical systems, and leading to EM of higher levels of complexity that give origin to teleology. I have argued that nonbeing thus understood cannot be classified as sheer nothingness, since constraints are always related and defined in reference to concrete existing entities, dynamic systems, and living organisms, which are susceptible to being constrained. I have also remained skeptical about Deacon's assigning to constraints (absences) a real causal character (see chapter 2, section 4.3). Apart from the difficulty of specifying the nature of such causation, I have shown that what is truly decisive about the existence of degrees of freedom in dynamical systems and organisms and about their possible reduction is both the irreducible formal principle—characteristic of each natural kind and its efficient causal dispositions—and natural teleology (the principle of final causation), which is also natural-kind specific.

I have also mentioned that—in response to my criticism—Deacon formulates, in one of his more recent publications, a rather surprising thesis that constraints, in fact, do not do any causal work. I think this line of reasoning brings him closer to the teaching of Aquinas, whose thought makes me conclude that, though informative about and indicative of the reduction of degrees of freedom, constraints do not actually cause it. This is the main point of my reinterpretation of Deacon's version of EM, which allows me to see his theory of a successive line of specific "absences" or "constraints"—in terms of both classical and new Aristotelianism—as referring to a successive line of specific states of "disposition" of primary matter to be informed by new (higher) substantial forms. This metaphysical refinement of Deacon's dynamical depth model of EM proved to be of crucial importance for the notion I have offered of divine action through the emergent processes in nature. I will turn to it one last time in the remaining section of this chapter.

5. THOMISTIC INTERPRETATION OF EMERGENCE VERSUS EMERGENTIST PANENTHEISM

My analysis of Aquinas's concept of divine action and its application in the theological interpretation of both the classical and Deacon's dynamical depth accounts of ontological EM shows that emergent phenomena can be regarded as a way of communication of God's goodness and perfection. In this final section I will indicate the main advantages of this new theological interpretation of EM over Peacocke's and Clayton's emergentist panentheism.

Most importantly, as I have already said in section 3.3 of this chapter, unlike the followers of the mereologically based emergentist panentheism of Peacocke, I do not have to "restrain" divine action to God's influence on the totality of the world (which assumes that divine action on entities and events on lower levels is somehow intermediated). Quite to the contrary, I see God as present and operative at all levels of complexity of matter, through the formal, efficient, and final aspects of the DC-based view of EM, as well as at all levels of the dynamical depth model of EM, which explains the growing complexity of entities and the natural processes they enter. This is important for the emphasis on the immanence of God, which, paradoxically, seems to be diminished in Peacocke's emergentist panentheism.

Moreover, I hold that my reference to (1) Aquinas's theory of primary and principal causation of God, working through secondary and instrumental agency of the creatures, (2) his explanation of divine transcendence and immanence, and (3) his plural notion of causation and distinction between *esse* and *essentia* answers several other problems challenging emergentist panentheism:

1. Conceiving God as the transcendent First Cause, acting through the natural secondary and instrumental causality of creatures, in reference to all four types of causality listed by Aristotle and the *per accidens* causality of chance and fortune, defines the "ontological interface" between God and the world. Unlike the interface suggested in emergentist panentheism, the one expressed in this conception is not physically "located," according to a strong meaning of this term, such that we would feel obliged to point toward the physical (empirically verifiable) boundary between God and the universe. Thus, this conception avoids the difficulties challenging Peacocke's idea of God acting on the totality of the world (world-as-a-whole).
2. Aquinas's emphasis on understanding divine action in terms of not only efficient but also formal and final causes—followed by the additional assertion that God causes primary matter as well—provides a sound theological description of God's causality in the universe. Aquinas's description is free from the difficulties challenging Peacocke's concept of the flow of information without an exchange of matter and energy and without physical forces. Whereas Peacocke's position seems to remain close to the physical—and, thus, empirically verifiable and mathematically describable—approach to causation, Aquinas's view is defined in clearly philosophical and metaphysical terms, which seems to be more appropriate for theological reasoning and argumentation. That such philosophical theology, developed in reference to classical and new Aristotelianism and Thomism, is not ignorant of the truth about nature revealed by natural sciences becomes apparent in the light of the project pursued in this volume.
3. Consequently, Aquinas's explanation of the God-world relation and divine action escapes the problem of the "causal joint conundrum," which remains unanswered in the emergentist panentheism of Peacocke. From the point of view of the Thomistic theory

of God's agency in the universe, the very question of the "causal joint" between God and the world is metaphysically ill conceived.

4. One of Peacocke's main arguments, when he speaks about the whole-part influence of God on the totality of the world, is his concern about divine action possibly abrogating natural laws and regulations. This univocal tendency in speaking about the causality of God is again corrected by Aquinas's philosophical and theological distinctions mentioned above. Once again, his reference to the primary and principal action of God—the source of all perfection, actuality, and being, acting "through" the secondary and instrumental causation of creatures as it is described in reference to all four Aristotelian types of causes, as well as through events attributed to chance and fortune—enables us to see how God acts immanently in all natural processes without interfering or canceling out causal powers of their participants. At the same time, this conception of God's action does not require introducing kenotic self-limitation of divine attributes on the part of God.

5. The same metaphysical background of the Thomistic interpretation of EM enables us to avoid the panentheistic metaphor of the world being "in" God, with all the ontological difficulties it generates. In the course of the investigation presented in this volume I have defended Aquinas's model of the God-world relationship, mediating between divine transcendence and immanence. Contrary to contemporary critics, I find it well balanced, relevant, and plausible, also in the context of contemporary science and philosophy.

6. Concerning the problem of physical and moral evil and the difficulty of emergentist panentheism treating it as a part of God's nature and thus struggling to reconcile it with the Creator's ultimate goodness, Aquinas's position once again seems to offer a feasible solution. His definition of evil as *privatio boni*—that is, the privation or lack of an actuality that should be present—and his description of evil occurrences and actions not as causes *per se* but merely (if they can be regarded causes at all) *per accidens* offer a reasonable foundation for theodicy free from the challenges faced by emergentist panentheism assuming that evil, in fact, does enter the very nature of God.

This list summarizes not only the constructive proposal developed in this chapter but also the entire project presented in this book. I hope that the Thomistic interpretation of divine action in the context of both the classical DC-based and Deacon's dynamical depth ("increased constraints") accounts of EM provides a viable alternative to Peacocke's and Clayton's emergentist panentheism.

Does it all mean, however, that the panentheistic turn is just another blind alley in contemporary philosophy of God and theology? Not necessarily. As notes Polish philosopher Jacek Wojtysiak, similar to Aristotelianism in its time, panentheism can reveal a perspective that goes beyond naturalism (secular or spiritual) and pantheism, which tend to characterize modern Western culture as attached to scientific discourse, distrustful of metaphysical speculation, thinking in spatial terms, and nature- and anthropocentric in its view of the reality. In other words, panentheism can successfully address the cognitive drama of modern human beings, who, shaped by the scientific mentality, when reflecting upon the mystery of the world, tend to narrow the horizon of being by assigning divine features to reality that is ontologically unspecified and impermanent.

Hence, even if one disagrees with the concept of the dipolar nature of God as described in panentheism, one should acknowledge that at least one of the two sides of the panentheistic metaphor and imagery directs our attention to the self-sufficient and absolutely perfect God, offering a reasonable and compelling argument in favor of his existence. All we need to do is "protect" this image of God from introducing into it dependency, temporality, or imperfection of any type, lest God's existence need an explanation similar to the one required in the case of all the contingent beings he creates. For only God, who is radically transcendent, unchangeable, omnipotent, and omniscient, and yet immanently present in the world he creates, explains the essence and existence of contingent entities of inanimate and animate nature. I hope to have shown and proved that classical theism, which construes and defends such an image of God, not only is still actual in the context of the contemporary science and philosophy of emergentism but also helps us develop their comprehensive understanding and interpretation.[116]

Conclusion

The existence of God, in so far as it is not self-evident to us, can be demonstrated from those of His effects which are known to us.
—Thomas Aquinas, *ST* I.2.2 co.

•

Our journey is coming to an end. It began with wonder over the beauty and the unique character of emergent phenomena observable in nature. The shortage of the scientific interpretative and explanatory categories needed for a proper description of such phenomena inspired a philosophical analysis of their fundamental characteristics. My critical evaluation of the two available accounts of ontological EM, developed recently in philosophy of biology and of science, inspired, in turn, the argument in favor of retrieving the plural notion of causality offered by Aristotle. Applied to emergentism in both its classical and contemporary dispositionalist versions, this notion allowed a new and more accurate reinterpretation and explanation of the common metaphysical ground of the numerous phenomena studied in systems biology and other branches of the natural and social sciences.

This metaphysical inquiry into the nature of emergents served as an introduction and a necessary background for the critical investigation of theology inspired by the wonder of EM, which was the object of my interest in the second part of the project. Critical about emergentist panentheism, developed within the circles of the theology-science dialogue, and in reference to my reinterpretation of both the classical DC-based and Deacon's dynamical depth accounts of ontological EM in terms of the classical and the new Aristotelianism, I have suggested explaining God's action in emergent phenomena in line with Aquinas's theology of divine action.

Beginning with Aquinas's understanding of God as the first and ultimate source of actuality and being and his definition of creation as dependence on God in being, I next explored the idea of God understood as the creator of all substantial forms (including exemplar causation of divine ideas) and as the ultimate source and goal of all teleology in nature. I believe that my analysis has shown and proved that God can be thus seen as the Creator who, although utterly transcendent (i.e., immutable, omniscient, omnipotent, infinite, eternal, and impassible), is nonetheless immanently present in all worldly events as the primary and principal agent, working through the secondary and instrumental actions of his creatures, including changes and events that we tend to attribute to chance. In this context, I saw both the classical DC-based and Deacon's dynamical depth ("increased constraints") models of EM as highlighting particular stages of one and the same process of communication of God's goodness and perfection at all levels of complexity in nature, perceived as an expression of God's actuality, which he wants to share abundantly with and bestow upon his creation.

Acknowledging the superiority of the dynamical depth ("increased constraints") account of EM over the classical DC-based theory in opening the way to see formal and final causes as permeating the whole reality of stable and changing entities—rather than being limited to higher levels of wholes, mereologically construed of their building parts—I have argued that Deacon's view of EM proves more promising for the theological interpretation of emergent phenomena. At the same time, my comparison of Aquinas's and Deacon's views on the metaphysical character and the role of nonbeing showed that even if nonbeing remains for Aquinas an important category in explaining human intellectual distinction and derivation of the many from the one, he does not ascribe to it an ontological and causal character. Nor does he think that it enters into the realm of exemplar ideas in the mind of God. Here Aquinas seems to correct Deacon's ascription of causal powers to absences (constraints), offering a better explanation of the ontological and causal aspects of the "engine" of EM, while leaving it open to a fruitful theological interpretation.

Consequently, having started with wonder over the beauty of emergent phenomena, we have finished our journey with the feeling of admiration and of awe in reference to God's divine action in nature, in and through emergent entities and dynamic processes. This fact confirms the truth of Aquinas's assurance that "the existence of God, in so far as it is not self-evident to us, can be demonstrated from those of His effects which are known to us."[1] I believe that the new theological reflection inspired by the theory of EM developed in this book offers a reasonable alternative to Peacocke's and Clayton's emergentist panentheism and contributes importantly to the ongoing dialogue between theology and science.

APPENDIX 1

I began the second part of this project with the intriguing reflection of Brierley, who speaks about the "quiet revolution" in Western theism over the last century—the revolution brought by the panentheistic understanding of the God-world relationship and its influence on the understanding of God's essence and divine attributes. My analysis of the historical development of panentheism in philosophical theism, offered in the third chapter, was limited to the main figures that seem to have influenced the most recent application of the panentheistic metaphor in the context of science-theology dialogue. In order to acknowledge that Brierley's placement of the recent turn toward panentheism in the category of revolution is not exaggerated, I list some of the most influential philosophers and theologians who are advocates of this version of theism, both historically and at present:[1]

1. Secular and Christian Neoplatonists such as Plotinus, Proclus, and Pseudo-Dionysius.
2. Medieval theologians and mystics such as John Scotus Eriugena, Eckhart, Nicholas of Cusa, Mechtild of Magdeburg, Julian of Norwich, and Jakob Böhme.
3. Renaissance thinkers such as Giordano Bruno, certain Cambridge Platonists (Ralph Cudworth, William Law, Henry More, John Smith, and Benjamin Whichcote), and Jonathan Edwards.
4. German idealists and Romantics such as Johann Gottlieb Fichte, Georg Wilhelm Friedrich Hegel, Gottfried Herder, Gotthold Lessing, Friedrich Wilhelm Joseph von Schelling, and Friedrich Schleiermacher.
5. Nineteenth-century influential figures in Germany, England, the United States, and France, such as Karl Krause, Isaak Dorner,

Gustav Theodor Fechner, Herman Lotze, Otto Pfleiderer, Samuel Taylor Coleridge, Thomas Hill Green, John and Edward Caird, James Ward, Andrew Seth Pringle-Pattison, Samuel Alexander, William Ralph Inge, Ralph Waldo Emerson, Charles Sanders Peirce, William James, Jules Lequier, Charles Renouvier, and Henri Bergson.

6. Twentieth-century philosophers and theologians from the Christian tradition who are often called panentheists by others, such as Martin Heidegger, Hans-Georg Gadamer, Nicolai Berdyaev, William Temple, John Robinson, John Macquarrie, Karl Rahner, Hans Küng, Pierre Teilhard de Chardin, Ian Barbour, Peter Berger, James Bethune-Baker, Dietrich Bonhoeffer, Martin Buber, Sergei Bulgakov, Rudolf Bultmann, Karl Heim, William Hocking, John Geddes MacGregor, John Polkinghorne, Albert Schweitzer, Paul Tillich, Ernst Troeltsch, Alan Watts, Paul Weiss, Wolfhart Pannenberg, and Alfred North Whitehead.

7. Twentieth-century liberation and ecological theologians such as James Cone, Gustavo Gutiérrez, Jan Luis Segundo, Leonardo Boff, Rosemary Ruether, Sallie McFague, and Matthew Fox.

8. Those contemporary theologians—coming mainly from an Anglo-American context—who openly identify themselves as panentheists: Charles Hartshorne, Norman Pittenger, Charles Birch, Schubert Ogden, John Cobb, James Will, Jim Garrison, David Pailin, Joseph Bracken, David Griffin, Jay McDaniel, Daniel Dombrowski, Anna Case-Winters, Alan Anderson, Marcus Borg, Philip Clayton, Scott Cowdell, Denis Edwards, Paul Fiddes, Donald Gelpi, Peter Hodgson, Christopher Knight, Jürgen Moltmann, Hugh Montefiore, Helen Oppenheimer, Arthur Peacocke, Piet Schoonenberg, Claude Stewart, and Kallistos Ware.

APPENDIX 2

In my analysis of the question of God's immanent presence in the universe in both classical theism and panentheism in chapter 4 (section 1.2) I referred to a crucial passage from Aquinas's *Summa theologiae* dedicated to this topic and quoted by Gregersen in the Blackfriars (Cambridge) translation and edition of the *Summa*.[1] Since the text differs in some important details from the standard Benzinger version, which seems to be more faithful to the original Latin text, I offer a short exegesis and comparison of both translations.

Latin original:

> Licet corporalia dicantur esse in aliquo sicut in continente, tamen spiritualia continent ea in quibus sunt, sicut anima continet corpus. Unde et Deus est in rebus sicut continens res. Tamen, per quandam similitudinem corporalium, dicuntur omnia esse in Deo, inquantum continentur ab ipso. (*ST* I.8.1 ad 2)

Blackfriars translation:

> That in which bodily things exist contains them, but immaterial things contain that in which they exist, as the soul contains the body. So God also contains things by existing in them. However, one does use the bodily metaphor and talk of everything being in God inasmuch as he contains them.

Benzinger translation:

> Although corporeal things are said to be in another as in that which contains them, nevertheless, spiritual things contain those things in

which they are; as the soul contains the body. Hence also God is in things as containing them; nevertheless, by a certain similitude to corporeal things, it is said that all things are in God; inasmuch as they are contained by Him.

The Blackfriars translation has "immaterial things" instead of "spiritual things," which is less precise in capturing Latin *spiritualia*. The first expression of God's immanence found in the passage is also better reflected by the Benzinger translation. Faithful to the original Latin *Deus est in rebus sicut continens res*, it begins by emphasizing God's presence in things ("God is in things"), then characterizes the mode of this presence ("as containing them"). The Blackfriars version reverses the order, expressing God's presence in things through the relation of containment ("God . . . contains things"), then specifying the mode of this containment ("by existing in them"). The first option better shows the transcendence of God, whose infinite being contains finite things. This perspective is less transparent in the Blackfriars translation, which merely states that God "contains things by existing in them," and thus does not refer to the infinitely transcendent nature of God's being.

Finally, the Benzinger translation captures better Aquinas's juxtaposition of the two alternative expressions of God's immanence in things created, which is the main point of the whole passage. Although Aquinas finds that the idea of God being present in all things "as containing them" in his being is the best way of describing God's immanence, he also accepts the alternative view of all things being in God. The juxtaposition of God's *being* in things and things *being* in God seems to be lost once we accept the alternative translation of *Deus est in rebus sicut continens res*, as we find it in the Blackfriars translation, which replaces "God is in things" with "God contains things." Consequently, God's *being* in things versus things *being* in God is replaced by God *containing* all things versus all things *being* in God.

NOTES

PREFACE

1. Two important terminological clarifications: (1) Speaking of God's relation to his creatures throughout the entire project, I will use interchangeably the terms "world" and "universe," because doing so has become a common practice among scholars working in the field of theology-science dialogue. However, I know that the terms have different meanings, which also depend on the context of the analysis. On the one hand, referring to the commonsense intuition, many people tend to associate "world" with the earth (including the atmosphere that surrounds it) and to associate "universe" with the entire cosmos (hence, on this interpretation the category of the universe is more inclusive). On the other hand, as notes Mullins, according to contemporary philosophy, "A universe is a spatiotemporal collection of physical objects that is not spatially, temporally, or causally related to any other such spatiotemporal collection of physical objects. A world is a maximally composible state of affairs. There is a possible world where God exists alone. There is a possible world where God and a universe exist. There is a possible world where God and a multiverse exist" (hence, on this interpretation the category of the world is more inclusive). See Richard T. Mullins, "The Difficulty with Demarcating Panentheism," *Sophia* 55, no. 3 (2016): 339. (2) I will also use interchangeably the phrases "theology-science dialogue" and "science-theology dialogue." The order of terms, as I understand it, does not imply superiority of one or the other discipline of knowledge. Nor does it define a methodological strategy of approaching natural sciences from the perspective of theology or vice versa.

2. See Mariusz Tabaczek, *Emergence: Towards a New Metaphysics and Philosophy of Science* (Notre Dame, IN: University of Notre Dame Press, 2019). Both projects originate from an interdisciplinary research in theology and science pursued by the author in 2011–16 during his doctoral studies at the Graduate Theological Union in Berkeley, California. They result from the subsequent reconsideration, deepening, and extending of the original material.

INTRODUCTION

1. Edwin Arthur Burtt, *The Metaphysical Foundations of Modern Science* (Mineola, NY: Dover, 2003; first published 1924), 99.

2. Divine concurrence in Thomistic terms does not mean that God's action and creaturely action add up to the effect of a particular change occurring in nature (so that the effect in question would causally depend partly on God and partly on some created causes). Rather, it depends fully on God's action and creaturely action, where the latter is defined as secondary and instrumental causation, dependent on the primary and principal causation of God.

3. Friedrich Schleiermacher, *The Christian Faith*, trans. H. R. Mackintosh and J. S. Stewart (Edinburgh: T&T Clark, 1928), 183.

4. In his *Jesus Christ and Mythology* Bultmann says that "in mythological thinking the action of God, whether in nature, history, human fortune, or the inner life of the soul, is understood as an action which intervenes between the natural, or historical, or psychological course of events; it breaks and links them at the same time. The divine causality is inserted as a link in the chain of the events which follow one another according to the causal nexus" (Rudolf Karl Bultmann, *Jesus Christ and Mythology* [New York: Charles Scribner's Sons, 1958], 61). Applying his method of demythologization, Bultmann states that "God's action generally, in nature and history, is hidden from the believer just as much as from the non-believer. But in so far as he sees what comes upon him here and now in the light of the divine word, he can and must take it as God's action" (ibid., 64). Consequently, "Only such statements about God are legitimate as express the existential relation between God and man. Statements which speak of God's actions as cosmic events are illegitimate. The affirmation that God is creator cannot be a theoretical statement about God as *creator mundi* in a general sense. The affirmation can only be a personal confession that I understand myself to be a creature which owes its existence to God" (ibid., 69). In other words, it is only based on one's personal experience of divine action in one's life that one may predicate actions of Creator God in nature. We have no other means to verify and objectify the notion of divine action.

5. Terrence W. Deacon, *Incomplete Nature: How Mind Emerged from Matter* (New York: W. W. Norton, 2012), 195.

6. As one shall see, the risk of panentheism collapsing into pantheism is related primarily to the difficulty in specifying the exact meaning of the preposition *en*, which is central in the definition of "panentheism." In fact, authors such as John Scotus Eriugena, Nicholas of Cusa, Spinoza, or Hegel are oftentimes classified among both panentheists and pantheists. The reason for this is that—contrary to the common opinion—pantheism is rarely defined in terms of a strict identity of God and the universe. As notes William Mander, "Such 'strict identity' is virtually impossible to define due to the extreme difficulty of stipulating what would count as an acceptable and what as an unacceptable sense, part, aspect, or element of difference." Hence,

"while some pantheists conceive of deity in mereological terms as the collection of things which make up the universe, many others have found this approach inadequate, maintaining that in some important sense 'the whole is greater than the sum of its parts'" (William Mander, "Pantheism," *Stanford Encyclopedia of Philosophy*, Spring 2020 ed., ed. Edward N. Zalta, https://plato.stanford.edu/archives/spr2020/entries/pantheism/, section 3). This argumentation opens the way to the logic of dialectical (relative) identity, or identity-in-difference, where God (or the divine aspect of reality) and the universe are thought to be simultaneously identical and different. This logic is characteristic of panentheism, which shows that the demarcation line between pantheism and panentheism is vague and porous.

7. See Étienne Gilson, *Methodical Realism* (San Francisco: Ignatius, 2011).

INTRODUCTION TO PART 1

1. See Sara Green, "Introduction to Philosophy of Systems Biology," in *Philosophy of Systems Biology: Perspectives from Scientists and Philosophers*, ed. Sara Green (Copenhagen: Springer, 2017), 1–23.

CHAPTER 1

1. Aristotle, *Meta.* 1.2.982b12–13 (unless otherwise noted, translations from Greek and Latin works are taken from the editions listed in the bibliography). Aristotle is echoing Socrates from Plato's *Theaetetus*, where we find him saying, "I see, my dear Theaetetus, that Theodorus had a true insight into your nature when he said that you were a philosopher, for wonder is the feeling of a philosopher, and philosophy begins in wonder" (Plato, *Theaetetus* 155d, in *Dialogues*, trans. Benjamin Jowett, vol. 4 [London: Oxford University Press, 1931]).

2. Another comprehensive list of (ontologically) emergent phenomena can be found in Michele Paolini Paoletti and Francesco Orilia, "Downward Causation: An Opinionated Introduction," in *Philosophical and Scientific Perspectives on Downward Causation*, ed. Michele Paolini Paoletti and Francesco Orilia (New York: Routledge, 2017), 9.

3. See Robert B. Laughlin, *A Different Universe: Reinventing Physics from the Bottom Down* (New York: Basic Books, 2005), 44–45. For an overarching view of EM, across various disciplines of science, see Harold J. Morowitz, *The Emergence of Everything: How the World Became Complex* (Oxford: Oxford University Press, 2002).

4. There is a vast literature and discussion on whether there ever was the scientific revolution of the seventeenth century. Many historians of science tend to reject the idea of a singular and discrete event, localized in time and space, that can

be pointed to as "the" scientific revolution. Rather, they speak about a number of complex changes in theory and practical methodology of explanation in natural philosophy and metaphysics, extended in time and space across Europe, in the advent of modernity. See, for instance, Steven Shapin, *The Scientific Revolution* (Chicago: University of Chicago Press, 1998).

5. Philosophy of science, by definition, finds its point of departure in the practice of natural science. Its main objectives include (1) offering a general synthetic view of results coming from various branches of natural science, (2) analyzing the methods of science and the nature of scientific progress and knowledge, and (3) bridging natural science with metaphysics (by addressing metaphysical questions raised by scientific evidence and theories). Philosophy of nature, by contrast, while not ignorant about natural science, emphasizes that, as such, the latter can study only certain aspects of the reality, based on the primitive and aided sensual experience. Reaching beyond scientific methodology, philosophy of nature claims to offer an original and unique approach to natural phenomena, addressing primary and prescientific questions concerning the species-individual structure of material things, stability and change in nature, quantitative and qualitative aspects of things, extension and intensity, space, motion, and time, determinism and indeterminism, chance and free will, etc. It analyzes these characteristic features of material beings in reference to the concepts of potency and act, unity and multiplicity, substance and accident, and so on. Thus, it is closely related to metaphysics and ontology. The latter, however, apply the same concepts not only to empirically verifiable material beings but also to entities that are immaterial.

6. As Stephen Gaukroger notes, "natural philosophy" designates a set of disciplines that we tend to classify—by our contemporary standards—as belonging to "hard science": physics, chemistry/alchemy, biology, and physiology. At the same time, it excludes some other disciplines like mathematics or medicine, which we would normally add to the same list. See Stephen Gaukroger, *The Emergence of a Scientific Culture: Science and the Shaping of Modernity, 1210–1685* (Oxford: Clarendon, 2006), 1–2n3.

7. In other words, we may say that practical-mathematical disciplines like mechanics were interested in explanations in terms of horizontal relations (those that do not move from one level of complexity of matter to another), whereas disciplines concerned with matter theory searched for vertical explanations (which did penetrate beyond phenomena, to discover what made them behave the way they did).

8. See Sophie Weeks, "The Role of Mechanics in Francis Bacon's Great Instauration," in *Philosophies of Technology: Francis Bacon and His Contemporaries*, ed. Claus Zittel, Gisela Engel, Romano Nanni, and Nicole C. Karafyllis (Leiden, Boston: Brill, 2008), 173–76.

9. Stephen Gaukroger, *The Collapse of Mechanism and the Rise of Sensibility: Science and the Shaping of Modernity, 1680–1760* (Oxford: Clarendon, 2010), 58. Gaukroger notes that as an outcome of the Cartesian turn in philosophy, Aristotelian

essentialism was replaced by mechanist micro-corpuscularianism (reductionist in nature). Both approaches are paradigm forms of the vertical explanation (operating in between the levels of complexity of matter). "The first requires that explanations take the form of demonstrations from 'natures' or essences of the bodies that 'underlie' the forms of behaviour of the body that arise from these essences.... [In] the micro-corpuscularian explanations . . . the crossing of level is not a qualitative one, but rather one of scale, and the idea of the realm of the *explanans* 'underlying' that of the macroscopic phenomenal realm is taken more literally" (ibid., 219). Hence, whereas in mechanics all explanations are necessarily horizontal, mechanists combined mechanics with micro-corpuscularianism to make mechanism's explanations vertical—by invoking microscopic events in order to explain macroscopic ones.

10. When we're referring to possible philosophical interpretations of the nature of science, it is important to distinguish between the contemporary (introduced in the nineteenth century and pragmatically oriented) definitions of "science" and "scientist," on the one hand, and the meaning of Latin *scientia*, which denotes a form of wisdom that derives from the systematic organization of material of different kinds. Unlike the former—defined in terms of a narrow, coherent, universal, and efficacious set of procedures leading to the growth of knowledge—the latter view of *scientia* designates a broad and diverse array of cultural, theoretical, and experimental (or rather experiential) practices, related to different modes of stability and change, aimed at understanding, explaining, and controlling of the natural world. See Gaukroger, *Emergence of a Scientific Culture*, 1–2n3.

11. See George Henry Lewes, *Problems of Life and Mind*, vol. 2 (London: Kegan Paul, Trench, Trübner & Co., 1875); John Stuart Mill, *A System of Logic* (New York and London: Harper & Brothers, 1846).

12. The term "emergence" derives from Medieval Latin *emergere* = rise out or up, bring forth, bring to light (e.g., from a liquid by virtue of buoyancy), which is opposite to Latin *mergere* = to dip, sink. It was also used in late Middle French (*émergence*, from the verb *émerger*). Over the last few decades "emergence" became very popular and found application in numerous scientific and unscientific contexts.

13. See Samuel Alexander, *Space, Time and Deity*, vol. 2 (London: Macmillan, 1920); Conwy Lloyd Morgan, *Emergent Evolution* (London: Williams & Norgate, 1923); Charlie Dunbar Broad, *The Mind and Its Place in Nature* (London: Routledge and Kegan Paul, 1925).

14. Ernest Nagel, *The Structure of Science: Problems in the Logic of Scientific Explanation* (New York: Harcourt, Brace & World, 1961), 352.

15. Although systems biology takes a nonreductionist approach to living systems, it does not resign from the achievements and tools of the more reductionist approach and mechanism-based explanations it provides. For more information on the nature of explanation in systems biology, see Pierre-Alain Braillard and Christophe Malaterre, eds., *Explanation in Biology: An Enquiry into the Diversity of Explanatory Patterns in the Life Sciences* (Dordrecht: Springer, 2015).

16. See Mario Bunge, *The Mind-Body Problem: A Psychobiological Approach* (Oxford and New York: Pergamon Press, 1980); Bunge, *Matter and Mind: A Philosophical Inquiry* (Dordrecht: Springer, 2010); a number of publications on this topic by Kim, especially Jaegwon Kim, *Philosophy of Mind*, 3rd ed. (Boulder, CO: Westview Press, 2011); Karl R. Popper and John Carew Eccles, *The Self and Its Brain* (Berlin/Heidelberg/London/New York: Springer, 1977); John Jamieson Carswell Smart, "Physicalism and Emergence," *Neuroscience* 6 (1981): 109–13; Roger Wolcott Sperry, "Discussion: Macro- versus Micro-Determinism," *Philosophy of Science* 53 (1986): 265–70.

17. Terrence W. Deacon, "Emergence: The Hole at the Wheel's Hub," in *The Re-Emergence of Emergence: The Emergentist Hypothesis from Science to Religion*, ed. Philip Clayton and Paul Davies (Oxford and New York: Oxford University Press, 2006), 122.

18. The suggested list of the main characteristics of the classical version of EM is related to similar analyses offered by Philip Clayton, *Mind and Emergence: From Quantum to Consciousness* (Oxford and New York: Oxford University Press, 2004), 60–62; Gregory R. Peterson, "Species of Emergence," *Zygon* 41 (2006): 692–95; Achim Stephan, "Emergence—a Systematic View on Its Historical Facets," in *Emergence or Reduction? Essays on the Prospects of Nonreductive Physicalism*, ed. Ansgar Beckermann, Hans Flohr, and Jaegwon Kim (Berlin and New York: de Gruyter, 1992), 27–39, 42–45.

19. Lewes, *Problems of Life and Mind*, 412.

20. Alexander, *Space, Time and Deity*, 2:45.

21. Carl Gillett, "Non-Reductive Realization and Non-Reductive Identity: What Physicalism Does Not Entail," in *Physicalism and Mental Causation: The Metaphysics of Mind and Action*, ed. Sven Walter and Heinz-Dieter Heckmann (Charlottesville, VA: Imprint Academic, 2003), 28.

22. See Hilary Putnam, "The Nature of Mental States," in *Mind, Language, and Reality: Philosophical Papers* (Cambridge: Cambridge University Press, 1975), 2:433–35.

23. Alexander, *Space, Time and Deity*, 2:46.

24. See Claus Emmeche, Simo Køppe, and Frederic Stjernfelt, "Explaining Emergence: Towards an Ontology of Levels," *Journal for General Philosophy of Science* 28 (1997): 105–13.

25. Broad, *Mind and Its Place*, 77–78.

26. Although the assertion that laws of nature are descriptive and not prescriptive seems to be predominant in the scientific circles, their metaphysical status is a subject of vivid debate in philosophy. See an introduction to it in John W. Carroll, "Laws of Nature," *Stanford Encyclopedia of Philosophy*, Fall 2016 ed., ed. Edward N. Zalta, https://plato.stanford.edu/archives/fall2016/entries/laws-of-nature/. Robert C. Koons and Timothy Pickavance in their comprehensive guide to metaphysics (*The Atlas of Reality* [Oxford: John Wiley & Sons, 2017], chapter 5, pp. 94–105) mention

three main positions in this conversation: (1) laws of nature ascribe powers to particular things; (2) laws of nature describe the unvarying sequences of events; (3) laws of nature attribute a special kind of relation, nomic necessitation, to ensembles of properties or universals. Nicholas Saunders, in the context of his analysis of the debate on the possibility of special divine action in the age of science, lists and discusses four possible accounts of the laws of nature: (1) the regulatory account (identifies laws with a constant conjunction view of regularities inspired by Hume—suggesting thus the primacy of events over laws); (2) the instrumentalist account (defines laws as descriptions which rational, although not necessarily human, minds impose on an external world of events—replacing thus laws of nature with laws of science); (3) the necessitarian account (states that physical laws ontologically determine which possibilities are open to the world and which are not—suggesting thus the primacy of the laws of nature to actual events); and (4) the irreducibly probabilistic account (assumes that some, or even all, laws of nature are irreducibly statistical in form—suggesting, in its most radical version, ontological priority of indeterminism in nature, which leads to denial of any strict causality). See Nicholas Saunders, *Divine Action and Modern Science* (Cambridge and New York: Cambridge University Press, 2002), 60–72.

27. Broad, *Mind and Its Place*, 55.

28. See Jaegwon Kim, "Making Sense of Emergence," *Philosophical Studies: An International Journal for Philosophy in the Analytic Tradition* 95 (August 1999): 8–9.

29. See ibid., 9–18.

30. Donald Thomas Campbell, "'Downward Causation' in Hierarchically Organized Biological Systems," in *Studies in the Philosophy of Biology*, ed. Francisco J. Ayala and Theodosius Dobzhansky (Berkeley and Los Angeles: University of California Press, 1974), 180.

31. See Tabaczek, *Emergence*, chapter 2.

32. Van Gulick brings both arguments together and states: "The challenge of those who wish to combine physicalism with a robustly causal version of emergence is to find a way in which higher-order properties can be causally significant without violating the causal laws that operate at lower physical levels. On one hand, if they override the micro-physical laws, they threaten physicalism. On the other hand, if the higher-level laws are merely convenient ways of summarizing complex micropatterns that arise in special contexts, then whatever practical cognitive value such laws may have, they seem to leave the higher-order properties without any real causal work to do" (Robert van Gulick, "Who's in Charge Here? And Who's Doing All the Work?," in *Evolution and Emergence: Systems, Organisms, Persons*, ed. Nancey C. Murphy and William R. Stoeger [Oxford and New York: Oxford University Press, 2007], 64–65).

33. The literal symbols used by Kim in the original version of his argument were adapted to my example of pain and escape reaction. For a more detailed version of the argument, see Kim, "Making Sense of Emergence," 31–33; Jaegwon Kim,

"Emergence: Core Ideas and Issues," *Synthese* 151 (2006): 557–58; Tabaczek, *Emergence*, 78–80.

34. Alwyn Scott rejects Kim's concept of diachronic DC, showing that from a nonlinear point of view the problem of circular causal loops and temporal framework of DC does not exist at all. According to Scott, an emergent structure does not pop into existence at time *t*. It "begins from an infinitesimal seed (noise) that appears at a lower level of description and develops through a process of exponential growth (instability). Eventually, this growth is limited by nonlinear effects, and a stable entity is established." Thus, what we deal with in cases of DC is a dynamic balance between upward and downward causation. See Alwyn C. Scott, *The Nonlinear Universe: Chaos, Emergence, Life* (Berlin, Heidelberg, New York: Springer, 2007), 287–88.

35. Erasmus Mayr, "Powers and Downward Causation," in *Philosophical and Scientific Perspectives on Downward Causation*, ed. Michele Paolini Paoletti and Francesco Orilia (New York: Routledge, 2017), 81. A detailed presentation of the major positions in this debate can be found in my *Emergence* (chapter 2).

36. Gillett, "Non-Reductive Realization," 42.

37. Ibid., 43.

38. See Alicia Juarrero, *Dynamics in Action: Intentional Behavior as a Complex System* (Cambridge, MA: MIT Press, 1999), 132–33. Arthur Peacocke uses "boundary conditions" language as well when defining whole-part constraint (his name for DC). See Arthur Peacocke, "God's Interaction with the World: The Implications of Deterministic 'Chaos' and of Interconnected and Independent Complexity," in *Chaos and Complexity: Scientific Perspectives on Divine Action*, ed. Robert J. Russell, Nancey Murphy, and Arthur Peacocke (Vatican: Vatican Observatory; Berkeley, CA: Center for Theology and the Natural Sciences, 1995), 273.

39. My argument assumes an Aristotelian moderate realist view of universals, which always finds them in things (*in rebus*), and never separate from them (in contrast to the Platonic realism of forms). However, even under the contemporary analytic substratum and bundle theories of entities, a similar argument can be made in favor of properties being instantiated in concrete entities.

40. Brian McLaughlin and Karen Bennett, "Supervenience," *Stanford Encyclopedia of Philosophy*, Winter 2018 ed., ed. Edward N. Zalta, https://plato.stanford.edu/archives/win2018/entries/supervenience/.

41. See Jaegwon Kim, "Supervenience as a Philosophical Concept," *Metaphilosophy* 21 (1990): 5–23.

42. See Paul Humphreys, "Emergence, Not Supervenience," in "Proceedings of the 1996 Biennial Meetings of the Philosophy of Science Association," part 2, "Symposia Papers," Supplement, *Philosophy of Science* 64 (1997): S341.

43. Hong Yu Wong, "The Secret Lives of Emergents," in *Emergence in Science and Philosophy*, ed. Antonella Corradini and Timothy O'Connor (New York: Routledge, 2010), 20.

44. Jaegwon Kim, "The Myth of Nonreductive Materialism," *Proceedings and Addresses of the American Philosophical Association* 63 (1989): 40. See also Kim, "Supervenience," 23–27.

45. Physical monism is ontologically weaker than physicalism, which tends to assume that at the end of the day all that exists is just physical elementary matter, which challenges EM and all other nonreductionist accounts of nature. It seems that emergentists opt for physical monism more than for physicalism. If they support the latter, they accept it in a metaphysically controversial form of NP.

46. See Jaegwon Kim, "'Downward Causation' in Emergentism and Nonreductive Physicalism," in *Emergence or Reduction? Essays on the Prospects of Nonreductive Physicalism*, ed. Ansgar Beckermann, Hans Flohr, and Jaegwon Kim (Berlin and New York: de Gruyter, 1992), 128–29.

47. See Robert van Gulick, "Reduction, Emergence, and the Mind/Body Problem," in Murphy and Stoeger, *Evolution and Emergence*, 65.

48. See Nancey C. Murphy, "Emergence and Mental Causation," in Clayton and Davies, *Re-Emergence of Emergence*, 228; Murphy, "Reductionism: How Did We Fall into It and Can We Emerge from It?," in Murphy and Stoeger, *Evolution and Emergence*, 26–27.

49. Menno Hulswit, "How Causal Is Downward Causation?," *Journal for General Philosophy of Science* 36 (2006): 284.

50. See Claus Emmeche, Simo Køppe, and Frederic Stjernfelt, "Levels, Emergence, and Three Versions of Downward Causation," in *Downward Causation: Mind, Bodies and Matter*, ed. Peter Bøgh Andersen, Claus Emmeche, Niels O. Finnemann, and Peder Voetmann Christiansen (Aarhus and Oxford: Aarhus University Press, 2000), 18–23.

51. See ibid., 26–31.

52. Ibid., 25.

53. Ibid.

54. Ibid., 17.

55. See Charbel Niño El-Hani and Antonio Marcos Pereira, "Higher-Level Descriptions: Why Should We Preserve Them?," in Andersen et al., *Downward Causation*, 134–35.

56. Michael Silberstein, "In Defence of Ontological Emergence and Mental Causation," in Clayton and Davies, *Re-Emergence of Emergence*, 218.

57. Alvaro Moreno and Jon Umerez, "Downward Causation at the Core of Living Organization," in Andersen et al., *Downward Causation*, 107.

58. Scott, *Nonlinear Universe*, 5–7.

59. See George F. R. Ellis, "Science, Complexity, and the Nature of Existence," in Murphy and Stoeger, *Evolution and Emergence*, 118–22.

60. See Terrence W. Deacon and Spyridon Koutroufinis, "Complexity and Dynamical Depth," *Information* 5 (2014): 404–23.

61. "I propose that we use the term *ententional* as a generic adjective to describe all phenomena that are intrinsically incomplete in the sense of being in relation to, constituted by, or organized to achieve something non-intrinsic" (Deacon, *Incomplete Nature*, 27).

62. Terrence Deacon, Alok Srivastava, and Augustus Bacigalupi, "The Transition from Constraint to Regulation at the Origin of Life," *Frontiers in Bioscience* 19 (2014): 947.

63. Terrence W. Deacon and Tyrone Cashman, "Teleology versus Mechanism in Biology: Beyond Self-Organization," in *Beyond Mechanism: Putting Life Back into Biology*, ed. Brian G. Henning and Adam C. Scarfe (Lanham, MD: Lexington Books, 2013), 290. See also Deacon and Koutroufinis, "Complexity and Dynamical Depth," 407–8; Deacon, *Incomplete Nature*, chapter 12.

64. Deacon, "Emergence," 124 (italics original).

65. Deacon, *Incomplete Nature*, 193.

66. Deacon, "Emergence," 143.

67. Deacon, *Incomplete Nature*, 194–95.

68. Ibid., 203.

69. Deacon and Koutroufinis, "Complexity and Dynamical Depth," 413.

70. The term "increased constraints emergentism" was proposed by O'Connor, in contrast with the classical DC-based view of EM, which he classifies as "closure-denying." The latter term is rather problematic when we take into account the view of those among the advocates of the classical account of ontological EM who claim that the reality of DC does not violate the rules of physical causal closure and physicalism. See Timothy O'Connor, "Philosophical Implications of Emergence," in *The Routledge Companion to Religion and Science*, ed. James W. Haag, Gregory R. Peterson, and Michael L. Spezio (New York: Routledge, 2012), 209.

71. Deacon, *Incomplete Nature*, 223, 226.

72. Deacon and Koutroufinis, "Complexity and Dynamical Depth," 413.

73. See Deacon, "Emergence," 127.

74. See Terrence W. Deacon, "The Hierarchic Logic of Emergence: Untangling the Interdependence of Evolution and Self-Organization," in *Evolution and Learning: The Baldwin Effect Reconsidered*, ed. Bruce H. Weber and David J. Depew (Cambridge, MA: MIT Press, 2003), 288; Deacon, "Emergence," 126–30.

75. See Deacon and Koutroufinis, "Complexity and Dynamical Depth," 413.

76. See Deacon, *Incomplete Nature*, 237–38; Deacon, "Emergence," 130; Deacon and Koutroufinis, "Complexity and Dynamical Depth," 414.

77. See Deacon and Koutroufinis, "Complexity and Dynamical Depth," 412, 414.

78. See Deacon, "Emergence," 130–37; Deacon, *Incomplete Nature*, 239–61.

79. Deacon, "Emergence," 136. See also Deacon, "Hierarchic Logic of Emergence," 293–97.

80. See Deacon and Koutroufinis, "Complexity and Dynamical Depth," 416–17.

81. Ibid., 417–18.

82. The term "autogenesis" has been used in biology to refer to spontaneous generation, abiogenesis, and the EM of eukaryotic cells. Deacon gives it a new meaning, because he uses it "for a very precise type of reciprocally interdependent interaction between different kinds of self-organizing process" (Deacon and Cashman, "Teleology versus Mechanism in Biology," 305n24).

83. Ibid., 300 (italics original). Deacon and Cashman emphasize that "in autogenesis, it is not just constituents that are joined in a reciprocally productive loop, but the constraints that each process generates, because each of these processes generates the boundary constraints that make the other process possible" (ibid., 299). See also ibid., 302; Deacon, *Incomplete Nature*, 305–8, 317; Deacon, "Emergence," 141.

84. See Deacon, *Incomplete Nature*, 315–19.

85. Deacon, "Reciprocal Linkage," 143; see also ibid., 137–38. In *Incomplete Nature* Deacon adds that "[the] retained foundation of reproduced constraints is effectively the precursor to genetic information (or rather, the general property that genetic information also exhibits). . . . Whether it is embodied in specific information-bearing molecules (as in DNA) or merely in the molecular interaction constraints of a simple autogenetic process, information is ultimately constituted by preserved constraints" (*Incomplete Nature*, 317–18).

86. See Deacon, *Incomplete Nature*, 271; Deacon and Cashman, "Teleology versus Mechanism in Biology," 300; Deacon, Srivastava, and Bacigalupi, "Transition from Constraint to Regulation," 952. At one point, Deacon adds that "what constitutes an autogenic 'self' cannot then be identified with any particular substrate, bounded structure, or energetic process. Indeed, in an important sense, the self that is created by the teleodynamics of autogens is only a virtual self, and yet is at the same time the locus of real physical and chemical influences" (*Incomplete Nature*, 311).

87. Deacon, *Incomplete Nature*, 35–36; Deacon, "Emergence," 114–17.

88. See Deacon, *Incomplete Nature*, 23.

89. See ibid., 34, 59, 210; Deacon, "Emergence," 113–14.

90. Deacon, *Incomplete Nature*, 230–31. We find a similar idea of form in Deacon's approval of the position of Emmeche et al., who argue to interpret DC as a case of formal causation. Deacon sees it as "the systemic 'geometric' position" within the dynamical network of the biomolecules building a living cell and generating specific higher constraints (see *Incomplete Nature*, 231–32). See also Deacon, *Incomplete Nature*, 212–13, 338.

91. See Deacon, Srivastava, and Bacigalupi, "Transition from Constraint to Regulation," 951–52, 954–55.

92. See Deacon, *Incomplete Nature,* 231, 234.

93. Ibid., 109.

94. Deacon, "Emergence," 143–44.
95. See Deacon, *Incomplete Nature*, 275.
96. Ibid., 140–41.
97. Deacon, Srivastava, and Bacigalupi, "Transition from Constraint to Regulation," 952.
98. See Deacon, *Incomplete Nature*, 22–24.
99. See ibid., 43.
100. Deacon, *Incomplete Nature*, 12, 14, 45; Deacon, "Emergence," 143–44.
101. See ibid., 119–20.
102. Deacon, *Incomplete Nature*, 203.
103. Deacon, "Emergence," 148.

CHAPTER 2

1. A much more thorough investigation of classical and the new Aristotelianism, as well as its application in EM theory, may be found in Tabaczek, *Emergence*, introduction, sections 2 and 3; chapter 3, sections 3 and 4; chapters 5, 6, and 7.

2. Andrew G. van Melsen argues that, due to the difficulty in deciding who has the greater competence in judging the character of changes in nature (the philosopher or the scientist), choosing change—substantial change in particular (see the next section)—as a starting point for the philosophy of nature developed in the contemporary setting endangers the independence and autonomy of that branch of philosophy. He thus suggests taking as a point of departure in philosophy of nature the rule of the species-individual structure of matter, which tells us that "any determination or form in matter . . . is realized in such a way that it is, in principle, not confined to just one particular individual material thing or event" (Andreas Gerardus Maria van Melsen, *The Philosophy of Nature* [Pittsburgh: Duquesne University Press, 1961], 118). Melsen claims that individuality and specificity (belonging to a general category of species) aspects of each entity/event are a fact beyond any possible doubt (scientific or philosophical) and refer to all matter—even if classifying a given entity/event as belonging to its proper natural kind is not always easy. He claims that the species-individual structure of matter indicates the distinction between different types of being and of change and also indicates the law of "composition" of substances of primary matter and substantial form (hylomorphism), which are the main points of Aristotle's philosophy of nature (see ibid., 107–53). While Melsen's proposition seems plausible and may lead to similar conclusions, I choose to follow the way followed originally by Aristotle.

3. The excerpts of Heraclitus's writings used in this chapter were translated by Richmond Lattimore and are found in Matthew Thompson McClure, *The Early Philosophers of Greece* (New York: Appleton-Century-Crofts, 1935), nos. 21–22, 119–28. In nos. 41 and 81 Heraclitus famously states, "You could not step twice in

the same rivers; for other and yet other waters are ever flowing on. . . . In the same rivers we step and we do not step. We are and we are not." See also Milton Charles Nahm, *Selections from Early Greek Philosophy* (Englewood Cliffs, NJ: Prentice-Hall, 1964), 76. Michael Dodds rightly notices that Heraclitus can be seen as a predecessor of Whitehead's philosophy of process (see Michael J. Dodds, *The Philosophy of Nature* [Oakland, CA: Western Dominican Province, 2010], 4).

4. See McClure, *Early Philosophers of Greece*, 145–49 (especially nos. 2, 4, 8). In no. 8 we read, "What *is* is without beginning, indestructible, entire, single, unshakable, endless; neither *has* it been nor *shall* it be, since now it *is*; all alike, single, solid. For what birth could you seek for it? Whence and how could it have grown? I will not let you say or think that it was from what is not; for it cannot be said or thought that anything is not. What need made it arise at one time rather than another, if it arose out of nothing and grew thence? So it must either be entirely, or not at all."

5. We know that Aristotle learned about Heraclitus from Plato, who portrayed him as denying any stability and identity of things over time. In *Cratylus* Plato says, "Heracleitus is supposed to say that all things are in motion and nothing at rest; he compares them to the stream of a river, and says that you cannot go into the same water twice" (Plato, *Cratylus* 402a, in *Dialogues*, trans. Benjamin Jowett, vol. 2 [London: Oxford University Press, 1931]). Some of the contemporary experts in ancient philosophy claim that Heraclitus was misunderstood in his time and subsequently neglected. They are convinced he was, in fact, trying to convey a much deeper insight: that it is precisely through change that identity is maintained. In other words, the continued existence of things depends precisely upon their continually changing: "Heraclitus's message seems to be that because the waters are constantly changing, the river is (at least relatively) constant" (Daniel W. Graham, "Heraclitus: Flux, Order, and Knowledge," in *The Oxford Handbook of Presocratic Philosophy*, ed. Patricia Curd and Daniel W. Graham [Oxford: Oxford University Press, 2008], 174). See also Charles H. Kahn, *The Art and Thought of Heraclitus: An Edition of the Fragments with Translation and Commentary* (Cambridge: Cambridge University Press, 1979).

6. "In all things there is a portion of everything except [*Nous*] mind; and there are things in which there is mind also. Other things include a portion of everything, but mind, is infinite and self-powerful and mixed with nothing, but it exists alone itself by itself" (Anaxagoras, excerpts in Nahm, *Selections from Early Greek Philosophy*, 141).

7. "And these [elements] never cease changing place continually, now being all united by Love into one, now each borne apart by the hatred engendered of Strife, until they are brought together in the unity of the all, and become subject to it" (excerpts in Nahm, *Selections from Early Greek Philosophy*, 118).

8. See Aristotle, *Phys.* 1.3.186a23–32.

9. Aristotle, *Meta.* 5.12.1019a19–20.

10. Aristotle, *Meta.* 9.1.1046a2.

11. Aristotle, *Meta.* 9.2.1046a35–1046b3.

12. Aristotle, *De gen. et corr.* 1.4.319b10–18. While distinguishing between accidents/accidental change and substances/substantial change, we must remember not to treat accidents as externally connected to substances (with no internal bond). Since the being of the former is intrinsically rooted in/dependent on the being of the latter, each accidental change implies a change of some determination of the substance in which it is grounded as well. And yet, such change of a substance in question differs from a substantial change sensu stricto in which one substance (S_1) ceases to be, while another substance (S_2) comes into being. Going back to Aristotle's example of a human body being "now healthy and now ill," the change that is being described, though accidental, affects the whole human person, who yet remains the same person.

13. Aristotle, *Phys.* 2.3.194b24–28.

14. Aristotle, *Meta.* 5.2.1013a24–29.

15. Aristotle, *Phys.* 1.7.191a8–12.

16. Aristotle, *Phys.* 1.9.192a25–33.

17. Aristotle, *Meta.* 7.3.1029a20–21, 24–25.

18. Aristotle, *Meta.* 9.7.1049a24. In Tredennick's translation: "If there is some primary stuff, which is not further called the material of some other thing, this is primary matter" (Aristotle, *The Metaphysics*, translated by Hugh Tredennick [London and Cambridge, MA: W. Heinemann, 1936]).

19. Commenting on the tendency to treat matter as substance, Aristotle notes that "both separability and 'thisness' are thought to belong chiefly to substance. And so form and the compound of form and matter would be thought to be substance, rather than matter" (*Meta.* 7.3.1029a29–30).

20. Aristotle, *Phys.* 2.3.194b26, in *Physica (Physics)*, trans. R. K. Gaye, in *The Basic Works of Aristotle*, ed. Richard McKeon (New York: The Modern Library, 2001), 213–394; *Meta.* 5.2.1013a27, in *Metaphysica (Metaphysics)*, trans. W. D. Ross, in *The Basic Works of Aristotle*, ed. Richard McKeon (New York: The Modern Library, 2001), 681–926.

21. Dodds, *Philosophy of Nature*, 25.

22. The language of eduction is an attempt of Aquinas to find a middle ground between, on the one hand, treating substantial form as external and coming to form things "from the outside" and, on the other, describing it as preexistent in a primitively actualized state in primary matter. Hence, he says that "motion is nothing else than the eduction of something from potentiality to actuality [educere aliquid de potentia in actum]" (*ST* I.2.3 co. [translation is mine]). If we can say that form preexists in primary matter, it does so as unactualized: "Every actuality of matter is educed from the potentiality of that matter [educi de potentia materiae]; for since matter is in potentiality to act, any act pre-exists in matter potentially" (*ST* I.90.2 ob. 2; see also ad 2). "Every form brought into being through the transmutation of matter is educed from the potentiality of matter [forma educta de potentia materiae]" (*SCG*

II.86.6). See also *ST* I.2.3 co.; I.4.1 ad 2; I.84.3 co.; III.8.3 co.; *SCG* II.45.3. Consequently, insofar as the effective cause brings out the form which it realizes in the potentiality of matter, it is said to induce (*inducere*) or introduce (*introducere*) form. See *Quod.* 7.4.9 ad 4.

23. Terence Irwin, *Aristotle's First Principles* (Oxford: Clarendon, 1988), 212.

24. Although in the logical works εἶδος (*eidos*) is usually understood as "species." The latter are also described in this part of the Aristotelian corpus as "types" or "genera" (γένη, *genē*), saying what a given thing is.

25. In the second book of the *Physics*, Aristotle explains briefly that the formal cause answers the question "why" something is the kind of thing it is, but without involving any motion (*Phys.* 2.7.198a17–18). In *Metaphysics* 1.3, he refers indirectly to the formal cause, defining it as the essence, the ultimate "why," and the ultimate principle: "Causes are spoken of in four senses. In one of these we mean the substance, i.e. the essence (for the 'why' is reducible finally to the definition, and the ultimate 'why' is a cause and principle)" (*Meta.* 1.3.983a26–29). See also *Meta.* 7.17.1041b11–32.

26. Irwin, *Aristotle's First Principles*, 100.

27. Some philosophers make an additional distinction between the primary substance (*substantia prima*)—that is, the concrete individual with its essence (e.g., a concrete puppy)—and the secondary substance (*substantia secunda*), that is, the abstract essence alone (e.g., the essence of "dogness"). Accordingly, they also speak about two kinds of substantial change: the one that modifies *substantia prima* with no alteration in *substantia secunda* (e.g., cutting a chunk of gold into two parts), and the one that modifies *substantia secunda*, when one species (type of substance) is superseded by another (e.g., dissolving gold in a mixture of nitric acid and hydrochloric acid in a molar ratio of 1:3 [*aqua regia*] turns gold into chloroauric acid [Au + HNO$_3$ + 4 HCl → HAuCl$_4$ + NO + 2 H$_2$O]). See Melsen, *Philosophy of Nature*, 128–30.

28. Aristotle, *Phys.* 2.3.194b29–195a2. A similar explanation can be found in *Meta.* 5.2.1013a29–1013b2. See also *Phys.* 2.7.198a18–20; *Meta.* 1.2.983a30–32.

29. Aristotle, *Phys.* 3.3.202a14.

30. Aristotle's original example of man being begotten by man and the sun as well is easily dismissed today due to its dependence on the outdated cosmology. However—philosophically speaking—it is not entirely implausible to see the energy emitted by the sun, forces of gravitation, and other general causal principles as contributing to educing particular forms from primary matter in processes of substantial change occurring in nature.

31. See, for instance, Aristotle, *De part. an.* 3.2.663b12–14; 4.5.679a25–30; *De gen. an.* 2.4.739b27–31; 3.4.755a17–30. Bostock lists a number of scholars claiming that "Aristotle would concede (at least for the sake of argument) that a complete materialist explanation might perhaps be available, and yet still insist that a teleological account was *also* needed." He suggests this "seems to be roughly the position that we

ourselves are in nowadays" (David Bostock, *Space, Time, Matter, and Form: Essays on Aristotle's Physics* [Oxford: Clarendon, 2006], 58, 60).

32. Bostock concludes we can speak about a "law of goodness" in Aristotle which assumes that "there is something that counts as good, namely what is good for the animal or plant concerned." Consequently "whatever parts a living thing needs, in order to live a life that is good for it, will for that very reason tend to be present in it (and therefore will grow as it grows). The law is limited in its application, of course, by the fact that the 'laws of matter' will only permit some kinds of parts to develop, and not others. . . . It is for him [Aristotle] a law that is basic and irreducible" (Bostock, *Space, Time, Matter, and Form*, 77–78).

33. Aristotle, *De part. an.* 2.1.646a31–34. Aristotle's profound conviction that "nature does nothing in vain" makes him think that there has to be a purpose for everything in nature.

34. Bostock, *Space, Time, Matter, and Form*, 71.

35. See also Bostock, *Space, Time, Matter, and Form*, 48–78; Allan Gotthelf, "Aristotle's Concept of Final Causality," *Review of Metaphysics* 30 (1976): 226–54.

36. Melsen notes that the notion of finality is widely present in natural science, which "is able to describe any motion [not only local motion but motion understood as any kind of change] in . . . two ways. Any one motion can be considered the result either of the acting forces or of tendencies expressed in the minimum principles [e.g., the principle of least action, which Russell has called "a law of cosmic laziness"]" (Melsen, *Philosophy of Nature*, 170). The latter strategy, referring to natural tendencies of entities/events, is clearly teleological in its nature.

37. Aristotle, *Phys.* 2.1.192b8–10.

38. Aristotle, *Phys.* 2.1.192b22–23.

39. See Aquinas, *In Meta.* 5.2.775; *De prin. nat.* 28.

40. Aquinas, *In Phys.* 5.11.246.

41. Aquinas, *De prin. nat.* 34. See also *In Phys.* 2.11.242.

42. Aristotle, *De part. an.* 1.1.639b14–16. Similar is the view of Aquinas when he says that "even though the end is the last thing to come into being in some cases, it is always prior in causality. Hence it is called the 'cause of causes,' because it is the cause of the causality of all causes" (*In Meta.* 5.3.782). See also *In Phys.* 2.5.186; *De prin. nat.* 29.

43. *De mixt. elem.* 17–18. It is important to remember that "mixture" in ancient and medieval philosophy often means a compound, i.e., a unified new entity informed by a new substantial form, and not merely a composite or a combination of elementary particles which, metaphysically speaking, is informed by an accidental form, which does not include a substantial change of the components.

44. Irwin, *Aristotle's First Principles*, 270.

45. Ibid., 276.

46. See Aristotle, *Phys.* 2.7.198a23–26.

47. See Aristotle, *Phys.* 2.3.195a26–34; 2.5.196b24–29.

48. See Aristotle, *Phys.* 2.3.195a34–195b6. Aristotle also uses a similar example of a house builder: *Phys.* 2.3.196b25–29. See also *Phys.* 2.3.195b24; 2.5.196b27–29.

49. Aristotle, *Phys.* 2.5.197a12–14.

50. "No incidental cause can be prior to a cause *per se*. Spontaneity and chance, therefore, are posterior to intelligence and nature. Hence, however true it may be that the heavens are due to spontaneity, it will still be true that intelligence and nature will be prior causes of this all and of many things in it besides" (Aristotle, *Phys.* 2.6.198a8–13).

51. Note that Aristotle has actually two terms for chance: (1) τύχη = luck or fortune (chance occurrences in causal situations related to human deliberating) and (2) a more general term ταὐτομάτον = chance (chance occurrences in causal situations related to an entity acting for its natural end [*telos*]).

52. "For Aristotle it is not legitimate to view the present condition of the world as the outcome of the interaction of chains of necessary causes, as many contemporary scientists and philosophers would hold. For Aristotle the human intellect can only trace back one chain of causes at a time, and will always have to stop the process when it reaches a free choice or a [sc. unusual] accidental cause, both of which introduce contingency into chains of causes, since the effect of free choices and [sc. unusual] accidents on the course of events is inherently unpredictable" (John Dudley, *Aristotle's Concept of Chance: Accidents, Cause, Necessity, and Determinism* [Albany: State University of New York Press, 2012], 323).

53. At the rise of the modern era Galileo famously states that the book of nature is "written in the language of mathematics" (Galileo Galilei, *The Assayer*, in *Discoveries and Opinions of Galileo*, trans. Stillman Drake [Garden City, NY: Doubleday, 1957], 238).

54. I offer a short overview of causal reductionism in modern science and philosophy in Tabaczek, *Emergence*, section 4 of the introduction. A thorough study of the same problem can be found in William A. Wallace, *Causality and Scientific Explanation*, vol. 2, *Classical and Contemporary Science* (Ann Arbor: University of Michigan Press, 1974).

55. David Hume, *A Treatise of Human Nature*, ed. L. A. Selby-Bigge (Oxford: Oxford University Press, 1978), 77, 266.

56. The following description is a summary of a detailed investigation of the main objectives and difficulties of all six neo-Humean views of causation and their more specific variants, offered in Tabaczek, *Emergence*, chapter 4. This investigation provides references to the most substantial primary sources. The list of important secondary sources that may be of interest to the reader includes Helen Beebee, Christopher Hitchcock, and Peter Menzies, eds., *The Oxford Handbook of Causation* (New York: Oxford University Press, 2009), especially parts 1–3; Menno Hulswit, *From Cause to Causation: A Peircean Perspective* (Dordrecht/Boston/London: Kluwer Academic, 2002), chapter 2; Ernest Sosa and Michael Tooley, eds., *Causation*

(Oxford and New York: Oxford University Press, 1993); Phyllis Illari and Federica Russo, *Causality: Philosophical Theory Meets Scientific Practice* (Oxford: Oxford University Press, 2014); John Losee, *Theories of Causality: From Antiquity to the Present* (New Brunswick, NJ, and London: Transaction Publishers, 2011).

57. Hume, *Treatise of Human Nature*, 172. Hume adds that this definition can be reformulated as saying that "a cause is an object precedent and contiguous to another, and so united with it, that the idea of the one determines the mind to form the idea of the other, and the impression of the one to form a more lively idea of the other" (ibid.).

58. David Hume, *An Enquiry concerning Human Understanding*, in Great Books of the Western World 35 (Chicago: Encyclopaedia Britannica, 1952), 477.

59. Molnar and Mumford suggest that both "dispositions" and "powers" can be classified among a number of other interrelated concepts such as "capacity" and "incapacity," "ability" and "liability," "skill," "aptitude," "propensity," "tendency," "potential," "amplitude," etc. See George Molnar, *Powers: A Study in Metaphysics*, ed. Stephen Mumford (Oxford and New York: Oxford University Press, 2003), 57; Stephen Mumford, *Dispositions* (Oxford and New York: Oxford University Press, 1998), 10.

60. See Alexander Bird, "Limitations of Power," in *Powers and Capacities in Philosophy: The New Aristotelianism*, ed. Ruth Groff and John Greco (New York: Routledge, 2013), 27. He also suggests replacing the term "powers" with "potency" and explains, "It is in the nature (essence) of a potency/power P that there is a specific dispositional character $D_{S,M}$ such that, necessarily, for any particular, x, that possesses P, $D_{S,M} x$ holds [x is disposed to manifestation M in response to stimulus S]" (ibid.).

61. Stephen D. Mumford, "Causal Powers and Capacities," in Beebee, Hitchcock, and Menzies, *Oxford Handbook of Causation*, 268.

62. See Stephen Mumford and Rani Lill Anjum, *Getting Causes from Powers* (Oxford and New York: Oxford University Press, 2011), 175–94.

63. Mumford states that dispositions are ascribed to at least three classes of things: objects, substances, and persons. In *Dispositions* he lists five examples of powers—fragility, belief, bravery, thermostats, and divisibility by 2—which shows the diversity and flexibility of dispositional ascriptions (see *Dispositions*, 5–11).

64. See Mumford, "Causal Powers and Capacities," 270–71.

65. See Molnar, *Powers*, 183–87.

66. Brian D. Ellis, *The Philosophy of Nature: A Guide to the New Essentialism* (Montreal and Ithaca, NY: McGill-Queen's University Press, 2002), 48.

67. Stephen Mumford and Rani Lill Anjum, "Causal Dispositionalism," in *Properties, Powers, and Structures: Issues in the Metaphysics of Realism*, ed. Alexander Bird, B. D. Ellis, and Howard Sankey (New York: Routledge, 2012), 104.

68. See Molnar, *Powers*, 194–99.

69. See Mumford and Anjum, *Getting Causes from Powers*, chapter 5, pp. 106–29; Robert C. Koons and Timothy Pickavance, *Metaphysics: The Fundamentals* (Oxford: Wiley-Blackwell, 2015), 69.

70. See Stephen Mumford and Rani Lill Anjum, "A Powerful Theory of Causation," in *The Metaphysics of Powers: Their Grounding and Their Manifestations*, ed. Anna Marmodoro (New York: Routledge, 2010), 153–55; *Getting Causes from Powers*, 143–48.

71. The dispositionalist approach to probability seems to follow the Aristotelian emphasis on the reference to *per se* causes in dealing with chance events (see section 1.6 above). Acknowledging the ontological character of chance (probabilistic) occurrences, dispositional metaphysics analyzes them in reference to kind-specific dispositions of entities and their manifestations, which are suppositional in nature.

72. A more detailed analysis of neo-Aristotelianism can be found in Tabaczek, *Emergence*, chapter 6.

73. Nancy Cartwright and John Pemberton, "Aristotelian Powers: Without Them, What Would Modern Science Do?," in *Powers and Capacities in Philosophy: The New Aristotelianism*, ed. Ruth Groff and John Greco (New York: Routledge, 2013), 93.

74. See Molnar, *Powers*, 61; John Heil, *From an Ontological Point of View* (Oxford: Clarendon, 2005), 222; Ullin T. Place, "Dispositions as Intentional States," in *Dispositions: A Debate*, by David Malet Armstrong, Charles Burton Martin, and Ullin T. Place (London: Routledge, 1996), 24.

75. David S. Oderberg, *Real Essentialism* (New York: Routledge, 2007), 137. Edward Feser rightly points out that the Aristotelian terminology of teleology and final causation is less vulnerable to the objection of panpsychism, which can be raised against the notion of physical intentionality. See Edward Feser, *Scholastic Metaphysics: A Contemporary Introduction* (Heusenstamm: Editiones Scholasticae, 2014), 104.

76. Pan-dispositionalism is followed by Mumford, Bird, Popper, Mellor, Shoemaker, and Rea. See more recent publications of Mumford, including Mumford, "Causal Powers and Capacities," 268; Mumford and Anjum, *Getting Causes from Powers*, 3–4; Stephen Dean Mumford, "The Power of Power," in Groff and Greco, *Powers and Capacities in Philosophy*, 12–13; Bird, "Limitations of Power"; Alexander Bird, "Monistic Dispositional Essentialism," in Bird, Ellis, and Sankey, *Properties, Powers, and Structures*, 11–26; Bird, "Causation and the Manifestation of Powers," in Marmodoro, *Metaphysics of Powers*, 35–41; Karl R. Popper, *The Logic of Scientific Discovery* (New York: Basic Books, 1959), 424; David Hugh Mellor, "Counting Corners Correctly," *Analysis* 42 (1982): 96–97; Mellor, "In Defense of Dispositions," *Philosophical Perspectives* 12 (1974): 283–312; Sydney Shoemaker, "Causal and Metaphysical Necessity," *Pacific Philosophical Quarterly* 79 (1998): 59–77; Shoemaker, "Causality and Properties," in *Time and Cause: Essays Presented to Richard Taylor*, ed. Richard Taylor and Peter van Inwagen (Dordrecht and Boston: Reidel, 1980), 109–35; Michael C. Rea, "Hylomorphism Reconditioned," *Philosophical Perspectives* 25 (2011): 344–45.

77. Pan-categoricalism is followed by Quine. See Willard V. O. Quine, *The Roots of Reference* (LaSalle, IL: Open Court, 1974).

78. Neutral monism was supported by Mumford in his earlier writings. See, for instance, Mumford, *Dispositions*, 192.

79. See Charles Burton Martin, *The Mind in Nature* (Oxford and New York: Clarendon, 2008); John Heil, "Powerful Qualities," in Marmodoro, *Metaphysics of Powers*, 58–72.

80. Property dualism is supported by Molnar, Ellis, and Lowe. See Molnar, *Powers*, chapter 10; Brian D. Ellis, "The Categorical Dimensions of Causal Powers," in Bird, Ellis, and Sankey, *Properties, Powers, and Structures*, 20; Ellis, "Causal Powers and Categorical Properties," in Marmodoro, *Metaphysics of Powers*, 138; Ellis, *Scientific Essentialism* (Cambridge and New York: Cambridge University Press, 2001), 217–18; Edward Jonathan Lowe, *The Four-Category Ontology: A Metaphysical Foundation for Natural Science* (New York: Oxford University Press, 2006), 139. At the same time, however, in agreement with some other powers theorists, Ellis acknowledges that at the bottom level of reality, as described in quantum physics, powers are groundless. Even if we discover in the future that present elementary particles are further complex and therefore reducible to yet more basic constituents, the latter will be classified as groundless dispositions (see Ellis, "Categorical Dimensions of Causal Powers," 24). His position may be thus classified as dispositional reductionism (categorical properties are real only at some levels of description) and contrasted with the categorical reductionism of Armstrong, who seems to claim that dispositional properties, though real, are nevertheless reducible to their categorical bases. See David Malet Armstrong, *A Materialist Theory of the Mind* (London and New York: Routledge & K. Paul, 1968), 86; Armstrong, "Defending Categoricalism," in Bird, Ellis, and Sankey, *Properties, Powers, and Structures*, 79–80.

81. Heil, *From an Ontological Point of View*, 221.

82. Note that this conversation has yet another level of complexity in reference to the most recent debate on the relevance and proper interpretation of hylomorphism in analytic metaphysics. I present and analyze this debate in chapter 6 of *Emergence*.

83. Even if dispositionalism is not simply a version or an expression of hylomorphism—and there is no easy transition from one position to the other—we have seen that it nonetheless becomes an advocate of the reality of essences. It seems that Aristotelian hylomorphism offers a better ground for essentialism than Locke's substratum theory of particulars or the neo-Humean "bundle" or "trope" theory of substance. Hence, I find it highly recommendable to those who embrace dispositionalism.

84. The emergent theory of mind itself seems to be problematic as long as it is interpreted in terms of a new property (mind) coming into being from other entity/ies (brain, neurons) that do not have it. It may be defended when understood with reference to Aristotle's hylomorphism, as a new property proper for an entity whose substantial form is the form of a human being.

85. See, for instance, Deacon, *Incomplete Nature*, 34. In Tabaczek, *Emergence*, chapter 3, section 3.1, I assert that Deacon, in fact, presents a much stronger metaphysical position concerning the nature of matter, in relation to the Copenhagen

interpretation of quantum mechanics. Nevertheless, because his model of dynamical depth is based on the analysis of micro- and macroscopic rather than quantum-level phenomena, I do not need to discuss his position here in more detail.

86. Among the advocates of different versions of the mereological and structural version of hylomorphism we find Kathrin Koslicki, Mark Johnston, Kit Fine, William Jaworski, and Robert Koons. See Kathrin Koslicki, "Aristotle's Mereology and the Status of Form," *Journal of Philosophy* 103 (2006): 715–36; Koslicki, *The Structure of Objects* (Oxford: Oxford University Press, 2010); Mark Johnston, "Hylomorphism," *Journal of Philosophy* 103 (2006): 652–98; Kit Fine, "Things and Their Parts," *Midwest Studies in Philosophy* 23 (1999): 61–74; William Jaworski, *Structure and the Metaphysics of Mind: How Hylomorphism Solves the Mind-Body Problem* (Oxford: Oxford University Press, 2016); Robert Koons, "Staunch vs. Faint-Hearted Hylomorphism: Toward an Aristotelian Account of Composition," *Res Philosophica* 91 (2014): 151–77.

87. Redefining the formal principle as a function of the behavior of the constituents of a dynamic system leaves us with the question about these constituents, their properties and tendencies to engage in processes described in reference to the geometric properties of the probability space. The question of formal principle simply goes back one level, but still remains unanswered. We could continue this research program going all the way down to the quantum level, always left with the same question of why the processes at the given level of complexity occur and why their participants tend to do what they do.

88. It is important to note that in one of his more recent papers, coauthored with Cashman—in response to an early draft of my first monograph (*Emergence: Towards a New Metaphysics and Philosophy of Science*)—Deacon seems to have changed his strategy radically. He said that his idea of "efficacy of absence" was just a "rhetorical trope," which actually "undermines the point that it intends to emphasize," and that "absences themselves don't do work, nor do they resist work" (Terrence W. Deacon and Tyrone Cashman, "Steps to a Metaphysics of Incompleteness," *Theology and Science* 14 [2016]: 419). Nonetheless, my analysis of Deacon's correction of his earlier views (see Tabaczek, *Emergence*, chapter 3, section 3.4) shows that he still finds it difficult to avoid using causal terms such as "allowing," "enabling," "channeling," or "opening up for" in reference to absences, which seems to suggest understanding them in terms of efficient causation.

89. See Aristotle, *Phys.* 2.3.195a11–14; *Meta.* 5.2.1013b11–16. Aristotle does not further develop his thought on this topic.

90. Remember the similar explanation of the causation of absences given in the context of my analysis of DVC in section 2.2.2 of this chapter.

91. Aristotle, *Phys.* 1.7.190b10–13.

92. This way of reasoning is radically foreign to Aristotle's understanding of final causation. Even if teleology can be regarded in his philosophy, together with efficient causation, as an extrinsic principle of living and nonliving things, it

nevertheless has an intrinsic aspect to it. Although the end or goal may be extrinsic to nonliving and living things, the directedness to it is intrinsic in each one of them, because it finds its source in their proper natures (defined in terms of formal causation).

INTRODUCTION TO PART 2

1. See Michael W. Brierley, "Naming a Quiet Revolution: The Panentheistic Turn in Modern Theology," in *In Whom We Live and Move and Have Our Being: Panentheistic Reflections on God's Presence in a Scientific World*, ed. Philip Clayton and Arthur Robert Peacocke (Grand Rapids, MI / Cambridge, UK: Eerdmans, 2004), 1–15.

2. As notes Gregersen, "Literally, pan-en-theism means that 'all' (Gk. pan) is 'in' God (Gk. Theos), but God is not exhausted by the world as a whole (G>W). As such panentheism attempts to steer a middle course between an acosmic theism, which separates God and World (G/W), and a pantheism that identifies God with the universe as a whole (G=W). Positively speaking, panentheists want to balance divine transcendence and immanence by preserving aspects of the former's claim of God's self-identity while embracing the latter's intimacy between God and Universe" (Niels Henrik Gregersen, "Three Varieties of Panentheism," in Clayton and Peacocke, *In Whom We Live and Move*, 19).

3. Brierley, "Naming a Quiet Revolution," 4.

4. The question concerning the extent to which panentheism departs from classical theism was addressed by Benedikt Göcke, who, contrary to Brierley and many other thinkers, went as far as to say that "panentheism and classical theism differ only as regards the modal status of the world. According to panentheism, the world is an intrinsic property of God—necessarily there is a world—and according to classical theism the world is an extrinsic property of God—it is only contingently true that there is a world. Therefore, as long as we do not have an argument showing that necessarily there is a world [and Göcke offers two such arguments, himself remaining neutral on whether one should follow any of them] panentheism is not an attractive alternative to classical theism" (Benedikt Paul Göcke, "Panentheism and Classical Theism," *Sophia* 52, no. 1 [2013]: 61). Göcke's assertion triggered an exchange of arguments between him and Raphael Lataster published in the philosophical journal *Sophia*. As we shall see, Lataster is right in clarifying (1) that at least for some panentheists "the universe *is* of the substance of God"; (2) that in the panentheistic scenario "the universe qua universe is as unnecessary as the universe in the theistic scenario"; (3) that panentheistic God, unlike God in classical theism, is both immutable and mutable; and (4) that panentheists for whom "the world is of the very substance of God" may not "be committed to a creation and especially the typically monotheistic or classical theistic concept of *creatio ex nihilo*" (Raphael Lataster,

"The Attractiveness of Panentheism—a Reply to Benedikt Paul Göcke," *Sophia* 53, no. 3 [2014]: 389–93). The remaining arguments in the Göcke-Lataster exchange may be found in Benedikt Paul Göcke, "Reply to Raphael Lataster," *Sophia* 53, no. 3 (2014): 397–400; Raphael Lataster, "Theists Misrepresenting Panentheism—Another Reply to Benedikt Paul Göcke," *Sophia* 54, no. 1 (2015): 93–98; Benedikt Paul Göcke, "Another Reply to Raphael Lataster," *Sophia* 54, no. 1 (2015): 99–102. For the further evaluation of Göcke's position, see Mullins, "Difficulty with Demarcating Panentheism," 338–42.

CHAPTER 3

1. See Charles Hartshorne and William L. Reese, eds., *Philosophers Speak of God* (New York: Humanity Books, 2000); John W. Cooper, *Panentheism—the Other God of the Philosophers: From Plato to the Present* (Grand Rapids, MI: Baker Academic, 2006). See also appendix 1; John Culp, "Panentheism," in *Stanford Encyclopedia of Philosophy*, Summer 2017 ed., ed. Edward N. Zalta, https://plato.stanford.edu/entries/panentheism/.

2. See Robert C. Whittemore, "The Meeting of East and West in Neglected Vedanta," *Dialogue & Alliance* 2 (1988): 41–44.

3. See Hartshorne and Reese, *Philosophers Speak of God*, 29–34.

4. Plato, *Timaeus* 30c.

5. Ibid., 34c.

6. See Cooper, *Panentheism*, 32–38; Hartshorne and Reese, *Philosophers Speak of God*, 38–57. Dirk Baltzly thinks that Plato's *Timaeus* is actually quite panentheistic:

> [It] describes a situation in which a cosmic god and a plurality of encosmic gods depend (beginninglessly) upon a creator god. . . . This view is best described as panentheistic. This is because panentheism is best understood as the position that asserts that there exist a creator god and a cosmos that are both somehow divine. In what sense the cosmos is divine—whether by being in the creator god or by being interpenetrated by the creator god or by being itself a god dependent upon the creator—is a matter to be specified by particular panentheists. Nonetheless, the creator god and the cosmos are not identical, as in pantheism. Plato's account in the *Timaeus* fits this description. Moreover, interpreted correctly, the *Timaeus* ascribes divinity to the soul–body composite that is the cosmos, as well as to the stars and planets within it. The notion that gods may be such soul–body composites sits uncomfortably alongside other passages in the Platonic corpus. Subsequent Platonists pursue a variety of strategies for resolving this tension. These include restricting the divine part of the cosmos to just that portion that has a particularly refined kind of matter, making the divinity of the cosmos a

second-hand one that results from the primary divinity of the World Soul, and simply paying lip service to the original Platonic text" (Dirk Baltzly, "Is Plato's *Timaeus* Panentheistic?," *Sophia* 49, no. 2 [2010], 214)

7. See Plotinus, *Enneads*, 7 vols., trans. A. H. Armstrong (Cambridge, MA: Harvard University Press, 1966–88), 1.3.4; 2.9.1; 5.4.1; 6.9.6.

8. Cooper, *Panentheism*, 41.

9. Plotinus, *Enneads* 4.3.22.

10. Cooper, *Panentheism*, 43.

11. See ibid., 44.

12. It is important to realize that Spinoza's notion of God has some important panentheistic overtones. Richard Mason notes that the term "God, or Nature" (Mason does not write it as many others do: "God-or-Nature")—the main argument allowing one to classify Spinozism as monistic pantheism—is actually seldom found and merely accidental in the *Ethics*. Moreover, he unveils an important inconsistency in Spinoza, who in the first part of his work denies that God is corporeal, whereas in the second part claims that he is an extended thing. Trying to understand properly the relation between God and the world, Mason refers to Edwin Curley's *Spinoza's Metaphysics: An Essay in Interpretation* and says that Spinoza did not understand the relation of particular things to God as the relation of predicate to its subject in a grammatical sense. Nor did he understand it as a relation of species to genus. He is of the opinion that Spinoza puts a major stress on causality. Therefore, modes are related to substance according to causality (substance is causally self-sufficient, whereas modes are not). Mason says that this approach distances Spinoza from pantheism and proves that he did not want to identify God with corporeal nature. Even if his theistic reflection does not establish a proper ontological difference between them, adds Mason, we can still consider Spinoza an important predecessor of the modern version of panentheism. See Richard Mason, *The God of Spinoza: A Philosophical Study* (Cambridge: Cambridge University Press, 1997), 28–32; Edwin M. Curley, *Spinoza's Metaphysics: An Essay in Interpretation* (Cambridge, MA: Harvard University Press, 1969).

13. See Johann Gottlieb Fichte, *Science of Knowledge (Wissenschaftslehre)*, ed. Peter Heath and John Lachs (New York: Appleton-Century-Crofts, 1970).

14. See Johann Gottlieb Fichte, "Faith," in *The Vocation of Man*, book 3, trans. Peter Preuss (Indianapolis: Hackett, 1987), 110.

15. Ibid., 107–8, 111.

16. See Alexandre Guilherme, "Fichte: Kantian or Spinozian? Three Interpretations of the Absolute I," *South African Journal of Philosophy* 29 (2010): 1–2, 5–7; Cooper, *Panentheism*, 93–94; Frederick C. Copleston, "Pantheism in Spinoza and the German Idealists," *Philosophy* 21 (1946): 47–49. Wilkens and Padgett note that Fichte's doctrine of the "Absolute I" as "a divine consciousness that contains the

world . . . eliminates any means of clearly speaking of the transcendence of God. Similarly, the idea of God as Creator becomes problematic. If the Absolute is infinite, all things are eternally found in it" (Steve Wilkens and Alan G. Padgett, *Christianity and Western Thought*, vol. 2, *Faith and Reason in the Nineteenth Century* [Downers Grove, IL: InterVarsity Press, 2000], 70).

17. See Karl Christian Friedrich Krause, *Vorlesungen über das System der Philosophie* (Göttingen: Dieterich'sche Buchhandlung, 1828).

18. Karl Christian Friedrich Krause, *Der zur Gewissheit der Gotteserkenntnis als des höchsten Wissenschaftsprinzips emporleitende Theil der Philosophie* (Prague: Tempsky, 1869), 10, quoted in Benedikt Paul Göcke, "On the Importance of Karl Christian Friedrich Krause's Panentheism," *Zygon* 48, no. 2 (2013): 366 (this and other quotations from Krause are in Göcke's translation).

19. Karl Christian Friedrich Krause, *Der Begriff der Philosophie* (Leipzig: Otto Schulze, 1893), 22.

20. Göcke quotes Krause's example of a grain of sand (which Krause adduces in defense of the alleged adequacy of his list of categories): "I think about a grain of sand. Although my recognition of this entity might be imperfect, I cannot fail . . . to think of it as possessing original unity, I cannot fail to think of it as possessing itselfness and wholeness, and also the unity of itselfness and wholeness. Furthermore, I distinguish the grain of sand as a whole from its parts, its internal constitution. . . . I cannot fail to think of the grain of sand as something positive which is directed upon itself in its act of being and thereby composes itself completely" (Krause, *Der zur Gewissheit*, 221).

21. Göcke notes that "the distinction between 'as such' and 'in itself' brings to mind the modern concept of emergence and applies to anything on any level of constitution. It also applies to the Absolute, that is, to the one infinite principle of being and recognition. Depending on how we conceive of the Absolute we either obtain the conclusion that everything is 'in' the Absolute or that the Absolute is 'outside of the world.' In this respect, Krause's theory of the relation between God and world puts us in mind of the fact that how we interpret things depends on how we understand them—whether *as such* or *in themselves*—while the things do not themselves change" (Göcke, "On the Importance," 371).

22. Göcke offers an overview of Krause's panentheism in Benedikt Paul Göcke, *The Panentheism of Karl Christian Friedrich Krause* (Berlin: Peter Lang, 2018), 33–44. He then analyzes Krause's philosophical theology in more detail in the following chapters of the same book. Our concern here is mainly God and his relation to the world (as part of Krause's philosophy of religion [Göcke evaluates its relevance in chapter 11 of his book]). Hence, I will not refer to Krause's psychology and his understanding of the transcendental constitution of the ego (as such and in itself), which Göcke describes in chapters 4 and 5, nor to his panpsychist philosophy of mind, and his philosophy of science, analyzed by Göcke in chapters 9, 12–13.

23. Göcke, "On the Importance," 372. He adds (ibid.) that in Krause's philosophy of religion the distinction between *Orwesen* and *Urwesen* answers the question "of whether God is an extramundane and the world is an extradivine being or not ... [since] through the distinction of Orwesen from himself as Urwesen one can see that God, as the One, identical, whole Essence, is neither out of nor above, nor next to, nor in the world, ... and that God as Urwesen is outside of and above the world, and the world outside of him as Urwesen" (Krause, *Vorlesungen über das System*, 401). See also Göcke, *Panentheism of Karl Christian Friedrich Krause*, 121–24, and 173–77, where he says that for Krause "there cannot be a *substantial* distinction" between God and the world, while, at the same time, they are distinguished "insofar as God is considered to be that in virtue of which the world exists and is what it is." Hence, in reference to Gregory R. Peterson's "Whither Panentheism?" (*Zygon* 36, no. 3 [2001]: 399) we should acknowledge that what Krause has in mind when thinking the God-world relationship is not merely their copresence ("weak panentheism"), but rather a sense of their identity, albeit in terms of part and whole ("strong panentheism").

24. Krause, *Der Zur Gewissheit*, 313. "*Orwesen* is not subordinated to any category, but instead, due to the unity of its being, is identical to each and every of its essentialities, to their unions, and to what distinguishes them. That is to say that according to Krause, in the fundamental intuition of God, God shows himself not only as an ordinary essence but as an essence that, as a whole, is identical with its essentialities" (Göcke, *Panentheism of Karl Christian Friedrich Krause*, 100). See also ibid., 100–112, on the relation of *Orwesen* to material and formal categories of the "analytic-ascending" part of science.

25. See Göcke, "On the Importance," 374.

26. Copleston, "Pantheism in Spinoza," 49–51; See also Friedrich Schelling, *System of Transcendental Idealism*, trans. Peter Heath (Charlottesville: University Press of Virginia, 1978), part 4, problem E, solution 3, parts 5 and 6; Alexandre Guilherme, "Schelling's *Naturphilosophie* Project: Towards a Spinozian Conception of Nature," *South African Journal of Philosophy* 29, no. 4 (2010): 373–81. Guilherme lists and describes six principal similarities between Schelling and Spinoza (see ibid., 381–86).

27. See Philip Clayton, "Panentheisms East and West," *Sophia* 49, no. 2 (2010): 183.

28. All quotations come from Cooper, *Panentheism*. See p. 100, with footnotes for references to Schelling's original works.

29. See Cooper, *Panentheism*, 94–104. Schelling uses classical Aristotelian and scholastic terminology when he refers to categories such as "eternity," "essence," "existence," "immutability," and the like. But he confronts them with the dialectic description of reality (based on categories such as change, dynamics, growth, and development).

30. Copleston, "Pantheism in Spinoza," 51.

31. See Georg Wilhelm Hegel, *The Phenomenology of Spirit*, translated by A. V. Miller (Oxford: Clarendon, 1977), 461–62. See also Andrew J. Reck, "Substance, Subject and Dialectic," *Tulane Studies in Philosophy* 9 (1960): 109–14; Timothy L. S. Sprigge, *The God of Metaphysics* (Oxford: Clarendon, 2006), 109–10.

32. "When man begins to philosophize, the soul must commence by bathing in this ether of the One Substance, in which all that man has held as true has disappeared; this negation of all that is particular, to which every philosopher must have come, is the liberation of the mind and its absolute foundation" (Georg Wilhelm Friedrich Hegel, *Lectures on the History of Philosophy*, vol. 3, trans. E. S. Haldane and Frances H. Simson [London: Routledge & Kegan Paul, 1955], 257–58).

33. Ibid., 3:283.

34. Ibid., 3:263–64, 282–83; Georg Wilhelm Friedrich Hegel, *The Logic of Hegel*, trans. William Wallace (Oxford: Clarendon, 1894), 317.

35. Hegel, *Lectures on the History of Philosophy*, 3:280–82; Hegel, *Logic of Hegel*, 89–90, 236–37; Efraim Shmueli, "Hegel's Interpretation of Spinoza's Concept of Substance," *International Journal for Philosophy of Religion* 1 (1970): 182–83.

36. See Hegel, *Lectures on the History of Philosophy*, 3:287–88; Shmueli, "Hegel's Interpretation," 183–84, 186; Reck, "Substance, Subject and Dialectic," 114–17.

37. See Hegel, *Lectures on the History of Philosophy*, 3:483, 499–501. Like Hegel, Clayton thus describes Fichte's insuperable dilemma: "Beginning with the finite ego leads (at best) to agnosticism about God.... Yet beginning with the creative act of an infinite ego leads to agnosticism about humans" (Philip Clayton, "Panentheism in Metaphysical and Scientific Perspective," in Clayton and Peacocke, *In Whom We Live and Move*, 79).

38. Hegel, *Lectures on the History of Philosophy*, 3:541–42; Cooper, *Panentheism*, 108.

39. See Hegel, *Logic of Hegel*, 147–152; Reck, "Substance, Subject and Dialectic," 120; James K. Feibleman, "Hegel Revisited," *Tulane Studies in Philosophy* 9 (1960): 22–31; Sprigge, *God of Metaphysics*, 114–55; Michael Forster, "Hegel's Dialectical Method," in *The Cambridge Companion to Hegel*, ed. Frederick C. Beiser (New York and Cambridge: Cambridge University Press, 1993), 130–70.

40. See Georg Wilhelm Hegel, *The Phenomenology of Mind*, trans. J. B. Baillie (New York: Dover, 2003), 80–88; Reck, "Substance, Subject and Dialectic," 124–31; Shmueli, "Hegel's Interpretation," 189–90; Copleston, "Pantheism in Spinoza," 52; Quentin Lauer, *Hegel's Concept of God* (Albany: State University of New York Press, 1982), 137–53. See also Nathan Rotenstreich, *From Substance to Subject: Studies in Hegel* (The Hague: Martinus Nijhoff, 1974).

41. Shmueli, "Hegel's Interpretation," 190. See also Reck, "Substance, Subject and Dialectic," 127–28; Lauer, *Hegel's Concept of God*, 158–61.

42. Joseph Prabhu sees Hegel as striving to overcome "a purely abstract and undialectical dichotomy ... between the secular and the theological" and calls him

a "secular theologian." He explains that "a secular theology demonstrates its commitment to secularity through three main affirmations: (1) the full reality and significance of this world, (2) the autonomy of the different fields of culture and knowledge including religion, and (3) the epistemological authority of reason and shared experience in determining the true and the real. Secularity, however, may be distinguished from secularism, which has a largely negative character, because it contends that these affirmations are trumped by three corresponding denials, namely the denial of (1) the existence of anything real or significant beyond this world, (2) the rationality and cultural authenticity of religion, and (3) sources of truth and models of reason other than those acknowledged by the empirical sciences. Thus, revelation, faith, 'reasons of the heart,' and the esprit de finesse, for example, are all ruled out of court in a reductionist secularism, of which Richard Dawkins in our time is a typical representative. It is clear that a secular theology wishes to affirm secularity but deny secularism" (Joseph Prabhu, "Hegel's Secular Theology," *Sophia* 49, no. 2 [2010]: 219).

43. See Copleston, "Pantheism in Spinoza," 53–54; Robert C. Whittemore, "Hegel as Panentheist," *Tulane Studies in Philosophy* 9 (1960): 136–40.

44. See Whittemore, "Hegel as Panentheist," 140–41; Copleston, "Pantheism in Spinoza," 55; Lauer, *Hegel's Concept of God*, 154–58.

45. Whittemore, "Hegel as Panentheist," 141.

46. Hegel, *Phenomenology of Mind*, 11.

47. See Whittemore, "Hegel as Panentheist," 147–62; Clark Butler, "Hegelian Panentheism as Joachimite Christianity," in *New Perspectives on Hegel's Philosophy of Religion*, ed. David Kolb (Albany: State University of New York Press, 1992), 139; Cooper, *Panentheism*, 113–16.

48. See Cooper, *Panentheism*, 117; Wilkens and Padgett, *Christianity and Western Thought*, 85–86; Butler, "Hegelian Panentheism as Joachimite Christianity," 137–38.

49. Georg Wilhelm Hegel, *Lectures on the Philosophy of Religion: One-Volume Edition; The Lectures of 1827*, ed. Peter C. Hodgson (Oxford and New York: Oxford University Press, 2006), 263.

50. Contrary to this conclusion, Prabhu claims that Hegel's idea of God is not panentheistic: "While panentheism is, from Hegel's perspective, an advance over both theism and pantheism, insofar as it insists on the transcendence, but not the separation, of God vis-à-vis his finite modalities; it still retains the notion of a personal God, even if still developing, and it is difficult to see how this can avoid the defect of externality, of a God ultimately set apart from all that he includes. Furthermore, the idea of finite things being contained in God ignores the fact that the finite is *aufgehoben* [at once preserved, negated, and transcended] in the Absolute. The first defect comes from its theistic feature, the second from its pantheistic one; so panentheism will not do either. All variants of theism are too 'substantial' for Hegel's taste. Hegel's idea of God, while bearing resemblances to neo-Platonic and also to some Christian

conceptions, is in fact a unique blend of the two" (Prabhu, "Hegel's Secular Theology," 224).

In the same article Prabhu adds that "Hegel's theology fits broadly ... within the category of nondualism and thus opens up a dialogue with the great nondual thinkers of the Eastern traditions from Sankara, Ramanuja, and Abhinavagupta in the Indian tradition to certain forms of Buddhist and Taoist thought" (ibid., 228). See also Loriliai Biernacki and Philip Clayton, eds., *Panentheism across the World's Traditions* (Oxford and New York: Oxford University Press, 2014). Mullins expresses skepticism about Biernacki and Clayton's project because it suggests considering under the banner of panentheism certain religious traditions that deny the existence of God. He sides with the strong criticism coming from Hutchings, who says that to commit oneself to panentheism is to commit oneself to anything. See Mullins, "Difficulty with Demarcating Panentheism," 326; Patrick Hutchings, "Postlude: Panentheism," *Sophia* 49, no. 2 (2010): 299.

51. See appendix 1 for the detailed list of panentheist philosophers and theologians in the nineteenth and twentieth centuries, and Cooper, *Panentheism*, 120–300, for more information about the most representative and influential figures among them.

52. Alfred North Whitehead, *Process and Reality* (New York: Free Press, 1979), 39.

53. See David R. Griffin, "Time in Process Philosophy," *KronoScope: Journal for the Study of Time* 1 (2001): 75–99; Griffin, *Whitehead's Radically Different Postmodern Philosophy: An Argument for Its Contemporary Relevance* (Albany: State University of New York Press, 2007), 106–38.

54. Terrence Deacon and Tyrone Cashman, "Eliminativism, Complexity, and Emergence," in Haag, Peterson, and Spezio, *Routledge Companion to Religion and Science*, 195.

55. See Whitehead, *Process and Reality*, especially chapters 2 and 3, pp. 18–36; Cooper, *Panentheism*, 166–72.

56. See Alfred North Whitehead, *Science and the Modern World* (New York: The Free Press, 1997; first published 1925), chapters 11–12; Cooper, *Panentheism*, 172–73.

57. See Alfred North Whitehead, *Religion in the Making* (New York: Macmillan, 1926); Cooper, *Panentheism*, 173–74.

58. Whitehead, *Process and Reality*, 343.

59. Ibid., 87–88, 343, 345.

60. Ibid., 88, 345.

61. Ibid., 345. Interestingly, Göcke notes that Krause's panentheism already implies negation of the doctrine of creation out of nothing and of the claim that God could have not created the world. Moreover, he says that the skepticism about *creatio ex nihilo* goes back even to Fichte, who saw the doctrine as the "absolute fundamental error of all false metaphysics and the doctrine of religion" (Johann Fichte,

"Die Anweisung zum seligen Leben, oder auch die Religionslehre," in *Fichtes Werke*, vol. 5, *Zur Religionsphilosophie*, ed. Immanuel H. Fichte [Berlin: de Gruyter, 1971], 479, quoted in Göcke, *Panentheism of Karl Christian Friedrich Krause*, 125, in his translation). Krause rejects *creatio ex nihilo* because the world according to him is not created at all (hence it cannot be created out of nothing). As the unity of all finite beings, the world is a part of the inner essentiality of *Orwesen*. Consequently, notes Göcke, "It is necessary . . . that *Orwesen* exist exactly when the world exists. Logically, therefore, the existence of the world is a necessary condition for the existence of *Orwesen*, as the non-existence of the world is sufficient condition for the nonexistence of *Orwesen*. . . . That the world necessarily exists when *Orwesen* exists, and that the world, for this reason, cannot be freely created by a God outside the world, implies that the world, as such, is eternal—of the same essence as *Orwesen* itself. Because *Orwesen* is eternal, the world is also eternal" (Göcke, *Panentheism of Karl Christian Friedrich Krause*, 125–26). Griffin embraces this position and says, "By saying that the world is in God, panentheism is distinguished from all forms of theism, according to which our world was created *ex nihilo* in such a way that the very existence of a realm of finite beings is wholly contingent upon a divine decision. Panentheism, by contrast, holds that the existence of the world is integral to the divine existence" (David R. Griffin, *Panentheism and Scientific Naturalism: Rethinking Evil, Morality, Religious Experience, Religious Pluralism, and the Academic Study of Religion* [Claremont, CA: Process Century Press, 2014], 13). As we shall see in section 3.4 below, contrary to Whitehead and Griffin, Clayton will argue that panentheism is consistent with *creatio ex nihilo*.

62. Whitehead, *Process and Reality*, 344.

63. See ibid., 222.

64. Ibid., 7, 21, 31, 88.

65. Ibid., 348–49, 351; See also Cooper, *Panentheism*, 176. When we consider Whitehead's version of panentheism, we should not ignore its possible relation to Hegel, whose thought might have reached and influenced Whitehead though British idealism. See Mariusz Tabaczek, "Hegel and Whitehead: In Search for Sources of Contemporary Versions of Panentheism in the Science–Theology Dialogue," *Theology and Science* 11 (2013): 143–61.

66. Charles Hartshorne, *Omnipotence and Other Theological Mistakes* (Albany: State University of New York Press, 1984), 94.

67. Charles Hartshorne, *The Divine Relativity: A Social Conception of God* (New Haven: Yale University Press, 1948), 138.

68. Ibid., 142.

69. Ibid., 122.

70. Hartshorne and Reese, *Philosophers Speak of God*, 22.

71. See Cooper, *Panentheism*, 181. In the following section of chapter 7 (pp. 185–90) Cooper investigates John Cobb and David Griffin's application of Whitehead

and Hartshorne's panentheism in specifically Christian theology. In their treatment of the "Divine Love" of process theology, in addition to offering arguments contrasting it with classical theism—similar to those developed by Hartshorne—they present it as "promoting enjoyment," understood primarily not as pleasure but as the experience of maximal actualization of an occasion's positive possibilities. Thus, they conclude that "in process thought, morality stands in the service of enjoyment" (John B. Cobb and David Ray Griffin, *Process Theology: An Introductory Exposition* [Louisville: Westminster John Knox, 1976], 56–57).

72. See Brierley, "Naming a Quiet Revolution," 5–12, with references to particular authors. See also the classification of the three main versions of panentheism proposed by Gregersen in "Three Varieties of Panentheism," 20–34; and Culp, "Panentheism," section 1.

73. See Philip Clayton, *Adventures in the Spirit: God, World, Divine Action* (Minneapolis: Fortress, 2008), 118–19.

74. Catherine Keller, *The Face of the Deep: A Theology of Becoming* (London and New York: Routledge, 2003), 219.

75. Göcke says that already for Krause, "although *Orwesen* is not subject to the flow of time, time is the form of the inner life of God, considered as such, and therefore the intrinsic constitution of *Orwesen*, that is, the nature of the Absolute in itself, with its separation and union of *Urwesen* and the world, is the temporal explication of what God, as such, is" (Göcke, *Panentheism of Karl Christian Friedrich Krause*, 181–82).

76. Once again, Göcke thinks this line of thinking is present already in Krause: "In Krause's panentheism the fact that there is a world is constitutive of the intrinsic nature of God as such, while at the same time, due to the freedom bestowed to the development of the world as such, it is not fixed how the history of the world will turn out to be. So, God requires a world, but not the world as it factually developed" (ibid., 182).

77. Whitehead, *Process and Reality*, 351.

78. See Ian G. Barbour, *Religion and Science: Historical and Contemporary Issues* (New York: HarperCollins, 1997), 285–87.

79. Ibid., 294.

80. See ibid., 322–25.

81. See Paul Davies, "Teleology without Teleology: Purpose through Emergent Complexity," in Clayton and Peacocke, *In Whom We Live and Move*, 96.

82. Ibid., 97.

83. Ibid.

84. Ibid., 104.

85. Ibid.

86. Ibid., 103. It is important to note that although being committed to a dipolar concept of God does not necessarily make one a panentheist, the emphasis on the

"two poles" in the essence of God is nevertheless common to those who follow this theological proposition.

87. See Arthur Peacocke, *Theology for a Scientific Age: Being and Becoming—Natural, Divine, and Human* (Minneapolis: Fortress, 1993), 25–80; Peacocke, *Paths from Science towards God: The End of All Our Exploring* (Oxford: Oneworld Publications, 2001), 39–90; Peacocke, "Articulating God's Presence in and to the World Unveiled by the Sciences," in Clayton and Peacocke, *In Whom We Live and Move*, 138–43.

88. Peacocke, "Articulating God's Presence," 139.

89. Ibid., 145.

90. Ibid., 146.

91. Ibid., 144.

92. Peacocke, *Theology for a Scientific Age*, 121.

93. Ibid., 122. On another occasion Peacocke states, "God cannot know definitely the precise outcome of any quantum event because God can only know that which is logically possible to know. . . . Ontological indeterminacy at the quantum level precludes such precise knowledge for God to have" (Arthur Peacocke, "Biological Evolution—a Positive Theological Appraisal," in *Evolutionary and Molecular Biology: Scientific Perspectives on Divine Action*, ed. Robert J. Russell, William R. Stoeger, and Francisco José Ayala (Vatican: Vatican Observatory; Berkeley, CA: Center for Theology and the Natural Sciences, 1998), 368–69n31.

94. Peacocke, *Theology for a Scientific Age*, 126; italics original. Referring to Paul Fiddes, Peacocke emphasizes once again: "There has been an increasing assent to the idea that it is possible 'to speak consistently of *a God who suffers eminently and yet is still God, and a God who suffers universally and yet is still present uniquely and decisively in the sufferings of Christ*'" (ibid., 127).

95. Ibid., 132; italics original.

96. Peacocke, "Articulating God's Presence," 150–51.

97. Ibid., 154. See also William Temple, *Nature, Man, and God* (London: Macmillan, 1934), chapter 19.

98. Peacocke, *Theology for a Scientific Age*, 371–72n75.

99. Clayton, "Panentheism in Metaphysical and Scientific Perspective," 91.

100. Ibid., 81–82.

101. Philip Clayton, *The Problem of God in Modern Thought* (Grand Rapids, MI / Cambridge, UK: Eerdmans, 2000), 488. See also ibid., 210, 235, 467–505. Cooper calls Clayton's theology a "neo-Schellingian emergent personalist panentheism" (see Cooper, *Panentheism*, 314).

102. See Clayton, *Problem of God*, 483–86; Philip Clayton, *God and Contemporary Science* (Grand Rapids, MI: Eerdmans, 1997), 93–96; "Panentheism in Metaphysical and Scientific Perspective," 82–83.

103. See Philip Clayton, "Kenotic Trinitarian Panentheism," *Dialog* 44, no. 3 (2005): 250–55.

104. See Clayton, "Panentheism in Metaphysical and Scientific Perspective," 84–91.

105. Polkinghorne's model originates in his *Belief in God in an Age of Science* (New Haven: Yale University Press, 1998).

106. Ignacio Silva, introduction to John C. Polkinghorne, ed., *The Work of Love: Creation as Kenosis* (Grand Rapids, MI / Cambridge, UK: Eerdmans, 2001), xii. Silva offers a critical evaluation of the development of Polkinghorne's thought in his article "John Polkinghorne on Divine Action: A Coherent Theological Evolution," *Science and Christian Belief* 24, no. 1 (2012): 19–30.

107. John C. Polkinghorne, *Science and the Trinity: The Christian Encounter with Reality* (New Haven: Yale University Press, 2004), 99. See also Polkinghorne, *Faith, Science and Understanding* (New Haven: Yale University Press, 2001).

108. John C. Polkinghorne, "Kenotic Creation and Divine Action," in *The Work of Love: Creation as Kenosis*, ed. John C. Polkinghorne (Grand Rapids, MI / Cambridge, UK: Eerdmans, 2001), 105. Note the transition from a purely informational to a both informational and energetic concept of divine action. Ignacio Silva notes that in his most recent reflection on divine action Polkinghorne becomes much more cautious, to the point of questioning the plausibility of his own model developed in the past: "In seeking to explore these possibilities [of God's action in the world], different people focused initially on different loci of intrinsic unpredictability, some looking to quantum indeterminacy and others to chaotic uncertainty. None of these attempted models should be taken with undue seriousness. They are what a physicist would call 'thought experiments'—attempts to explore and try out ideas in a simplified way, rather than purporting to be complete solutions to the problem of divine action" (John C. Polkinghorne, *Theology in the Context of Science* [New Haven and London: Yale University Press, 2009], 78).

109. John C. Polkinghorne, *Science and Christian Belief: Theological Reflections of a Bottom-Up Thinker* (London: SPCK, 1994), 64.

110. Polkinghorne, *Science and the Trinity*, 98.

111. Ibid., 101.

112. Ibid., 108. In his article written for one of the volumes in the series published jointly by the Vatican Observatory and the Center for Theology and the Natural Sciences, Polkinghorne subscribes even more openly to a dipolar (time/eternity) theism developed within process thought. He speaks not only about God not knowing the future but also about the *kenosis* (emptying) of divine omnipotence and omniscience. See John C. Polkinghorne, "The Laws of Nature and the Laws of Physics," in *Quantum Cosmology and the Laws of Nature: Scientific Perspectives on Divine Action*, ed. Robert J. Russell, Nancey C. Murphy, and C. J. Isham (Vatican: Vatican Observatory; Berkeley, CA: Center for Theology and the Natural Sciences, 1999), 438–39.

113. Polkinghorne, *Science and the Trinity*, 109. See Daniel Day Williams, *The Spirit and the Forms of Love* (New York: Harper & Row, 1968); Paul S. Fiddes, *The Creative Suffering of God* (Oxford: Oxford University Press, 1988).

114. "I do not accept panentheism as a present theological reality, but I do affirm *the eschatological hope of a sacramental panentheism* [the same term that Peacocke applies to the present] as the character of the new creation" (Polkinghorne, *Science and the Trinity*, 166).

115. Ibid., 165.

116. Ibid., 115–16. See also Polkinghorne, *Science and Christian Belief*, 168; Cooper, *Panentheism*, 315–17.

117. Ursula Goodenough and Terrence W. Deacon, "The Sacred Emergence of Nature," in *The Oxford Handbook of Religion and Science*, ed. Philip Clayton and Zachary Simpson (Oxford: Oxford University Press, 2006), 865–67.

118. Ibid., 867.

119. Ibid.

120. Although Sherman and Deacon do not explain further the category of "self-creation," it becomes clear from the context of their reasoning that what they have in mind is the idea that all purposeful actions of humans do not have any "non-wordly" origins, i.e., that they "arise spontaneously from a universe devoid of any such property" (Jeremy Sherman and Terrence W. Deacon, "Teleology for the Perplexed: How Matter Began to Matter," *Zygon* 42 [2007]: 896).

121. See ibid., 897–900.

122. Ibid., 899.

123. Ibid., 899–900. In our private conversation Deacon acknowledged he is more in favor of the apophatic tradition than of the religious naturalism of Goodenough, or Sherman's naturalism, which seems to be devoid of religion.

124. Willem B. Drees, *Creation: From Nothing until Now* (New York: Routledge, 2002), 25.

125. Davies, "Teleology without Teleology," 104.

126. Alexander, *Space, Time and Deity*, 362.

127. Ibid., 361–62, 365, 397.

128. Niels Henrik Gregersen, "Emergence: What Is at Stake for Religious Reflection?," in Clayton and Davies, *Re-Emergence of Emergence*, 290. See Harold Morowitz, "Emergence of Transcendence," in *From Complexity to Life: On the Emergence of Life and Meaning*, ed. Niels Henrik Gregersen (New York: Oxford University Press, 2003), 177–86.

129. In his *Process and Reality*, Whitehead refers to Alexander twice. First, defining "becoming" as "creative advance into novelty," he states that "every ultimate actuality embodies in its own essence what Alexander terms 'a principle of unrest,' namely, its becoming" (28). Second, describing "actual entity" as "an act of experience arising out of data ... a process of 'feeling' the many data," he says that "this use of the term 'feeling' has a close analogy to Alexander's use of the term 'enjoyment'" (40–41). Although Whitehead does not mention Alexander in his chapter "God and the World," Alexander's idea of "becoming God" is certainly

close to Whitehead's dipolar concept of deity. On the influence of British idealism on Whitehead, see Tabaczek, "Hegel and Whitehead," 149–50.

130. Alexander, *Space, Time and Deity*, 357.

131. Ibid., 358.

132. Arthur Peacocke, "Emergent Realities with Causal Efficacy: Some Philosophical and Theological Applications," in Murphy and Stoeger, *Evolution and Emergence*, 278.

133. Arthur Peacocke, "God's Interaction with the World," *Studies in Science and Theology* 3 (1995): 146.

134. Ibid.

135. Arthur Peacocke, "Emergence, Mind, and Divine Action: The Hierarchy of the Sciences in Relation to the Human Mind-Brain-Body," in Clayton and Davies, *Re-Emergence of Emergence*, 274–75. See also Peacocke, *Theology for a Scientific Age*, 161, 164.

136. Peacocke, *Paths from Science towards God*, 109.

137. Peacocke, "Emergence, Mind, and Divine Action," 274–75.

138. Peacocke, "Emergent Realities with Causal Efficacy," 279. Special divine action is usually contrasted with general (objective) divine action. Saunders thus defines both terms: "General Divine Action (GDA): Those actions of God that pertain to the whole of creation universally and simultaneously. These include actions such as the initial creation and the maintenance of scientific regularity and the laws of nature by God. Special Divine Action (SDA): Those actions of God that pertain to a *particular* time and place in creation as distinct from another. This is a broad category and includes the traditional understanding of 'miracles,' the notion of particular providence, response to intercessionary prayer, God's personal actions, and some forms of religious experience" (Saunders, *Divine Action and Modern Science*, 21).

139. Peacocke, "Emergence, Mind, and Divine Action," 276.

140. Gregersen, "Emergence," 295.

141. Philip Clayton, "Emergence from Quantum Physics to Religion: A Critical Appraisal," in Clayton and Davies, *Re-Emergence of Emergence*, 307.

142. Philip Clayton, "Toward a Constructive Christian Theology of Emergence," in Murphy and Stoeger, *Evolution and Emergence*, 327.

143. Clayton, "Emergence from Quantum Physics to Religion," 319.

144. James W. Haag, "Emergence and Christian Theology," in Haag, Peterson, and Spezio, *Routledge Companion*, 218.

145. Gordon D. Kaufman, "A Religious Interpretation of Emergence: Creativity as God," *Zygon* 42 (2007): 919. See also ibid., 917.

146. Ibid., 916.

147. Ibid., 918.

148. Gordon D. Kaufman, *In the Beginning . . . Creativity* (Minneapolis: Fortress, 2004), 55.

149. See ibid., 75–100; "Religious Interpretation of Emergence," 917–18.

150. Although identifying God with the "mystery of creativity" seems to save Kaufman's position as theistic, it remains unclear whether the term in question describes an ontological place/role for God, or just names an aspect of the natural world that we tend to call "God."

151. Kaufman, "Religious Interpretation of Emergence," 926.

152. Ibid.

CHAPTER 4

1. The last metaphor listed below—the one saying that God is totally dependent on, or coterminous with, the cosmos (i.e., they have the same boundaries or extent in time and space)—is clearly pantheistic. One might object that pantheism is monistic whereas panentheism defends ontological pluralism. However, the truth is that unlike Spinoza, who believed in substance monism, the majority of pantheists are de facto pluralists. They believe the universe contains many things. Its unity might be defined (1) ontologically, yet not necessarily in terms of (a) substance monism. It can be interpreted naturistically in reference to (b) ordering principle(s) and/or force(s) or (c) the fact of all things being grounded in "being." It might also be expounded in terms of either (2) identity of origin of all things or (3) teleological identity of their ultimate destiny or purpose, i.e., oneness with or full expression of deity. See Mander, "Pantheism," section 4; Michael P. Levine, *Pantheism: A Non-Theistic Concept of Deity* (London and New York: Routledge, 1994), 36–46.

2. See Michael W. Brierley, "The Potential of Panentheism for Dialogue between Science and Religion," in Clayton and Simpson, *Oxford Handbook of Religion and Science*, 636–41.

3. Philip Clayton, "Panentheism Today: A Constructive Systematic Evaluation," in Clayton and Peacocke, *In Whom We Live and Move*, 253. Clayton does not provide any reference to Oord's original list.

4. See Peterson, "Whither Panentheism?," 399–400.

5. See ibid., 402–3. The idea of the substantial unity of God and the world has important consequences. It seems to suggest the divinity of the universe, defined either distributively (each thing in the cosmos is divine) or collectively (the cosmos as a whole is divine). This brings panentheism to pantheism. One might object that unlike in classical theism and panentheism, divinity in pantheism does not have an aspect characterized in terms of being rational, "minded," and/or personal. It is usually defined in terms of the phenomenological content of a numinous experience that is historically and culturally pervasive in relation to human encounter with the beauty and harmony of nature. If this is true, then pantheism can be classified as fostering a nontheistic concept of deity, which certainly differentiates it from both classical theism and panentheism. At the same time, what seems to bring together pantheism and

panentheism is their limitation of the transcendence of the divine to its merely epistemological dimension. This limitation becomes true assuming that (1) the pantheistic emphasis of the ultimate immanence of divine unity does not entail the denial of that unity's epistemological transcendence, and (2) the panentheistic "in" expresses substantial unity of God and the created universe, i.e., that panentheism truly denies ontological transcendence of the divine (God).

  6. Lataster, "Attractiveness of Panentheism," 390–91. However, he does not support his statement with references to Christian panentheists who would think this way. Rather, he refers his reader to Barua's article which points toward those (e.g., Augustine) who remained rather skeptical about the idea that God can be regarded the soul of the world (the body), because it comes close to pantheism. See Ankur Barua, "God's Body at Work: Rāmānuja and Panentheism," *International Journal of Hindu Studies* 14, no. 1 (2010): 2–3.

  7. Göcke, *Panentheism of Karl Christian Friedrich Krause*, 177–78.

  8. Krause, *Der zur Gewissheit*, 307–8 (quoted and translated in Göcke, *Panentheism of Karl Christian Friedrich Krause*, 178–79). In support of this argument Krause provides two examples recalled by Göcke: (1) the sun in relation to nature and (2) human ego in relation to God. However, although both examples do avoid spatial-empirical interpretation of the relation among the entities (realities) in question, they may be challenged on their indirect and hidden assumption that each pair of *relata* belongs to one and the same ontological order. Hence, they still do not express nor explain the meaning of the panentheistic "in." Consider the following quotations from Krause, in which he develops these examples:

> (1) So, we say some finite natural entity, e.g. the sun, might be in nature. This contains the following compound thoughts: the sun is a finite entity. It is a part of a higher whole, nature. In its essentialities, the sun is similar to nature. But the sun is bounded. And the limits separate the sun from the whole of nature, but also unite it with it. Further, this limit is only the limit of this sun. The whole of nature, as a whole, is not also bounded or circumscribed by the limit of the sun. All this is what we want to say when we assert that the sun is in nature. (Krause, *Vorlesungen über das System*, 307, translated in Göcke, *Panentheism of Karl Christian Friedrich Krause*)

> (2) Similarly, if it is asserted that the ego, or any finite rational being, is in God, then this claim means the following: God is also the ego, also all the egos, but only as a part. Not the whole of God is a finite ego or all the finite egos. It is further thought that the ego is of the essentiality of Essence, so that the ego is also a self-same and whole essence, as God is, but finite and limited, not infinite and unconditioned like God. (ibid., 307–8, translated in Göcke, *Panentheism of Karl Christian Friedrich Krause*)

  9. Peacocke, "Articulating God's Presence," 145–46.

10. Clayton, "Panentheism in Metaphysical and Scientific Perspective," 88.

11. Philip Clayton, "The Panentheistic Turn in Christian Theology," *Dialog* 38, no. 4 (1999): 290.

12. Clayton, "Panentheisms East and West," 185. On the criticism of Clayton's demarcation of panentheism and pantheism see Mullins, "Difficulty with Demarcating Panentheism," 337–38. David Griffin, following the principles of process philosophy, arrives at a conclusion similar to Clayton's position. In reference to the mind/brain dependency (interaction), where he understands mind as "*numerically distinct* from the brain" whereas "the mind and the brain cells are *not ontologically different in kind*," he claims that the proposed doctrine, "which can be called *nondualistic interactionism*, also provides an analogy for understanding the interaction of God and the world" (David Ray Griffin, "Panentheism: A Postmodern Revelation," in Clayton and Peacocke, *In Whom We Live and Move*, 44–45). I find the assumption of an ontological unity in kind of God and the world gravitating towards pantheism.

13. Hartshorne, *Divine Relativity*, 143–44.

14. See Owen C. Thomas, "Problems in Panentheism," in Clayton and Simpson, *Oxford Handbook of Religion and Science*, 653–55. Thomas rightly notes that in process theism God's knowledge is neither complete nor of the contemporary world. He prehends actual occasions only after they reach their satisfaction and have perished. Furthermore, because God is not the Creator *ex nihilo*, the totality of the world, along with "eternal objects" and "creativity," seems to be ultimately unexplained.

15. Bracken sees "Each of the three divine persons [as] a personally ordered society of divine actual occasions" presiding "from moment to moment over an unlimited field of activity proper to itself. Yet because each of these fields of activity is infinite in scope, it necessarily coalesces with the others to form a common field of activity structured by the interrelated decisions of the divine persons from moment to moment" (Joseph A. Bracken, "Whitehead and Roman Catholics: What Went Wrong?," *American Journal of Theology & Philosophy* 30, no. 2 [2009]: 159). Bracken's view of the Trinity is inspired by Colin E. Gunton's *The One, the Three and the Many: God, Creation and the Culture of Modernity; The 1992 Bampton Lectures* (Cambridge and New York: Cambridge University Press, 1993).

16. See Culp, "Panentheism," section 4.

17. Joseph A. Bracken, "Panentheism and the Classical God-World Relationship: A Systems-Oriented Approach," *American Journal of Theology & Philosophy* 36, no. 3 (October 31, 2015): 222. Toward the end of the article Bracken embraces the kenotic aspect of process theism and states,

> If God and the world are thus involved in one and the same comprehensive life-system, then both God and the world are constrained in their respective modes of operation vis-à-vis one another. That is, the three divine persons find themselves limited in their dealings with their creatures by the constraints proper to the mode of operation of the cosmic process both as a whole and in its multiple

constituent subsystems. The divine persons cannot force but only persuade the finite subjects of experience (actual entities) within each of these subsystems proper to the cosmic process as a whole. (224–25)

18. Clayton, "Panentheism in Metaphysical and Scientific Perspective," 83.

19. John B. Cobb, "Review of Clayton and Peacocke," *Theology and Science* 3, no. 2 (2005): 241.

20. Gregersen, "Three Varieties of Panentheism," 23.

21. Aquinas, *ST* I.8.1 co.

22. See Aquinas, *ST* I.8.3 co. Georg Gasser agrees that the best way of speaking about God's omnipresence is to refer to his nonfundamental occupational relation to contingent creatures by presence, by power, and by essence. He also claims,

> Such an account helps to interpret the classical distinction between general divine action, which "merely" conserves creation, and special divine action, which "interferes" purposively in the course of creation, as one kind of action differing only in terms of agentive intensity. Special divine action is "special" because we experience divine presence in a more intense manner due to God's particular activity. General divine action, instead, we do not conceive of as anything extraordinary because we do experience the same degree of God's presence in virtue of the same conserving action all the time. (Georg Gasser, "God's Omnipresence in the World: On Possible Meanings of 'En' in Panentheism," *International Journal for Philosophy of Religion* 85, no. 1 [2019]: 59–60)

Before he reaches this conclusion, Gasser—in reference to Hud Hudson's "Omnipresence," in *The Oxford Handbook of Philosophical Theology*, ed. Thomas P. Flint and Michael C. Rea (Oxford: Oxford University Press, 2011), 199–216—analyzes the idea of divine immanence defined as (1) God's ubiquitous entending in space, i.e., God being located wholly and entirely at the most inclusive region of space—say, the cosmos—and wholly at its proper subregions; (2) God having among divine attributes the attribute of absolute space, which serves as a "receptacle" for physical space, which comes into existence with creation; and (3) God's entending in space immaterially. See Gasser, "God's Omnipresence," 48–57.

23. Aquinas, *ST* I.8.1 ad 2, as found in Gregersen, "Three Varieties of Panentheism," 23–24. I will further analyze the God-world / soul (mind)-body analogy in section 1.5 of this chapter. It needs to be noted that Gregersen uses the Blackfriars (Cambridge) translation and edition of the *Summa theologiae*, which differs in some important details from the standard Benzinger version, which is more faithful to the original Latin text. See appendix 2.

Owen Thomas adds that Karl Barth has an even stronger doctrine of God's immanence. After criticizing various historical versions of panentheism, Barth states,

> Now the absoluteness of God strictly understood in this sense means that God has the freedom to be present with that which is not God, to communicate

Himself and unite Himself with the other and the other with Himself, in a way which utterly surpasses all that can be effected in regard to reciprocal presence, communication and fellowship between other beings. . . . God . . . is free to be immanent, free to achieve a uniquely inward and genuine immanence of His being in and with the being which is distinct from Himself. (Karl Barth, *Church Dogmatics*, vol. 2, *The Doctrine of God, Part 1: The Knowledge of God* [Edinburgh: T&T Clark, 1957], 313, quoted in Thomas, "Problems in Panentheism," 656)

24. Gregersen, "Three Varieties of Panentheism," 24.

25. Aquinas, *ST* I.13.7 co.

26. See Michael J. Dodds, *Unlocking Divine Action: Contemporary Science and Thomas Aquinas* (Washington, DC: Catholic University of America Press, 2012), 171. He refers to William Lane Craig, *God, Time, and Eternity* (Dordrecht: Kluwer, 2001), 61, 78; David Tracy, *Blessed Rage for Order: The New Pluralism in Theology* (New York: Seabury Press, 1975), 177.

27. Gregersen, "Three Varieties of Panentheism," 20.

28. Ibid., 22, 24.

29. Müller, *Glauben, Fragen, Denken*, vol. 3, *Selbstbeziehung und Gottesfrage* (Münster: Aschendorf, 2010), 744, quoted in Göcke, *Panentheism of Karl Christian Friedrich Krause*, 180.

30. Göcke, *Panentheism of Karl Christian Friedrich Krause*, 182.

31. Clayton, "Panentheism in Metaphysical and Scientific Perspective," 83. Edgar Towne, in reference to several articles in the volume on panentheism edited by Clayton and Peacocke, notes that although the mutual dependence of God and creatures may be assumed by the majority of the contemporary Western theologians, it is not so willingly welcomed by Eastern Orthodox thinkers.

So Ware can say that "the penetration of the world by the uncreated energies does not enrich God, as he is in himself, but it certainly enriches the creation in its relation to the creator" (p. 167). Nesteruk, senior lecturer in mathematics at the University of Portsmouth, interprets panentheism in terms of the nature-hypostasis distinction. "The reciprocity of the Divine and the created is ultimately initiated and held by the person of the Logos of God. [This is] one-sided and entirely determined by the Logos himself" (p. 176). Similarly, Louth, professor of patristic and Byzantine studies at the University of Durham, discussing panentheism in terms of the divine logoi and energies, says, "there is no sense in which God may be said to be affected by the cosmos itself" (p. 184). Like Edwards and Bracken, Nesteruk and Louth locate the divine interaction with the universe within the divine Trinity. Gregersen rightly sees dipolar theism as an incompatible metaphysical option, which relates God and world in such a way there is real ontological identity, to whose view other panentheisms appear to be equivocal. (Edgar A. Towne, "The Variety of Panentheisms," *Zygon* 40, no. 3 [2005]: 784)

Towne refers to the following essays in Clayton and Peacocke, *In Whom We Live and Move*: Kallistos Ware, "God Immanent yet Transcendent: The Divine Energies according to Saint Gregory Palamas" (157–68); Alexei V. Nesteruk, "The Universe as Hypostatic Inherence in the Logos of God: Panentheism in the Eastern Orthodox Perspective" (169–83); Andrew Louth, "The Cosmic Vision of Saint Maximos the Confessor" (184–96).

32. Dodds, *Unlocking Divine Action*, 171, 169. See also Michael J. Dodds, "Ultimacy and Intimacy: Aquinas on the Relation between God and the World," in *Ordo Sapientiae et Amoris: Hommage au Professeur Jean-Pierre Torrell, O.P.*, ed. Carlos-Josaphat Pinto de Oliveira (Fribourg, Switzerland: Editions Universitaires, 1993), 211–27.

33. Dodds, *Unlocking Divine Action*, 172. See also Matthew R. McWhorter, "Aquinas on God's Relation to the World," *New Blackfriars* 94, no. 1049 (2013): 3–19; Thomas Gerard Weinandy, *Does God Change? The Word's Becoming in the Incarnation* (Still River, MA: St. Bede's Publications, 1984), 184.

34. Michael J. Dodds, *The Unchanging God of Love: Thomas Aquinas and Contemporary Theology on Divine Immutability* (Washington, DC: Catholic University of America Press, 2008), 169.

35. Brian J. Shanley, *The Thomist Tradition* (Dordrecht: Kluwer Academic, 2002), 59.

36. One of the important aspects and consequences of the panentheists' understanding of God's immanence is their interpretation of *creatio continua* in terms of the ongoing creative action of God, which he shares with the creatures. In section 3.2 of chapter 3 we saw Davies speaking of a God who decided to "give a vital, cocreative role to nature itself" (Davies, "Teleology without Teleology," 104). In the same chapter (section 4.3) Clayton suggested we should consider God as the "'co-creator' with finite agents" (Clayton, "Emergence from Quantum Physics to Religion," 307). Peacocke, similarly, speaks of God continuously creating through the processes of the natural order, working from inside the universe. He thinks "creation goes on all the time and is not just a one-off event" (Peacocke, *Theology for a Scientific Age*, 170). He sees God as "creating at every moment of the world's existence through perpetually giving creativity to the very stuff of the world" (Peacocke, "Articulating God's Presence," 144). Such claims are rather problematic from the point of view of the tradition of classical theism. If we define creation as bringing entities into existence *ex nihilo*, we must acknowledge that such an act requires an infinite power. Hence, notes Aquinas, only God can create (see *ST* I.45.5 co.). Consequently, the incessant processes of changes in nature described by panentheists should be perceived as important aspects of God's governance (*gubernatio*) of the created universe, rather than creation (*creatio ex nihilo*). The same classical theological tradition

acknowledges that creation finds its logical continuation in God's keeping all created things in existence (*conservatio a nihilo*), which may, properly speaking, be called *creatio continua*. But if creatures participate both in God's governance and in God's keeping things in existence causally, they do so acting as secondary and instrumental causes under the primary and principal causality of God, and not as sharing (univocally) in the creative action of God per se. I will say more on the relation of divine and creaturely causation from the point of view of classical theism in chapter 5, section 2.3. See Mariusz Tabaczek, "Pantheism and Panentheism," in *T&T Clark Handbook of the Doctrine of Creation*, ed. Jason Goroncy (London and New York: Bloomsbury T&T Clark, forthcoming).

37. Clayton, "Panentheism in Metaphysical and Scientific Perspective," 83.

38. See, respectively, William A. Christian, *An Interpretation of Whitehead's Metaphysics* (New Haven: Yale University Press, 1959), chapter 18, "God and the World: Transcendence and Immanence"; Burton Z. Cooper, *The Idea of God: A Whiteheadian Critique of St. Thomas Aquinas' Concept of God* (The Hague: Martinus Nijhoff, 1974), 102; Palmyre M. F. Oomen, "God's Power and Almightiness in Whitehead's Thought," *Open Theology* 1 (2015): 287–89.

39. Culp, "Panentheism," section 5.

40. See Alexander S. Jensen, *Divine Providence and Human Agency: Trinity, Creation and Freedom* (Burlington, VT: Ashgate, 2014), 131; Philip Clayton, "Creation Ex Nihilo and Intensifying the Vulnerability of God," in *Theologies of Creation: Creatio Ex Nihilo and Its New Rivals*, ed. Thomas Jay Oord (New York: Routledge, 2014), 27.

41. See Charles Hartshorne, *Man's Vision of God and the Logic of Theism* (Chicago: Willet, Clark, 1941), chapter 5.

42. In its rejection of the notion of *creatio ex nihilo* panentheism comes close to pantheism, which—as notes Levine—"claims only that Unity exists and explains what constitutes it. It need not explain its existence or claim an explanation is possible" (Levine, *Pantheism*, 187). Because panentheistic unity is divine, it is regarded as infinite, metaphysically perfect, necessarily existent and eternal. Hence, no notion of creation is necessary. Creation may be left behind, "along with the theistic concept of deity generally, as anthropocentric and anthropomorphic" (ibid., 194).

43. Barbour, *Religion and Science*, 295–96.

44. Gene Reeves lists four formative elements in the creation of actual occasions in Whitehead: past occasions, eternal objects, creativity, and God. He refers to Cobb arguing that God has always the decisive reason that each new occasion becomes. Barbour's position is similar. Nevertheless, Reeves's argument raises the same objection. If creativity is "the ultimate" in Whitehead, the claim that God is still the decisive factor is nonetheless questionable. See Gene Reeves, "God and Creativity," in *Explorations in Whitehead's Philosophy*, ed. Lewis S. Ford and George Louis Kline (New York: Fordham University Press, 1983), 239–51.

45. Davies, "Teleology without Teleology," 101.

46. Ibid., 102–3.
47. Ibid., 103, 105, 108.
48. Ibid., 108.
49. Arthur Peacocke, "God's Action in the Real World," *Zygon* 26 (1991): 460–61.
50. Clayton, *God and Contemporary Science*, 260.
51. Gregory Vlastos, "Organic Categories in Whitehead," *Journal of Philosophy* 34 (1937): 253–54. Unlike that of Hegel, Whitehead's dialectic is heterogeneous. Vlastos says that in the philosophy of organism, thesis is material (physical), whereas antithesis is ideal (conceptual). Because of this heterogeneity it is not the case either that the second dialectical stage can be generated from the first or that the third can be generated from the second. Thus, in this version of dialectics one finds no space for an internal contradiction, which is essential for homogenous dialectic.
52. Peacocke, "Articulating God's Presence," 151–52.
53. Gregersen, "Three Varieties of Panentheism," 32–33.
54. "Soul" and "mind" seem to be used interchangeably in this context, which is rather unfortunate. Nevertheless, I will follow the terminology found in the writings of panentheists, without introducing any further (and necessary) ontological distinctions between these two phenomena (entities).
55. See *ST* I.8.1 ad 2.
56. Clayton, "Panentheism in Metaphysical and Scientific Perspective," 83–84.
57. See chapter 3, section 3.3.
58. See Warren Brown, *Whatever Happened to the Soul? Scientific and Theological Portraits of Human Nature*, ed. Nancey Murphy and H. Newton Malony (Minneapolis: Fortress, 1998); Niels Henrik Gregersen, Ulf Görman, and Willem B. Drees, eds., *The Human Person in Science and Theology* (Edinburgh: T&T Clark International, 2003).
59. The contemporary views of the mind/body (brain) dependency once again raise the above-mentioned question concerning the reversal of the relation of containment (from Aquinas's idea of soul containing body to the panentheistic idea of the immaterial principle contained within the material [bodily] principle). Gregory Peterson notes that "in modern physicalist anthropology, the mind itself is bodily in character in the sense that it arises out of the operations of the brain (ignoring for the moment that the human brain functions properly only in a social context). By analogy, God (mind/brain) is in and part of the world (body). But the reverse is not true; the body is not in the brain. Furthermore, God (mind/brain) acts on the world (body), but the world (body) cannot be said to be in God (mind/brain)" (Peterson, "Whither Panentheism?," 402).
60. Apart from the challenges coming from contemporary neuroscience and philosophy of mind, Joseph Bracken states that the metaphor in question lacks clarity about the freedom and self-identity of the creatures in their relation to God. Similar is the complaint of Anna Case-Winters, who finds the metaphor perceiving soul (God)

as dominating the body (world) and failing to recognize the world as a unified organism. See Joseph A. Bracken, "The Issue of Panentheism in the Dialogue with the Non-Believer," *Studies in Religion/Sciences Religieuses* 21, no. 2 (1992): 211; Anna Case-Winters, "Toward a Theology of Nature: Preliminary Intuitions," *Religiologiques* 11 (1995): 251, 254.

61. Gregersen, "Three Varieties of Panentheism," 20. Mullins notes that the soul (mind)/body//God/world analogy does not help "to demarcate panentheism since pantheists can easily say that the universe is God's body. Further, classical theists like T. J. Mawson (Tim Mawson, "God's Body," *Heythrop Journal* 47, no. 2 [2006]: 171–81) also affirm that the universe is God's body" (Mullins, "Difficulty with Demarcating Panentheism," 335). He refers to Swinburne's five conditions that need to be met for a mind to be embodied (see Richard Swinburne, *The Coherence of Theism* [Oxford: Oxford University Press, 1977], 102–4) and asserts, "Surely something more substantive needs to be in place to demarcate panentheism from theism than the mere definition of embodiment. If mere omnipresence is all it takes for the universe to be God's body, then all forms of theism logically entail panentheism. Thus, there would be no difference between theism and panentheism, and as such this demarcation fails" (Mullins, "Difficulty with Demarcating Panentheism," 336).

62. This view seems to be identical with pantheistic belief, which—defined "from below"—"arises when the things of this world excite a particular sort of religious reaction in us. We feel, perhaps, a deep *reverence for* and *sense of identity with* the world in which we find ourselves. . . . To think of oneself as part of a vast interconnected scheme may give one a sense of being 'at home in the universe.' Here ecological thinking may come to the fore; like the individual creatures in a complex ecosystem, small but vital contributors to a larger whole, we too may be thought to have our place in the connected whole that is Nature" (Mander, "Pantheism," sections 2 and 9).

63. Such is certainly the position of Aquinas, based on his reading of Pseudo-Dionysius. On the other hand, in Orthodox Christianity apophatic theology is often taught as superior to cataphatic theology (e.g., the fourth-century Cappadocian Fathers, Maximus the Confessor, or—more recently—Vladimir Lossky). Some claim that Pseudo-Dionysius thought the same. Nonetheless, whether they would remain skeptical about any positive assertion concerning God's nature to the extent that Deacon and Cashman are skeptical remains questionable.

64. Davies, "Teleology without Teleology," 99–100.

65. Ibid., 100.

66. Kaufman, "Religious Interpretation of Emergence," 917.

67. Ibid., 927.

68. Ibid.

69. Robert J. Russell, *Cosmology from Alpha to Omega: The Creative Mutual Interaction of Theology and Science* (Minneapolis: Fortress, 2008), 136.

70. Peacocke, *Theology for a Scientific Age*, 164. Ignacio Silva rightly notes—in reference to Polkinghorne, who uses the same concept of divine action through a nonenergetic input of information, "a pattern-forming influence"—that the problem of the proponents of this idea is their univocal predication of God's action in the universe. Unlike classical theists—who, when referring to God the same categories of four Aristotelian causes, always pointed toward "the similarities and dissimilarities natural and divine causality have"—Polkinghorne and Peacocke seem to think about divine input of information in terms of God's raising information's level on Shannon's scale, which renders God a cause among causes. See Silva, "John Polkinghorne on Divine Action," 27–28.

71. In his *Unlocking Divine Action*, Dodds shows that the univocal approach to divine action is common to many theologians contributing to the science-theology dialogue. See 153–59.

CHAPTER 5

1. I have already mentioned this in section 2.4 of chapter 4. In section 2 of chapter 3 I noted that panentheists do not seem to distinguish between God's existence and the existence proper to creatures, which may be regarded as one of the major points of divergence between their position and the one of classical theism.

2. Aquinas, *ST* I.4.2 co. "In Him essence does not differ from existence" (I.3.4 co.). "Since therefore God is subsisting being itself, nothing of the perfection of being can be wanting to Him" (I.44.1 co.) "God alone is actual being through His own essence, while other beings are actual beings through participation, since in God alone is actual being identical with His essence" (*SCG* III.66.7). See also *ST* I.4.3 ad 4; I.104.1 co.; *In I Sent.* 37.1.1 co.; *Q. de ver.* 5.8 ad 9; *SCG* III.65.3; *Super De causis* 24. On the meaning of *ipsum esse subsistens*, see Rudi A. te Velde, *Participation and Substantiality in Thomas Aquinas* (Leiden, New York, Cologne: Brill, 1995), 119–25.

3. Dodds compares Aristotle's and Aquinas's understandings of potency and act to show that

> to Aristotle, act (substantial form) is a determining principle. For Thomas, act (*esse*) is, in itself, an unlimited or boundless principle. In creatures, of course, neither substantial form nor *esse* exists apart from its corresponding principle of potency. Both Thomas and Aristotle recognize, however, that there is a being that is "pure act" apart from all potency. The "pure act" Aristotle attributes to this being, however, is the determinate perfection of pure substantial form. The "pure act" Aquinas envisions is the boundless perfection of pure *esse*. For Thomas, as for Aristotle, pure act is the immovable summit of all perfection. In

Aristotle, this is the immovable mover. For Aquinas, it is the God of revelation. (Dodds, *Unchanging God of Love*, 129–30)

See also Aristotle, *Meta.* 12.6.1071b20; Aquinas, *SCG* I.16; *Q. de pot.* 7.2 ad 9; *ST* I.3.1 co.; I.3.4 co.

4. "Creation is not change" (Aquinas, *ST* I.45.2 ad 2). "The proper effect of God creating is what is presupposed to all other effects, and that is absolute being" (I.45.5 co.). "Creation in the creature is only a certain relation to the Creator as to the principle of its being" (I.45.3 co.). "Being is the most common first effect and more intimate than all other effects: wherefore it is an effect which it belongs to God alone to produce by his own power" (*Q. de pot.* 3.7 co.). "The being of every creature depends on God, so that not for a moment could it subsist, but would fall into nothingness were it not kept in being by the operation of the Divine power" (*ST* I.104.1 co.). "God is the cause not indeed only of some particular kind of being, but of the whole universal being" (I.103.5 co.). On the unity of *creatio ex nihilo* and divine *conservatio* of things see, Rudi A. te Velde, *Aquinas on God: The "Divine Science" of the Summa Theologiae* (Aldershot: Ashgate, 2006), 125. On being as the proper effect of God, see te Velde, *Participation and Substantiality*, 176–83.

5. "Two things pertain to the care of providence—namely, the 'reason of order,' which is called providence and disposition; and the execution of order, which is termed government. Of these, the first is eternal, and the second is temporal" (Aquinas, *ST* I.22.1 ad 2). "[A] thing's ultimate perfection consists in the attainment of its end. Therefore it belongs to the Divine goodness, as it brought things into existence, so to lead them to their end: and this is to govern" (I.103.1 co.). "Things are ordered to the ultimate end which God intends, that is, divine goodness, not only by the fact that they perform their operations, but also by the fact that they exist, since, to the extent that they exist, they bear the likeness of divine goodness which is the end for things. . . . Therefore, it pertains to divine providence that things are preserved in being" (*SCG* III.65.2). See also *In I Sent.* 39.2.1 co.; *ST* I.22.3 co.; I.45.5 co.; I.104.1 co.; *SCG* III.77.2.

6. Aquinas, *SCG* II.9.4. "Furthermore, an action that is not the substance of the agent is in the agent as an accident in its subject; and that is why action is reckoned as one of the nine categories of accident. But nothing can exist in God in the manner of an accident. Therefore, God's action is not other than His substance and His power" (*SCG* II.9.5). See also *ST* I.30.2 ad 3.

7. Aquinas, *ST* I.25.2 ad 2; see also *In IX De div. nom.* 2.232–59. As an agent God is not contained in any genus. Hence, creatures cannot participate in the likeness of God according to the same specific or generic formality, but only analogically. And because action follows being, God must differ from every other being, not only in *esse* but also in *actio*.

8. Aquinas, *SCG* III.101.1. On the perfection of God and God as creator, see Étienne Gilson, *The Christian Philosophy of St. Thomas Aquinas*, trans. L. K. Shook

(New York: Random House, 1956), 110–29. He also describes divine simplicity in *The Elements of Christian Philosophy* (New York: New American Library, 1963), 121–35.

9. In the case of separated intellectual substances like angels, their form is endowed with a determined nature subsisting in itself, apart from primary matter. And yet, as spiritual substances, angels are not identical with God, and their nature remains a mixture of potency and act. The unique and individual form of each angel (making it a separate species) is related to its act of being as potency to act. Thus, as a simple intelligence and pure form, free from any matter, an angel has still only a limited amount of being, and his *essentia* is not the same as his *esse*. As Aquinas states:

> Although there is no composition of matter and form in an angel, yet there is act and potentiality. And this can be made evident if we consider the nature of material things which contain a twofold composition. The first is that of form and matter, whereby the nature is constituted. Such a composite nature is not its own existence but existence is its act. Hence the nature itself is related to its own existence as potentiality to act. Therefore if there be no matter, and supposing that the form itself subsists without matter, there nevertheless still remains the relation of the form to its very existence, as of potentiality to act. And such a kind of composition is understood to be in the angels. (*ST* I.50.2 ad 3)

See also *SCG* II.50; Gilson, *Christian Philosophy of St. Thomas Aquinas*, 165.

10. Dodds illustrates the relation between the form and the act of existence (*esse*) through an analogy: "Form is a principle of being in a substance in the same way transparency is a principle of illumination in the air. As the transparency of the air makes the air a suitable subject for receiving and transmitting light, so the form of a substance makes the substance a suitable subject for receiving the act of existing (*esse*). And as the transparency of the air, apart from the act of the sun's illumination, is darkness, so the form of a substance, apart from the act of existing (*esse*), is nonbeing" (Dodds, *Unchanging God of Love*, 128). See Aquinas, *SCG* II.54.5.

11. "The third error is that of David of Dinant, who most absurdly [*stultissime*] taught that God was primary matter" (Aquinas, *ST* I.3.8 co.). See also *In I Sent.* 34.1.2 co.

12. Aquinas, *ST* I.44.2 co. "Although matter as regards its potentiality recedes from likeness to God, yet, even in so far as it has being in this wise, it retains a certain likeness to the divine being" (I.14.11 ad 3). "Since God is the efficient, the exemplar and the final cause of all things, and since primary matter is from Him, it follows that the first principle of all things is one in reality" (I.44.4 ad 4). "Primal matter has a likeness to God in so far as it has a share of being. For even as a stone, as a being, is like God, although it has no intelligence as God has, so primal matter in so far as it has being and yet not actual being, is like God. Because being is, so to say, common to potentiality and act" (*Q. de pot.* 3.1 ad 12).

13. See Aquinas, *In I Sent.* 36.2.3 ad 2; *Q. de ver.* 3.5; *ST* I.7.2 ad 3; John F. Wippel, *The Metaphysical Thought of Thomas Aquinas: From Finite Being to Uncreated Being* (Washington, DC: Catholic University of America Press, 2000), 322–26.

14. Aquinas, *In Phys.* 1.15.135. "All created things, so far as they are beings, are like God as the first and universal principle of all being" (*ST* I.4.3 co.). See also *SCG* III.19.4.

15. Aquinas, *Quod.* 8.2. Aquinas adopts teaching on divine ideas from Augustine, who presents his most comprehensive account of them in the forty-sixth of his *Eighty-Three Different Questions* (*De Diversis Quaestionibus LXXXIII*). After identifying Plato as the first who used the term "Ideas" (although not the first who grasped the divine reality signified by them), Augustine analyzes other names given to the ideas: "forms" (*formae*), "species" (*species*), and "reasons" (*rationes*). He then characterizes ideas as unchangeable and fixed, eternal and existing always in the same state, contained in the Divine Intelligence. After stating that it is by participation in ideas that whatever is exists in whatever manner it does exist in and elucidating how rational souls come to know the ideas (through an act of divine illumination), Augustine presents his argument for both the existence and the multiplicity of the divine ideas in the mind of God. He concludes by saying it would be "sacrilegious" to suggest that God had to look to something outside himself to get the pattern for what he was going to create. Augustine's teaching on divine ideas served as the chief patristic source on the topic throughout the entire Middle Ages.

16. In my investigation of divine ideas as exemplar causes in Aquinas I am indebted to Gregory T. Doolan and his excellent study: *Aquinas on the Divine Ideas as Exemplar Causes* (Washington, DC: Catholic University of America Press, 2008). Although the doctrine posits forms apart from matter, Doolan emphasizes that it does not contradict the Aristotelian position. In reference to *Q. de ver.* 3.1 ad 4, he reminds us that "natural forms cannot exist immaterially of themselves, but they can acquire an immateriality from the one in whom they exist. This is evident with our own intellects in which they exist in an immaterial way, and so such forms can also exist in an immaterial way in the divine intellect" (Doolan, *Aquinas on the Divine Ideas*, 82).

17. "By ideas are understood the forms of things, existing apart from the things themselves" (Aquinas, *ST* I.15.1 co.). See also Aquinas, *Q. de ver.* 3.3; *In I Sent.* 36.2.1. In a broader sense an idea can also pertain to purely speculative knowledge, as the notion (*ratio*) or likeness (*similitude*) of a thing.

18. Aquinas, *Super De causis* 14. See also *Q. de ver.* 3.1; *Quod.* 8.1.2; *In V De div. nom.* 3.665; *ST* I.35.1 ad 1.

19. Aquinas, *Q. de ver.* 2.5 co.

20. Aquinas, *ST* I.15.3 co.

21. Aquinas, *In V De div. nom.* 3.665.

22. Aquinas, *ST* I.15.3 co.

23. It is important to note that although in *Q. de ver.* Aquinas claims that the term "exemplar" belongs both to knowledge that is actually practical (i.e., leads to an

actual realization and action) and virtually practical (i.e., can but does not have to be realized), beginning from *In De div. nom.* he holds that only actually practical ideas can be called exemplars.

24. See Aquinas, *In I Sent.* 35.1.2; *Q. de ver.* 2.1; 2.3; *SCG* I.44; *ST* I.44.3.
25. See Aquinas, *Q. de ver.* 3.1; *SCG* I.49; *ST* I.15.1 co.
26. See Aquinas, *ST* I.14.5 co.
27. See Aquinas, *In Meta.* 1.11.178–79; 1.17.259; Doolan, *Aquinas on the Divine Ideas*, 61–64. Doolan analyzes the argument of those who claim that Aquinas's Fourth Way (from degrees of perfection) is principally influenced not by his Aristotelianism but rather by his Platonism, and thus becomes one more argument in favor of divine exemplarism. In their opinion,

> the Fourth Way attempts to prove that God exists by proving that he is the cause of all transcendental and simple perfections, and it attempts to do this by focusing on two types of causality: exemplarism and efficiency. In the first stage of the Fourth Way, Thomas argues for the existence of a *maxime ens* that he presents as an exemplar for all the lesser degrees of perfection that approach it. In the second stage, he argues that this absolute maximum must be the efficient cause of all of the perfections that approach it in likeness. It must be granted that he never explicitly mentions either of these two types of causality by name, but scholars are generally agreed that the argument is nonetheless referring to both. (Ibid., 70–71)

And yet Doolan notes that, unlike the arguments from teleology, divine similitude, and divine self-knowledge, which refer to the exemplars in God's intellect of all the things that he makes, the argument in the Fourth Way refers to the exemplarism of the one divine essence acting as a "natural exemplar" for perfections that can be possessed by created entities in diverse degrees (e.g., goodness, truth, unity, etc.). See ibid., 75–80.

28. "Ontologically there is but one exemplar of all things, which is God. As the fullness of being (*esse*), the divine essence is imitable in diverse ways. Still, when Thomas addresses the subject of divine ideas, it is not simply to this imitability that he is referring. Rather, for him a divine idea consists in God's *knowing* his essence as imitable in these diverse ways. It is this knowledge that constitutes an idea. Since these ways are themselves diverse, so too is God's knowledge and, hence, his ideas" (Doolan, *Aquinas on the Divine Ideas*, 116). A composite thing has a perfect idea in God by reason of its intrinsic substantial form. Primary matter has an idea only imperfectly (as it is unknowable in itself). Accidents fall short of having a perfect idea in God as well, but to the extent they do imitate God's essence, the divine essence is their idea. God knows pure possibilia not as existing in themselves in any way, but as existing solely in his divine power. Privation does not have an idea in God at all. Evil cannot have one either, as it is a privation, named from its lack of form.

29. Aquinas, *ST* I.44.3 co.

30. Doolan, *Aquinas on the Divine Ideas*, 90. See also Aquinas, *Q. de ver.* 3.2; 2.4 ad 2.

31. Thomas speaks about *forma exemplaris* in *In I Sent.* 18.1.5; *Q. de ver.* 3.1; *Quod.* 4.1; 8.2; *ST* I.44.3 co.

32. See Aquinas, *In I Sent.* 36.2.1; *Q. de ver.* 3.1.

33. See Aquinas, *Q. de ver.* 3.1, where Aquinas teaches that the form can be understood in three ways: (1) as that "by which" (*a qua*) a thing is formed, as it proceeds from the form of an agent; (2) as that "according to which" (*secundum quam*) something is formed, i.e., a part of a composite (e.g., soul as the form of a man or the figure of a statute as the form of bronze); and (3) as that "in regard to which" (*ad quam*) something is formed. Only the last refers to the exemplar form, in imitation of which something is made. Moreover, something can imitate form according to the intention of an agent or accidentally. Only the former applies to exemplar forms, with an additional requirement that an agent predetermines the end for himself.

34. Doolan suggests that in his early commentary on the *Sentences* Thomas goes even further, treating exemplarism "as being, at least in part, a type of efficient causality" (Doolan, *Aquinas on the Divine Ideas*, 33–34n76). He sees it as Aquinas's youthful view, which he comes to abandon later in his career. Doolan bases his theory on the term *causa efficiens exemplaris* (efficient exemplar cause), which appears in several places in Aquinas's *In I Sent.* However, a closer analysis of the context in which Aquinas uses the term in question shows that he refers it to divine attributes (or divine names) rather than God's exemplar ideas. For instance, in *In I Sent.* 1.4.2 we read: "Opera divina possunt comparari ad divina attributa sicut ad causam efficientem exemplarem; et hoc modo sapientia creaturae est a sapientia Dei, et esse creaturae ab esse divino, et bonitas a bonitate; et sic loquitur Boetius" (God's works can be compared to divine attributes as to efficient exemplar causes; and in this way the wisdom of a creature is from God's wisdom, and the being of a creature from God's being, and goodness from God's goodness, as Boethius says). In *In I Sent.* 10.1.5 ad 4 we find Aquinas stating: "Omnia attributa divina sunt principium productionis per modum efficientis exemplaris; sicut bonitatem omnia bona imitantur, et essentiam omnia entia, et sic de aliis" (All divine attributes are principles of production in the manner of efficient exemplars, as all good things imitate [God's] goodness, and all beings [imitate God's] essence). Similar is the meaning of *principium effectivum exemplari* (effective exemplar principle) in *In I Sent.* 19.5.2 ad 3, of *modus communis efficiens exemplaris* (common mode of efficient exemplars) in *In I Sent.* 38.1.1 co., and of *causalitas efficiens exemplaris* (causality of efficient exemplars) in *In I Sent.* 8.1.3 ad 2 (English translation of excerpts from *In I Sent.* is my own). In all cases listed here Aquinas speaks about the efficient causality of exemplars in terms of divine attributes (or divine names), and not in terms of divine exemplar ideas. It thus seems more appropriate to hold that if exemplars play a role in efficient causality, it is possible only through their relation to the will of the divine agent, whose action realizes exemplar ideas.

35. See Aquinas, *ST* I.21.2; *Q. de pot.* 6.3 ad 3. When he considers the example of an artisan, Aquinas states that "an artisan's knowledge manifests and his will intends an end, commanding the act through which the work is produced" (*In I Sent.* 38.1.1).

36. Aquinas, *Q. de ver.* 2.10 ad sed contra 1.

37. Aquinas, *ST* I.14.11 ad 1. We must note that divine exemplar ideas are not reproduced in creatures in the same way they exist in God, but according to the mode that the nature of creatures allows. Thus, even if the divine ideas are themselves immaterial, they nonetheless produce material things. See *Q. de ver.* 2.10 ad 1.

38. See Aquinas, *Q. de ver.* 2.5; Doolan, *Aquinas on the Divine Ideas*, 163.

39. See Doolan, *Aquinas on the Divine Ideas*, 127–29, 163–65. Doolan acknowledges that Aquinas introduces the distinction between *forma totius* and *forma partis* in a different context, in his commentary on the *Metaphysics* and in *De ente et essentia*, and never applies that distinction to divine ideas. But Doolan claims that the distinction can be used and is insightful with reference to divine exemplars. He reminds us, however, that we must remember that "unlike a *forma totius* that exists in *our* intellects, which corresponds only to the essence of a species . . . divine forms correspond to the essences of individuals" (*Aquinas on the Divine Ideas*, 165).

40. On medieval occasionalists and their critiques, see John Marenbon, "The Medievals," in Beebee, Hitchcock, and Menzies, *Oxford Handbook of Causation*, 40–54; William E. Carroll, "Creation and Science in the Middle Ages," *New Blackfriars* 88 (2007): 680–86; John Henry and Mariusz Tabaczek, "Causation," in *Science and Religion: A Historical Introduction*, ed. Gary B. Ferngren (Baltimore: Johns Hopkins University Press, 2017), 379–82; Ignacio Silva, "Divine Action and Thomism: Why Thomas Aquinas's Thought Is Attractive Today," *Acta Philosophica* 25 (2016): 66–67.

41. See Aquinas, *Q. de pot.* 3.8; 3.1 ad 12; *SCG* III.69.

42. See Aquinas, *In I Sent.* 7.1.1 ad 3; *Q. de ver.* 5.8 ad 8; *Q. de pot.* 5.1; *ST* I.104.1.

43. Aquinas, *ST* I.105.5 co.

44. Aquinas, *ST* I.22.3 ad 2. See also *ST* I.19.6 ad 3; I.19.8 co.; I.23.5 co.; I.105.5 ad 2; I-II.10.4 ad 2; Gregory T. Doolan, "The Causality of the Divine Ideas in Relation to Natural Agents in Thomas Aquinas," *International Philosophical Quarterly* 44 (2004): 407; Gilson, *Christian Philosophy of St. Thomas Aquinas*, 176, 182–84; te Velde, *Participation and Substantiality*, 170–75.

45. Aquinas, *SCG* III.70.8. "Just as it is not unfitting for one action to be produced by an agent and its power, so it is not inappropriate for the same effect to be produced by a lower agent and God: by both immediately, though in different ways" (*SCG* III.70.5). See also Doolan, "Causality of the Divine Ideas," 408–9.

46. Aquinas, *ST* I.4.3 co.

47. Aquinas, *SCG* III.67.5. See also *ST* I.21.4 co.; I.36.3 ad 4; *Q. de ver.* 5.9 ad 10; *Q. de pot.* 3.7 co.

48. Following Aristotle and his cosmology—which placed universal causality in the sun and, partially, in the stars—when he spoke about the distinction between primary and secondary causes, Aquinas used the example of lower bodies, which "act through the power of the celestial bodies" (*SCG* III.67.5). Although he would also mention the example of the lower artisans, who "work in accord with the direction of the top craftsman" (ibid.), he developed more the former argument (the standard example is a man being begotten by man and by the sun). See *SCG* III.69.24; *ST* I.118.1 ad 3; *Q. de pot.* 3.8 ad 15. See also my comment on the same example in Aristotle above, chapter 2, section 1.4, n. 30. The distinction between primary and secondary causation was rejected by Duns Scotus, who claimed that to explain the contingency of things in the world, we must say that the First Cause itself causes contingently. See Marenbon, "Medievals," 44.

49. Aquinas, *Q. de ver.* 27.4 co.; 27.4 ad 8. See also *ST* III.62.1 co.; III.62.1 ad 2; *In III Sent.* 18.1.1 ad 4; *SCG* III.147.6; *ST* I.45.5 co.; III.19.1 co.; 62.4 co.; 66.5 ad 1.

50. See Dodds, *Unlocking Divine Action*, 193–94.

51. Aquinas, *SCG* III.66.4. See also III.67.1; II.21; III.66.1–3; III.66.5; *Q. de pot.* 3.7 ad 3; 3.7 ad 16; 5.1 co.; *ST* I.45.5 co.; I.104.1 co. Wippel notes that

> for Thomas, whenever a new substance is efficiently caused by a natural or created agent, that agent's causation applies both to the act of being itself (*esse*) of the new substance and to a particular determination of esse as realized in that substance. Causation of the particular determination (this or that kind of form) is owing to the created efficient cause insofar as it operates by its own inherent power as a principal cause. Causation of the act of being itself (*esse*) is assigned to it as an instrumental cause acting with the power of God and to God himself as the principal cause of the same. From this it follows that one should not maintain that Thomas denies that created causes can efficiently cause the act of existing or the act of being, at least in the process of bringing new substances into being. (John F. Wippel, "Thomas Aquinas on Creatures as Causes of Esse," *International Philosophical Quarterly* 40 [2000]: 213)

See also Doolan, "Causality of the Divine Ideas," 400–408; Étienne Gilson, *Thomism: The Philosophy of Thomas Aquinas* (Toronto: Pontifical Institute of Medieval Studies, 2002), 210–12.

52. Silva presents similar accounts of the same typology in three articles: Ignacio Silva, "Thomas Aquinas Holds Fast: Objections to Aquinas within Today's Debate on Divine Action," *Heythrop Journal* 48 (2011): 5–7; Silva, "Revisiting Aquinas on Providence and Rising to the Challenge of Divine Action in Nature," *Journal of Religion* 94 (2014): 280–85; Silva, "Divine Action and Thomism," 71–74.

53. Aquinas, *Q. de pot.* 3.7 co.

54. Ibid.

55. Silva, "Thomas Aquinas Holds Fast," 6.

56. Aquinas, *Q. de pot.* 3.7 co.

57. Ibid. In other words, "God is the cause of every action, inasmuch as every agent is an instrument of the divine power operating" (ibid.).

58. Commenting on this aspect of divine action, te Velde says: "The effect of being, *esse*, Thomas frequently says, belongs solely to God according to his own power. This is a well-known but often misinterpreted statement. Being is not simply poured in by God from above in all particular effects of natural causes as their common actualization. Although God's power is immediately related to the being of things, which is its formal effect, this immediacy does not mean that the divine gift of being remains extrinsic to the effects of the natural agents, as if being were exclusively God's effect; it is on the contrary by the immediacy (intimacy) of God's operation that every other agent is mediated with the being-in-act of its effect and thus constituted in its proper action. And it is by reflection on this mediation that it appears to us that being must be attributed to God as the effect proper to his universal power. This means that God gives being by causing every other agent to give being in a particular way, adapted to its particular power" (te Velde, *Participation and Substantiality*, 176–77).

59. Silva speaks about "founding" (in his first paper on this topic from 2011 he uses the term "static") and "dynamic" "moments." I find it more appropriate to speak about "founding" and "dynamic" aspects of efficient divine action in the world.

60. I am following Silva's argumentation presented in "Thomas Aquinas Holds Fast," 5–9; "Revisiting Aquinas on Providence," 281.

61. Thomas F. Tracy, "Special Divine Action and the Laws of Nature," in *Scientific Perspectives on Divine Action: Twenty Years of Challenge and Progress*, ed. Robert J. Russell, Nancey C. Murphy, and William R. Stoeger (Vatican: Vatican Observatory; Berkeley, CA: Center for Theology and the Natural Sciences, 2008), 257.

62. See te Velde, *Participation and Substantiality*, 165–66; Silva, "Revisiting Aquinas on Providence," 281n10.

63. Stoeger also mentions God's particular application of a secondary cause in cases of special divine action: "God [is] not only acting as primary cause to maintain secondary causes in existence, but possibly working through secondary causes to produce an effect God desires, a special or particular effect, outside of the ordinary pattern of what we would expect" (William R. Stoeger, "Describing God's Action in the World in Light of Scientific Knowledge of Reality," in Russell, Murphy, and Peacocke, *Chaos and Complexity*, 254, 256).

64. William E. Carroll, "Divine Agency, Contemporary Physics, and the Autonomy of Nature," *Heythrop Journal* 49 (2008): 592.

65. In one of his articles Carroll does seem to pay attention—although indirectly—to the first of the two dynamic aspects of efficient divine action in the world, when he quotes a passage from *ST* I.105.5 co. in which Aquinas mentions God as one who "moves things to operate, as it were applying their forms and powers to operation, just as the workman applies the axe to act." See William E. Carroll, "Aquinas on Creation and the Metaphysical Foundations of Science," paper presented at

the annual Thomistic Institute sponsored by the Jacques Maritain Center, University of Notre Dame, July 23, 1998, accessed July 30, 2019, https://maritain.nd.edu/jmc/ti98/carroll.htm.

66. Dodds, *Unlocking Divine Action*, 190.

67. See ibid., 190–99.

68. Nancey Murphy, "Divine Action in the Natural Order: Buridan's Ass and Schrödinger's Cat," in Russell, Murphy, and Peacocke, *Chaos and Complexity*, 333.

69. John C. Polkinghorne, *Science and Theology: An Introduction* (Minneapolis: Fortress, 1998), 86. On another occasion Polkinghorne says, "It is not clear to me what is gained by so apophatic an account of God's action. In the end, the answer seems to be 'God only knows'" (Polkinghorne, "The Metaphysics of Divine Action," in Russell, Murphy, and Peacocke, *Chaos and Complexity*, 150).

70. Peacocke, *Theology for a Scientific Age*, 148–49.

71. Clayton, *God and Contemporary Science*, 177.

72. Polkinghorne, *Science and Theology*, 86.

73. Murphy, "Divine Action in the Natural Order," 333. Ignacio Silva offers a similar account, listing most of the charges mentioned here and then attempting to answer them in "Thomas Aquinas Holds Fast," 2–5; "Divine Action and Thomism," 79–82.

74. Polkinghorne, *Science and Theology*, 86.

75. Austin Marsden Farrer, *Faith and Speculation* (London: Adam and Charles Black, 1967), 159.

76. Ibid., 66.

77. Ibid., 62.

78. I will say more about the distinction between *per se* and *per accidens* causes in the section dedicated to chance and fortune. On divine action and evil, see Dodds, *Unlocking Divine Action*, 236–43; Ignacio Silva, "Providence, Contingency, and the Perfection of the Universe," *Philosophy, Theology and the Sciences* 2 (2015): 153.

79. Aquinas, *ST* I.105.6 ad 1. See also *SCG* III.100.2.

80. Dodds, *Unlocking Divine Action*, 253.

81. Ibid., 198. "[God] governs things inferior by superior, not on account of any defect in His power, but by reason of the abundance of His goodness; so that the dignity of causality is imparted even to creatures" (Aquinas, *ST* I.22.3 co.). See also *Q. de ver.* 5.8 ad 11; *SCG* III.70.7. In the same vein, answering the question concerning causation of divine exemplars, we may speak about intermediary exemplar causes: "Just as the divine power, the first agent, does not exclude the action of the natural power, so neither does the first exemplar form, which is God, exclude the derivation of forms from other lower forms whose action produces forms like themselves" (*Q. de ver.* 3.8 ad 17). See also *ST* I.65.4 ad 2; I.104.1 co.; I-II.6.7 ad 1.

82 Aquinas, *ST* I.103.2 co. That Aquinas is thinking here about the end of the whole universe becomes clear from the title of the article, as well as the last sentence

of the *respondeo*, in which he speaks about "the end of the whole universe" (*finis totius universi*). Hence, the term "the end of all things" (*finis rerum*) in the earlier part of the same passage should be interpreted as referring to the entire universe as well.

83. Aquinas, *ST* I.6.4 co. "God moves as the object of desire and apprehension" (I.105.2 ad 2). See also *In I Sent.* 34.1.2 co.; *ST* I.6.1 co.; I.6.1 ad 2; I-II.109.6 co.

84. Aquinas, *ST* I.6.1 ad 2.

85. Aquinas, *ST* I.44.4 ad 3. "Thus then does God work in every worker, according to these three things. First as an end. For since every operation is for the sake of some good, real or apparent; and nothing is good either really or apparently, except in as far as it participates in a likeness to the Supreme Good, which is God; it follows that God Himself is the cause of every operation as its end" (*ST* I.105.5 co.). See also Étienne Gilson, *The Spirit of Mediaeval Philosophy* (New York: Charles Scribner's Sons, 1940), 75; Corey L. Barnes, "Natural Final Causality and Providence in Aquinas," *New Blackfriars* 95 (2014): 349–61.

86. Dodds, *Unlocking Divine Action*, 181.

87. *ST* I.44.4 co. See also *ST* I.25.2 co.; Dodds, *Unlocking Divine Action*, 181–82. Although we find certain similarities between Aristotle and Aquinas in their understanding of the first source of teleology in nature, Étienne Gilson notices that Aquinas's God differs significantly from Aristotle's first mover, which moves only by the love which it excites, does not create *ex nihilo*, and cannot breathe love into contingent beings. The last is proper for the Christian God, who is the Creator. Consequently, although the unmoved mover does not refuse to be loved, the God of Aquinas loves and is love himself. He "shares" his perfection and goodness with each creature, and each creature groans for fulfillment in God (See Gilson, *Spirit of Mediaeval Philosophy*, 75).

88. See Aquinas, *SCG* III.39; *ST* I.63.9; *Q de ver.* 3.1 co.

89. Aquinas, *In meta.* 6.3.1210. See also *SCG* III.99; Silva, "Providence, Contingency," 143–48; Ignacio Silva, "Thomas Aquinas on Natural Contingency and Providence," in *Abraham's Dice: Chance and Providence in the Monotheistic Traditions*, ed. Karl W. Giberson (New York: Oxford University Press, 2016), 162–65.

90. Aquinas, *Q. de ver.* 5.9 ad 12.

91. See Polkinghorne, "Metaphysics of Divine Action," 151–55; Russell, *Cosmology from Alpha to Omega*, chapters 4–5 (pp. 110–211); Murphy, "Divine Action in the Natural Order," 340–43; Thomas F. Tracy, "Creation, Providence, and Quantum Chance," in *Quantum Mechanics: Scientific Perspectives on Divine Action*, ed. Robert J. Russell, Philip Clayton, Kirk Wegter-McNelly, and John Polkinghorne (Vatican: Vatican Observatory; Berkeley, CA: Center for Theology and the Natural Sciences, 2001), 239–50. Polkinghorne argues in favor of divine action through indeterministic aspects of chaotic occurrences in nature. Robert Russell develops a comprehensive argument in favor of divine action at the level of quantum indeterminacy. He introduces the category of noninterventionist objective divine action (NIODA) based on the Copenhagen interpretation of quantum mechanics. He sees God as

acting in particular quantum events to bring, indirectly, some specific effects on the macroscopic level—effects that can be classified as cases of both general and special divine providence. In other words, divine action in individual quantum events may be described in terms of an ontological (divine) entanglement (see Culp, "Panentheism," section 1). Russell's ideas receive support from Ian Barbour, Philip Clayton, Nancey Murphy, and Thomas Tracy. For a Thomistic critical response to this view, see Dodds, *Unlocking Divine Action*, 53–56, 63–71, 119–20, 126–34, 136–38, 140–47; Ignacio Silva, "A Cause among Causes? God Acting in the Natural World," *European Journal for Philosophy of Religion* 7, no. 4 (2015): 99–114; Carroll, "Divine Agency, Contemporary Physics," 582–95. Theories of divine action referring to chaotic and quantum events are discussed in two volumes of the series published by the Vatican Observatory and the Ayala Center for Theology and the Natural Sciences (CTNS) in Berkeley, California (Russell, Murphy, and Peacocke, *Chaos and Complexity*; Robert J. Russell, Philip Clayton, Kirk Wegter-McNelly, and John Polkinghorne, eds., *Quantum Mechanics: Scientific Perspectives on Divine Action* [Vatican: Vatican Observatory; Berkeley, CA: Center for Theology and the Natural Sciences, 2001]). Another critical account of both theories was offered by Saunders (*Divine Action and Modern Science*, chapters 5–7, pp. 94–206). Although he finds Peacocke's whole-part model of divine action the most promising at this stage of the debate (ibid., chapter 8, pp. 207–13), Saunders concludes that even if we can "continue to assert that God is active in the physical world . . . we should also note that much of the traditional account of God's activity cannot hold up against our modern understanding of science." At the same time, he encourages theologians "to press on with these seemingly intractable problems and in the process become more scientifically aware," striving to understand "the claims made by science, the assumptions behind the laws of nature, and the philosophy of determinism" (ibid., 216). It is important to note that—unlike at least some of the authors contributing to the above-mentioned edited volumes copublished by the Vatican Observatory and the CTNS—Saunders does not refer to the classical Aristotelian-Thomistic account of divine action.

92. Aquinas, *SCG* III.74.3.

93. See Aquinas, *In Phys.* 2.7–10.198–238, especially 218. Along the way of his argumentation Aquinas raises an interesting question, asking whether everything that happens due to a *per accidens* cause should be regarded as a chance event, to which he answers that such events must occur rarely, for "what is always or frequently joined to the effect falls under the intention itself" (2.8.214). Moreover, we should also remember that "among the per accidens causes, some are nearer [to the *per se* cause] and others are more remote. Those which are more remote seem less to be causes" (ibid., 2.9.221). See also *ST* I.116.1 co.; II-II.95.5 co.; *SCG* III.86.12; III.94.2; *In Peri herm.* 1.14.11.

94. Aquinas, *SCG* III.94.11. See also *ST* I.19.8 co.; *Q. de ver.* 23.5 co.

95. Aquinas, *ST* I.103.7 ad 2. "The very fact that an element of chance is found in . . . things proves that they are subject to government of some kind. For unless

corruptible things were governed by a higher being, they would tend to nothing definite, especially those which possess no kind of knowledge. So nothing would happen unintentionally; which constitutes the nature of chance" (*ST* I.103.5 ad 1).

96. Aquinas, *ST* I.22.2 ad 1.

97. Dodds, *Unlocking Divine Action*, 220, 223–25.

98. In the case of human beings, the situation is considerably different, because their substantial forms (immortal souls) are not simply educed from the potentiality of primary matter but are created *ex nihilo* in an act of God's direct special divine action, exercised by the Creator whenever a new human being comes into existence. God creates each new human soul when the primary matter it is about to inform is properly disposed by the parents to receive it. This emphasizes even more the fact that all substantial forms are rooted in the ultimate actuality and perfection of God's essence (which is identical with his existence). At the same time, such origin of each human soul does not have to be considered miraculous since, by definition, miracles are unusual and rare, whereas the coming into being of a new human is a quite regular and common occurrence. The creation of each new human soul at the moment of conception of a new human being may be thus regarded as an effect of a direct and special, but not miraculous, agency of God. We might say that it belongs to human nature that its new exemplars come into existence precisely in this way. In other words, because God is the author of nature, his direct special divine action in the coming into existence of each new human being may be regarded as natural.

99. Once again, it is important to remember that although creaturely *esse* has its source in God's *esse*, it is not identical with it. For the *esse* of a given entity is an act that is proportionate to the essence of that entity but never identical with it (unlike God's *esse*, which is identical with his *essentia*).

100. Even if among created things some can be called exemplars of others because of their likeness to those other things (according to the same species or according to the analogy of some kind of imitation), their exemplarism has its source in one divine essence. See Doolan, *Aquinas on the Divine Ideas*, 19, 102–3. Doolan holds that, for Aquinas, divine exemplarism plays an integral role in the theory of participation. In reference to the twofold divine exemplarism, he notes that a finite being participates in divine nature according to its many perfections inasmuch as the finite being has being, life, goodness, and the like. By contrast, a finite being participates in only one divine exemplar, because it has a particular mode of being, i.e., a determinate nature. The latter does not go without qualification, since divine ideas—sensu stricto—are not themselves participated in but "are rather the 'participabilities' of the likeness of the divine nature as it is known by God, that is, they are his knowledge of the ways in which the likeness of his essence can be participated" (ibid., 249).

101. This question naturally refers to the DC-based ontological EM as well.

102. My analysis of Aquinas's position on the importance of nonbeing for the explanation of the plurality of beings in nature is based on an important study by John F. Wippel, "Thomas Aquinas on the Distinction and Derivation of the Many

from the One: A Dialectic between Being and Nonbeing," *Review of Metaphysics* 38 (1985): 563–90.

103. See Aquinas, *Super De Trin.* 4.2. See also *Meta.* 10.3.1054a22.

104. Aquinas, *Super De Trin.* 4.2 co.

105. Aquinas does not restrict his analysis to complete entities and substances only. Creatures that do not enjoy complete being in themselves (such as primary matter) or any substantial form or a universal do not fall short of the divine simplicity by being composed, but because they are potentially divisible or divisible *per accidens* or because they can enter into composition with something else (*componible alteri*). "Prime matter, for example, insofar as it enters into Thomas's account of individuation, is capable of being divided only insofar as it is subject to quantity, or designated by quantity. A given substantial form can be divided or distinguished from other forms of the same type or species by being received in different instances of quantified matter. Moreover, prime matter and substantial form enter into composition with one another" (Wippel, "Thomas Aquinas on the Distinction," 575–76). Finally, even in the case of creatures whose essence is not composed of matter and form (such as an angel), their essence principle enters into composition with their corresponding *esse* principle and vice versa, which makes them fall short of God's simplicity as well. See Aquinas, *In I Sent.* 8.5.1; *Q. de ver.* 2.3; *SCG* II.52, where Thomas shows that in created intellectual substances essence (*quod est*) and existence (*esse*) differ.

In his *Treatise on Separate Substances*, Aquinas further specifies his position on the nature of nonbeing: "If, therefore, when I say 'non-being,' the effect is to remove only the 'to be' in act, the form, considered in itself, is non-being but sharing in 'to be.' But if 'non-being' removes not only the 'to be' in act but also the act or the form through which something shares in 'to be,' then, in this sense, matter is non-being, whereas a subsistent form is not non-being but an act which is a form that can participate in the ultimate act which is the 'to be'" (*De sub. sep.* 8.44).

106. Wippel, "Thomas Aquinas on the Distinction," 582.

107. Vincent P. Branick, "The Unity of the Divine Ideas," *New Scholasticism* 42 (1968): 199.

108. Wippel, "Thomas Aquinas on the Distinction," 584. He discusses the relative status of nonbeing in the last section of the article, on pp. 585–90.

109. Doolan, *Aquinas on the Divine Ideas*, 109–10.

110. Aquinas, *Q. de ver.* 3.2 co. Other passages quoted by Branick, Wippel, and Doolan express similar ideas:

> Since a created thing imperfectly imitates the divine essence, it happens that different things imitate it in different ways; yet every one of them has been produced according to a likeness of the divine essence.... Therefore, it is the common character of all things in so far as it is the one thing which all things imitate; but it is the proper character of this or that thing inasmuch as things

imitate it in different ways. In this way the divine essence causes proper knowledge of each and every thing, for it is the proper intelligible character of all. (Aquinas, *Q. de ver.* 2.4 ad 2)

Now, every form, both proper and common, considered as positing something, is a certain perfection; it includes imperfection only to the extent that it falls short of true being. The intellect of God, therefore, can comprehend in His essence that which is proper to each thing by understanding wherein the divine essence is being imitated and wherein each thing falls short of its perfection. (*SCG* I.54.4)

See also *Quod.* 8.1.1 co.

111. Doolan, *Aquinas on the Divine Ideas*, 110.
112. Branick, "Unity of the Divine Ideas," 199.
113. Aquinas, *In Phys.* 1.14.124, 127.
114. We must remember that, according to Aquinas, *De prin. nat.* 9–13, the source of potentiality (primary or secondary, i.e., proximate matter) is only one of the three principles of change. The other two principles are form (substantial or accidental), which is a principle of actuality, and privation. Out of these three principles, the first two are *per se* principles of change, whereas the third (privation) can be called a principle of change merely *per accidens*. At the same time, privation is a necessary, and not merely a predicamental, accident of matter (both primary and proximate), and thus cannot be separated from it. Moreover, whereas matter and form are principles in both being and in coming to be of the newly generated thing or its accidental feature, privation is only a principle in coming to be, and not in the being of a thing (see my comments on the nature of privation in chapter 2, section 4.3). Later on, in *De prin. nat.* 22–23, Aquinas claims that we may classify efficient and final causes as *per se* principles as well. He defines "principle" as something first (*primum*) from which something else, a posterior, begins or takes its origin, and he defines "cause" as something first from which the existence of something else follows (a narrower category). Thus, we can list four *per se* principles, which are also causes (matter, form, agency, end), and one *per accidens* principle (which can also be called a *per accidens* cause), i.e., privation.
115. Wippel claims that in his understanding of nonbeing Aquinas strives to avoid the two extremes of defining it as an absolute nothingness or reducing it to a mode of *esse*. He thinks Aquinas finds an aspect of nonbeing in essence, which is not identical with its existence principle (*esse*). But because essence stands in relation to *esse* as potency to act, this aspect of nonbeing cannot be regarded as absolute nonbeing. Thus, contrary to Branick—who speaks about absolute nonbeing entering God's intellection (see "Unity of the Divine Ideas," 199)—Wippel concludes that nonbeing in Aquinas's metaphysics is defined as relative (see Wippel, "Thomas Aquinas on the Distinction," 585–90). In support of his claim he quotes Aquinas, *Q. de pot.* 3.5 ad 2; 3.1 ad 17, but neither of these passages uses the term "relative nonbeing" in reference

to essences of things. This is not surprising in the context of my analysis offered in the main text, which showed that for Aquinas essence and primary matter can be regarded as principles of potentiality, but not as nonbeing. Thus, it seems more relevant in this context to use Aquinas's own categories, classifying nonbeing as a *per accidens* principle of novelty and of change and contrasting it with essence as being in potency to receive existence (*esse*) and with primary matter as being in potency to be informed by substantial form—both existence and substantial form being defined by Aquinas as *per se* principles of the coming to be of new entities. This approach proves even more appropriate once we realize that the very term "relative nonbeing," used by Wippel, is rather foreign to Aquinas's vocabulary.

116. See Jacek Wojtysiak, "Panenteizm," in *Filozofia Boga: Część II, Odkrywanie Boga* [Philosophy of God: Part 2, Discerning God], ed. Stanisław Janeczek and Anna Starościc (Lublin: Wydawnictwo KUL, 2017), 506–13.

CONCLUSION

1. Aquinas, *ST* I.2.2 co.

APPENDIX 1

1. My classification is based on Brierley, "Naming a Quiet Revolution," 2–4 (see bibliographical references to the works of many thinkers in footnotes); Hartshorne and Reese, *Philosophers Speak of God*; Cooper, *Panentheism*.

APPENDIX 2

1. Thomas Aquinas, *Summa theologiae*, vol. 2, *Existence and Nature of God (Ia. 2–11)*, trans., with introduction, notes, appendices, and glossary, by Timothy McDermott, O.P., with additional appendices by Thomas Gilby, O.P. [Blackfriars (Cambridge) edition] (Cambridge: Cambridge University Press, 2006).

BIBLIOGRAPHY

Alexander, Samuel. *Space, Time and Deity*. Vol. 2. London: Macmillan, 1920.
Andersen, Peter Bøgh, Claus Emmeche, Niels O. Finnemann, and Peder Voetmann Christiansen, eds. *Downward Causation: Mind, Bodies and Matter*. Aarhus and Oxford: Aarhus University Press, 2000.
Aquinas, Thomas. *Commentarium in Aristotelis libros Peri hermeneias*. Turin and Rome: Marietti, 1955. Translated by Jean T. Oesterle as *Commentary on Aristotle's On Interpretation*. In *Aristotle: On Interpretation; Commentary by St. Thomas and Cajetan* (Milwaukee: Marquette University Press, 1962).
———. *De ente et essentia*. In *Opera omnia iussu Leonis XIII P. M. edita*, 43:131–57. Rome: Editori di San Tommaso, 1976. Translated by Joseph Bobik as *Aquinas on Being and Essence: A Translation and Interpretation* (Notre Dame, IN: University of Notre Dame Press, 1965).
———. *De mixtione elementorum ad magistrum Philippum de Castro Caeli*. In *Opera omnia iussu Leonis XIII P. M. edita*, 43:315–81. Rome: Editori di San Tommaso, 1976. Translated by Joseph Bobik as *Aquinas on Matter and Form and the Elements: A Translation and Interpretation of the De Principiis Naturae and the De Mixtione Elementorum of St. Thomas Aquinas* (Notre Dame, IN: University of Notre Dame Press, 1998).
———. *De principiis naturae*. In *Opera omnia iussu Leonis XIII P. M. edita*, 43:39–47. Rome: Typographia polyglotta, 1976. Translated by Robert P. Goodwin as *The Principles of Nature*. In *Selected Writings of St. Thomas Aquinas* (New York: Bobbs-Merrill, 1965), 7–28.
———. *De substantiis separatis*. In *Opuscula philosophica*. Turin and Rome: Marietti, 1954.
———. *In librum beati Dionysii De divinis nominibus expositio*. Turin and Rome: Marietti, 1950.
———. *In Metaphysicam Aristotelis commentaria*. Turin and Rome: Marietti, 1926. Translated by John Rowan as *Commentary on The Metaphysics of Aristotle*, 2 vols. (Chicago: Regnery Press, 1961).
———. *In octo libros Physicorum Aristotelis expositio*. Turin and Rome: Marietti, 1965. Translated by Richard J. Blackwell, Richard J. Spath, and W. Edmund

Thirlkel as *Commentary on Aristotle's Physics* (Notre Dame, IN: Dumb Ox Books, 1999).

———. *Quaestiones disputatae de potentia Dei*. Turin and Rome: Marietti, 1965. Translated by English Dominican Fathers as *On the Power of God* (Westminster, MD: Newman Press, 1952).

———. *Quaestiones disputatae de veritate*. Vol. 22/1–3 of *Opera Omnia*. Rome: Typographia polyglotta, 1972–76. Translated by Robert W. Mulligan, S.J., et al. as *Truth*, 3 vols. (Albany, NY: Preserving Christian Publications, 1993).

———. *Quaestiones quodlibetales*. Turin and Rome: Marietti, 1949.

———. *Scriptum super Libros Sententiarum Magistri Petri Lombardi Episcopi Parisiensis*. Edited by P. Mandonnet. Paris: P. Lethielleux, 1929–47.

———. *Summa contra gentiles*. 3 vols. Turin and Rome: Marietti, 1961. Translated by Anton C. Pegis et al. as *On the Truth of the Catholic Faith: Summa Contra Gentiles*, 4 vols. (Garden City, NY: Image Books, 1955–57).

———. *Summa theologiae*. Rome: Editiones Paulinae, 1962. Translated by the Fathers of the English Dominican Province as *Summa Theologica*, 3 vols. (New York: Benzinger Bros., 1946).

———. *Summa theologiae*. Vol. 2, *Existence and Nature of God (Ia. 2–11)*. Translated, with introduction, notes, appendices, and glossary, by Timothy McDermott, O.P. Additional appendices by Thomas Gilby, O.P. [Blackfriars (Cambridge) edition]. Cambridge: Cambridge University Press, 2006.

———. *Super Boetium De Trinitate*. In *Opera Omnia iussu Leonis XIII P. M. edita*, 50:75–171. Rome and Paris: Éditions du Cerf, 1992. Translated and annotated by Armand A. Maurer as *Saint Thomas Aquinas, Faith, Reason, and Theology: Questions I–IV of His Commentary on the "De Trinitate" of Boethius* (Toronto: Pontifical Institute of Medieval Studies, 1987).

———. *Super librum De causis expositio*. Edited by H. D. Saffrey. Fribourg and Leuven: Société Philosophique, 1954. Translated and annotated by Vincent A. Guagliardo, O.P., Charles R. Hess, O.P., and Richard C. Taylor as *Commentary on the Book of Causes* (Washington, DC: Catholic University of America Press, 1996).

Aristotle. *De generatione et corruptione (On Generation and Corruption)*. Translated by Harold H. Joachim. In *The Basic Works of Aristotle*, edited by Richard McKeon, 465–531. New York: The Modern Library, 2001.

———. *Generation of Animals*. Translated by A. L. Peck. London and Cambridge, MA: Harvard University Press, 1943.

———. *Generation of Animals*. Translated by Arthur Platt. In *The Complete Works of Aristotle: The Revised Oxford Translation*, vol. 1, edited by Jonathan Barnes, 1111–1218. Princeton: Princeton University Press, 1984.

———. *Metaphysica (Metaphysics)*. Translated by W. D. Ross. In *The Basic Works of Aristotle*, edited by Richard McKeon, 681–926. New York: The Modern Library, 2001.

———. *The Metaphysics*. Translated by Hugh Tredennick. London and Cambridge, MA: W. Heinemann, 1936.

———. *Meteorology*. Translated by W. Webster. In *The Complete Works of Aristotle: The Revised Oxford Translation*, vol. 1, edited by Jonathan Barnes, 555–625. Princeton: Princeton University Press, 1984.

———. *Parts of Animals*. Translated by William Ogle. In *The Complete Works of Aristotle: The Revised Oxford Translation*, vol. 1, edited by Jonathan Barnes, 994–1086. Princeton: Princeton University Press, 1984.

———. *Physica (Physics)*. Translated by R. K. Gaye. In *The Basic Works of Aristotle*, edited by Richard McKeon, 213–394. New York: The Modern Library, 2001.

———. *The Physics*. Translated by Philip H. Wicksteed and Francis M. Cornford. London and New York: W. Heinemann, 1929.

Armstrong, David Malet. "Defending Categoricalism." In *Properties, Powers, and Structures: Issues in the Metaphysics of Realism*, edited by Alexander Bird, B. D. Ellis, and Howard Sankey, 27–33. New York: Routledge, 2012.

———. *A Materialist Theory of the Mind*. London and New York: Routledge & K. Paul, 1968.

Baltzly, Dirk. "Is Plato's *Timaeus* Panentheistic?" *Sophia* 49, no. 2 (2010): 193–215.

Barbour, Ian G. *Religion and Science: Historical and Contemporary Issues*. New York: HarperCollins, 1997.

Barnes, Corey L. "Natural Final Causality and Providence in Aquinas." *New Blackfriars* 95 (2014): 349–61.

Barth, Karl. *Church Dogmatics*, vol. 2, *The Doctrine of God, Part 1: The Knowledge of God*. Translated and edited by G. W. Bromiley and T. F. Torrance. Edinburgh: T&T Clark, 1957.

Barua, Ankur. "God's Body at Work: Rāmānuja and Panentheism." *International Journal of Hindu Studies* 14, no. 1 (2010): 1–30.

Beebee, Helen, Christopher Hitchcock, and Peter Charles Menzies, eds. *The Oxford Handbook of Causation*. Oxford and New York: Oxford University Press, 2009.

Biernacki, Loriliai, and Philip Clayton, eds. *Panentheism across the World's Traditions*. Oxford and New York: Oxford University Press, 2014.

Bird, Alexander. "Causation and the Manifestation of Powers." In *The Metaphysics of Powers: Their Grounding and Their Manifestations*, edited by Anna Marmodoro, 35–41. New York: Routledge, 2010.

———. "Limitations of Power." In *Powers and Capacities in Philosophy: The New Aristotelianism*, edited by Ruth Groff and John Greco, 25–47. New York: Routledge, 2013.

———. "Monistic Dispositional Essentialism." In *Properties, Powers, and Structures: Issues in the Metaphysics of Realism*, edited by Alexander Bird, B. D. Ellis, and Howard Sankey, 11–26. New York: Routledge, 2012.

Bostock, David. *Space, Time, Matter, and Form: Essays on Aristotle's Physics*. Oxford: Clarendon, 2006.

Bracken, Joseph A. "The Issue of Panentheism in the Dialogue with the Non-Believer." *Studies in Religion/Sciences Religieuses* 21, no. 2 (1992): 207–18.

———. "Panentheism and the Classical God-World Relationship: A Systems-Oriented Approach." *American Journal of Theology & Philosophy* 36, no. 3 (October 31, 2015): 207–25.

———. "Whitehead and Roman Catholics: What Went Wrong?" *American Journal of Theology & Philosophy* 30, no. 2 (2009): 153–67.

Braillard, Pierre-Alain, and Christophe Malaterre, eds. *Explanation in Biology: An Enquiry into the Diversity of Explanatory Patterns in the Life Sciences*. Dordrecht: Springer, 2015.

Branick, Vincent P. "The Unity of the Divine Ideas." *New Scholasticism* 42 (1968): 171–201.

Brierley, Michael W. "Naming a Quiet Revolution: The Panentheistic Turn in Modern Theology." In Clayton and Peacocke, *In Whom We Live and Move*, 1–15.

———. "The Potential of Panentheism for Dialogue between Science and Religion." In *The Oxford Handbook of Religion and Science*, edited by Philip Clayton and Zachary Simpson, 635–51. Oxford: Oxford University Press, 2006.

Broad, Charlie Dunbar. *The Mind and Its Place in Nature*. London: Routledge and Kegan Paul, 1925.

Brown, Warren. *Whatever Happened to the Soul? Scientific and Theological Portraits of Human Nature*. Edited by Nancey Murphy and H. Newton Malony. Minneapolis: Fortress, 1998.

Bultmann, Rudolf Karl. *Jesus Christ and Mythology*. New York: Charles Scribner's Sons, 1958.

Bunge, Mario. *Matter and Mind: A Philosophical Inquiry*. Dordrecht: Springer, 2010.

———. *The Mind-Body Problem: A Psychobiological Approach*. Oxford and New York: Pergamon Press, 1980.

Burtt, Edwin Arthur. *The Metaphysical Foundations of Modern Science*. Mineola, NY: Dover, 2003. First published 1924.

Butler, Clark. "Hegelian Panentheism as Joachimite Christianity." In *New Perspectives on Hegel's Philosophy of Religion*, edited by David Kolb, 131–42. Albany: State University of New York Press, 1992.

Campbell, Donald Thomas. "'Downward Causation' in Hierarchically Organized Biological Systems." In *Studies in the Philosophy of Biology*, edited by Francisco J. Ayala and Theodosius Dobzhansky, 179–86. Berkeley and Los Angeles: University of California Press, 1974.

Carroll, John W. "Laws of Nature." In *Stanford Encyclopedia of Philosophy*. Fall 2016 ed. Edited by Edward N. Zalta. https://plato.stanford.edu/archives/fall2016/entries/laws-of-nature/.

Carroll, William E. "Aquinas on Creation and the Metaphysical Foundations of Science." Paper presented at the annual Thomistic Institute sponsored by the

Jacques Maritain Center, University of Notre Dame, July 23, 1998. Accessed July 30, 2019. https://maritain.nd.edu/jmc/ti98/carroll.htm.

———. "Creation and Science in the Middle Ages." *New Blackfriars* 88 (2007): 678–89.

———. "Divine Agency, Contemporary Physics, and the Autonomy of Nature." *Heythrop Journal* 49 (2008): 582–602.

Cartwright, Nancy, and John Pemberton. "Aristotelian Powers: Without Them, What Would Modern Science Do?" In *Powers and Capacities in Philosophy: The New Aristotelianism*, edited by Ruth Groff and John Greco, 93–112. New York: Routledge, 2013.

Case-Winters, Anna. "Toward a Theology of Nature: Preliminary Intuitions." *Religiologiques* 11 (1995): 249–67.

Christian, William A. *An Interpretation of Whitehead's Metaphysics*. New Haven: Yale University Press, 1959.

Clayton, Philip. *Adventures in the Spirit: God, World, Divine Action*. Minneapolis: Fortress, 2008.

———. "Conceptual Foundations of Emergence Theory." In Clayton and Davies, *Re-Emergence of Emergence*, 1–31.

———. "Creation Ex Nihilo and Intensifying the Vulnerability of God." In *Theologies of Creation: Creatio Ex Nihilo and Its New Rivals*, edited by Thomas Jay Oord, 17–30. New York: Routledge, 2014.

———. "Emergence from Quantum Physics to Religion: A Critical Appraisal." In Clayton and Davies, *Re-Emergence of Emergence*, 303–22.

———. *God and Contemporary Science*. Grand Rapids, MI: Eerdmans, 1997.

———. "Kenotic Trinitarian Panentheism." *Dialog* 44, no. 3 (2005): 250–55.

———. *Mind and Emergence: From Quantum to Consciousness*. Oxford and New York: Oxford University Press, 2004.

———. "Panentheism in Metaphysical and Scientific Perspective." In Clayton and Peacocke, *In Whom We Live and Move*, 73–91.

———. "Panentheisms East and West." *Sophia* 49, no. 2 (2010): 183–91.

———. "Panentheism Today: A Constructive Systematic Evaluation." In Clayton and Peacocke, *In Whom We Live and Move*, 249–64.

———. "The Panentheistic Turn in Christian Theology." *Dialog* 38, no. 4 (1999): 289–93.

———. *The Problem of God in Modern Thought*. Grand Rapids, MI / Cambridge, UK: Eerdmans, 2000.

———. "Toward a Constructive Christian Theology of Emergence." In Murphy and Stoeger, *Evolution and Emergence*, 315–44.

Clayton, Philip, and Paul Davies, eds. *The Re-Emergence of Emergence: The Emergentist Hypothesis from Science to Religion*. Oxford and New York: Oxford University Press, 2006.

Clayton, Philip, and Arthur Robert Peacocke, eds. *In Whom We Live and Move and Have Our Being: Panentheistic Reflections on God's Presence in a Scientific World*. Grand Rapids, MI / Cambridge, UK: Eerdmans, 2004.

Cobb, John B. "Review of Clayton and Peacocke." *Theology and Science* 3, no. 2 (2005): 240–42.

Cobb, John B., and David Ray Griffin. *Process Theology: An Introductory Exposition*. Louisville: Westminster John Knox, 1976.

Cooper, Burton Z. *The Idea of God: A Whiteheadian Critique of St. Thomas Aquinas' Concept of God*. The Hague: Martinus Nijhoff, 1974.

Cooper, John W. *Panentheism—the Other God of the Philosophers: From Plato to the Present*. Grand Rapids, MI: Baker Academic, 2006.

Copleston, Frederick C. "Pantheism in Spinoza and the German Idealists." *Philosophy* 21 (1946): 42–56.

Craig, William Lane. *God, Time, and Eternity*. Dordrecht: Kluwer, 2001.

Culp, John. "Panentheism." In *Stanford Encyclopedia of Philosophy*. Summer 2017 ed. Edited by Edward N. Zalta. https://plato.stanford.edu/entries/panentheism/.

Curley, Edwin M. *Spinoza's Metaphysics: An Essay in Interpretation*. Cambridge, MA: Harvard University Press, 1969.

Davies, Paul. "Teleology without Teleology: Purpose through Emergent Complexity." In Clayton and Peacocke, *In Whom We Live and Move*, 95–108.

Deacon, Terrence W. "Emergence: The Hole at the Wheel's Hub." In Clayton and Davies, *Re-Emergence of Emergence*, 111–50.

———. "The Hierarchic Logic of Emergence: Untangling the Interdependence of Evolution and Self-Organization." In *Evolution and Learning: The Baldwin Effect Reconsidered*, edited by Bruce H. Weber and David J. Depew, 273–308. Cambridge, MA: MIT Press, 2003.

———. *Incomplete Nature: How Mind Emerged from Matter*. New York: W. W. Norton, 2012.

———. "Reciprocal Linkage between Self-Organizing Processes Is Sufficient for Self-Reproduction and Evolvability." *Biological Theory* 1 (2006): 136–49.

Deacon, Terrence W., and Tyrone Cashman. "Eliminativism, Complexity, and Emergence." In *The Routledge Companion to Religion and Science*, edited by James W. Haag, Gregory R. Peterson, and Michael L. Spezio, 193–205. New York: Routledge, 2012.

———. "Steps to a Metaphysics of Incompleteness." *Theology and Science* 14 (2016): 401–29.

———. "Teleology versus Mechanism in Biology: Beyond Self-Organization." In *Beyond Mechanism: Putting Life Back into Biology*, edited by Brian G. Henning and Adam C. Scarfe, 287–308. Lanham, MD: Lexington Books, 2013.

Deacon, Terrence W., and Spyridon Koutroufinis. "Complexity and Dynamical Depth." *Information* 5 (2014): 404–23.

Deacon, Terrence, Alok Srivastava, and Augustus Bacigalupi. "The Transition from Constraint to Regulation at the Origin of Life." *Frontiers in Bioscience* 19 (2014): 945–57.

Dodds, Michael J. *The Philosophy of Nature*. Oakland, CA: Western Dominican Province, 2010.

———. "Ultimacy and Intimacy: Aquinas on the Relation between God and the World." In *Ordo Sapientiae et Amoris: Hommage au Professeur Jean-Pierre Torrell, O.P.*, edited by Carlos-Josaphat Pinto de Oliveira, 211–27. Fribourg, Switzerland: Editions Universitaires, 1993.

———. *The Unchanging God of Love: Thomas Aquinas and Contemporary Theology on Divine Immutability*. Washington DC: Catholic University of America Press, 2008.

———. *Unlocking Divine Action: Contemporary Science and Thomas Aquinas*. Washington DC: Catholic University of America Press, 2012.

Doolan, Gregory T. *Aquinas on the Divine Ideas as Exemplar Causes*. Washington, DC: Catholic University of America Press, 2008.

———. "The Causality of the Divine Ideas in Relation to Natural Agents in Thomas Aquinas." *International Philosophical Quarterly* 44 (2004): 393–409.

Drees, Willem B. *Creation: From Nothing until Now*. New York: Routledge, 2002.

Dudley, John. *Aristotle's Concept of Chance: Accidents, Cause, Necessity, and Determinism*. Albany: State University of New York Press, 2012.

El-Hani, Charbel Niño, and Antonio Marcos Pereira. "Higher-Level Descriptions: Why Should We Preserve Them?" In Andersen et al., *Downward Causation*, 118–42.

Ellis, Brian D. "The Categorical Dimensions of Causal Powers." In *Properties, Powers, and Structures: Issues in the Metaphysics of Realism*, edited by Alexander Bird, B. D. Ellis, and Howard Sankey, 11–26. New York: Routledge, 2012.

———. "Causal Powers and Categorical Properties." In *The Metaphysics of Powers: Their Grounding and Their Manifestations*, edited by Anna Marmodoro, 133–42. New York: Routledge, 2010.

———. *The Philosophy of Nature: A Guide to the New Essentialism*. Montreal and Ithaca, NY: McGill-Queen's University Press, 2002.

———. *Scientific Essentialism*. Cambridge and New York: Cambridge University Press, 2001.

Ellis, George F. R. "Science, Complexity, and the Nature of Existence." In Murphy and Stoeger, *Evolution and Emergence*, 113–40.

Emmeche, Claus, Simo Køppe, and Frederic Stjernfelt. "Explaining Emergence: Towards an Ontology of Levels." *Journal for General Philosophy of Science* 28 (1997): 83–119.

———. "Levels, Emergence, and Three Versions of Downward Causation." In Andersen et al., *Downward Causation*, 13–34.

Farrer, Austin Marsden. *Faith and Speculation*. London: Adam and Charles Black, 1967.
Feser, Edward. *Scholastic Metaphysics: A Contemporary Introduction*. Heusenstamm: Editiones Scholasticae, 2014.
Feibleman, James K. "Hegel Revisited." *Tulane Studies in Philosophy* 9 (1960): 16–49.
Fichte, Johann Gottlieb. "Die Anweisung zum seligen Leben, oder auch die Religionslehre." In *Fichtes Werke*, vol. 5, *Zur Religionsphilosophie*, edited by Immanuel H. Fichte. Berlin: de Gruyter, 1971.
———. "Faith." In *The Vocation of Man*, book 3, translated by Peter Preuss, 67–123. Indianapolis: Hackett, 1987.
———. *Science of Knowledge (Wissenschaftslehre)*. Edited by Peter Heath and John Lachs. New York: Appleton-Century-Crofts, 1970.
Fiddes, Paul S. *The Creative Suffering of God*. Oxford: Oxford University Press, 1988.
Fine, Kit. "Things and Their Parts." *Midwest Studies in Philosophy* 23 (1999): 61–74.
Forster, Michael. "Hegel's Dialectical Method." In *The Cambridge Companion to Hegel*, edited by Frederick C. Beiser, 130–70. New York and Cambridge: Cambridge University Press, 1993.
Galileo Galilei. *The Assayer*. In *Discoveries and Opinions of Galileo*, translated by Stillman Drake, 229–80. Garden City, NY: Doubleday, 1957.
Gasser, Georg. "God's Omnipresence in the World: On Possible Meanings of 'En' in Panentheism." *International Journal for Philosophy of Religion* 85, no. 1 (2019): 43–62.
Gaukroger, Stephen. *The Collapse of Mechanism and the Rise of Sensibility: Science and the Shaping of Modernity, 1680–1760*. Oxford: Clarendon, 2010.
———. *The Emergence of a Scientific Culture: Science and the Shaping of Modernity, 1210–1685*. Oxford: Clarendon, 2006.
Gillett, Carl. "Non-Reductive Realization and Non-Reductive Identity: What Physicalism Does Not Entail." In *Physicalism and Mental Causation: The Metaphysics of Mind and Action*, edited by Sven Walter and Heinz-Dieter Heckmann, 23–49. Charlottesville, VA: Imprint Academic, 2003.
Gilson, Étienne. *The Christian Philosophy of St. Thomas Aquinas*. Translated by L. K. Shook. New York: Random House, 1956.
———. *Elements of Christian Philosophy*. New York: New American Library, 1963.
———. *Methodical Realism*. San Francisco: Ignatius, 2011.
———. *The Spirit of Mediaeval Philosophy*. New York: Charles Scribner's Sons, 1940.
———. *Thomism: The Philosophy of Thomas Aquinas*. Toronto: Pontifical Institute of Medieval Studies, 2002.

Göcke, Benedikt Paul. "Another Reply to Raphael Lataster." *Sophia* 54, no. 1 (2015): 99–102.

———. "On the Importance of Karl Christian Friedrich Krause's Panentheism." *Zygon* 48, no. 2 (2013): 364–79.

———. "Panentheism and Classical Theism." *Sophia* 52, no. 1 (2013): 61–75.

———. *The Panentheism of Karl Christian Friedrich Krause*. Berlin: Peter Lang, 2018.

———. "Reply to Raphael Lataster." *Sophia* 53, no. 3 (2014): 397–400.

Goodenough, Ursula, and Terrence W. Deacon. "The Sacred Emergence of Nature." In *The Oxford Handbook of Religion and Science*, edited by Philip Clayton and Zachary Simpson, 853–71. Oxford: Oxford University Press, 2006.

Gotthelf, Allan. "Aristotle's Concept of Final Causality." *Review of Metaphysics* 30 (1976): 226–54.

Graham, Daniel W. "Heraclitus: Flux, Order, and Knowledge." In *The Oxford Handbook of Presocratic Philosophy*, edited by Patricia Curd and Daniel W. Graham, 169–88. Oxford: Oxford University Press, 2008.

Green, Sara. "Introduction to Philosophy of Systems Biology." In *Philosophy of Systems Biology: Perspectives from Scientists and Philosophers*, edited by Sara Green, 1–23. Copenhagen: Springer, 2017.

Gregersen, Niels Henrik. "Emergence: What Is at Stake for Religious Reflection?" In Clayton and Davies, *Re-Emergence of Emergence*, 279–302.

———. "Three Varieties of Panentheism." In Clayton and Peacocke, *In Whom We Live and Move*, 19–35.

Gregersen, Niels Henrik, Ulf Görman, and Willem B. Drees, eds. *The Human Person in Science and Theology*. Edinburgh: T&T Clark International, 2003.

Griffin, David R. *Panentheism and Scientific Naturalism: Rethinking Evil, Morality, Religious Experience, Religious Pluralism, and the Academic Study of Religion*. Claremont, CA: Process Century Press, 2014.

———. "Panentheism: A Postmodern Revelation." In Clayton and Peacocke, *In Whom We Live and Move*, 36–47.

———. "Time in Process Philosophy." *KronoScope: Journal for the Study of Time* 1 (2001): 75–99.

———. *Whitehead's Radically Different Postmodern Philosophy: An Argument for Its Contemporary Relevance*. Albany: State University of New York Press, 2007.

Guilherme, Alexandre. "Fichte: Kantian or Spinozian? Three Interpretations of the Absolute I." *South African Journal of Philosophy* 29 (2010): 1–16.

———. "Schelling's *Naturphilosophie* Project: Towards a Spinozian Conception of Nature." *South African Journal of Philosophy* 29 (2010): 373–90.

Gulick, Robert van. "Reduction, Emergence, and the Mind/Body Problem." In Murphy and Stoeger, *Evolution and Emergence*, 40–73.

———. "Who's in Charge Here? And Who's Doing All the Work?" In Murphy and Stoeger, *Evolution and Emergence*, 74–87.
Gunton, Colin E. *The One, the Three and the Many: God, Creation and the Culture of Modernity; The 1992 Bampton Lectures*. Cambridge and New York: Cambridge University Press, 1993.
Haag, James W. "Emergence and Christian Theology." In *The Routledge Companion to Religion and Science*, edited by James W. Haag, Gregory R. Peterson, and Michael L. Spezio, 213–22. New York: Routledge, 2012.
Hartshorne, Charles. *The Divine Relativity: A Social Conception of God*. New Haven: Yale University Press, 1948.
———. *Man's Vision of God and the Logic of Theism*. Chicago: Willet, Clark, 1941.
———. *Omnipotence and Other Theological Mistakes*. Albany: State University of New York Press, 1984.
Hartshorne, Charles, and William L. Reese, eds. *Philosophers Speak of God*. New York: Humanity Books, 2000.
Hegel, Georg Wilhelm Friedrich. *Lectures on the History of Philosophy*. Vol. 3. Translated by E. S. Haldane and Frances H. Simson. London: Routledge & Kegan Paul, 1955.
———. *Lectures on the Philosophy of Religion: One-Volume Edition; The Lectures of 1827*. Edited by Peter C. Hodgson. Oxford and New York: Oxford University Press, 2006.
———. *The Logic of Hegel*. Translated by William Wallace. Oxford: Clarendon, 1874.
———. *The Phenomenology of Mind*. Translated by J. B. Baillie. New York: Dover, 2003.
———. *The Phenomenology of Spirit*. Translated by A. V. Miller. Oxford: Clarendon, 1977.
Heil, John. *From an Ontological Point of View*. Oxford: Clarendon, 2005.
———. "Powerful Qualities." In *The Metaphysics of Powers: Their Grounding and Their Manifestations*, edited by Anna Marmodoro, 58–72. New York: Routledge, 2010.
Henry, John, and Mariusz Tabaczek. "Causation." In *Science and Religion: A Historical Introduction*, edited by Gary B. Ferngren, 377–94. Baltimore: Johns Hopkins University Press, 2017.
Hudson, Hud. "Omnipresence." In *The Oxford Handbook of Philosophical Theology*, edited by Thomas P. Flint and Michael C. Rea, 199–216. Oxford: Oxford University Press, 2011.
Hulswit, Menno. *From Cause to Causation: A Peircean Perspective*. Dordrecht/Boston/London: Kluwer Academic, 2002.
———. "How Causal Is Downward Causation?" *Journal for General Philosophy of Science* 36 (2006): 261–87.

Hume, David. *An Enquiry concerning Human Understanding.* In Great Books of the Western World 35, 451–509. Chicago: Encyclopaedia Britannica, 1952.

———. *A Treatise of Human Nature.* Edited by L. A. Selby-Bigge. Oxford: Oxford University Press, 1978.

Humphreys, Paul. "Emergence, Not Supervenience." In "Proceedings of the 1996 Biennial Meetings of the Philosophy of Science Association," part 2, "Symposia Papers." Supplement, *Philosophy of Science* 64 (1997): S337–45.

Hutchings, Patrick. "Postlude: Panentheism." *Sophia* 49, no. 2 (2010): 297–300.

Illari, Phyllis, and Federica Russo. *Causality: Philosophical Theory Meets Scientific Practice.* Oxford: Oxford University Press, 2014.

Irwin, Terence. *Aristotle's First Principles.* Oxford: Clarendon, 1988.

Jaworski, William. *Structure and the Metaphysics of Mind: How Hylomorphism Solves the Mind-Body Problem.* Oxford: Oxford University Press, 2016.

Jensen, Alexander S. *Divine Providence and Human Agency: Trinity, Creation and Freedom.* Burlington, VT: Ashgate, 2014.

Johnston, Mark. "Hylomorphism." *Journal of Philosophy* 103 (2006): 652–98.

Juarrero, Alicia. *Dynamics in Action: Intentional Behavior as a Complex System.* Cambridge, MA: MIT Press, 1999.

Kahn, Charles H. *The Art and Thought of Heraclitus: An Edition of the Fragments with Translation and Commentary.* Cambridge: Cambridge University Press, 1979.

Kaufman, Gordon D. *In the Beginning . . . Creativity.* Minneapolis: Fortress, 2004.

———. "A Religious Interpretation of Emergence: Creativity as God." *Zygon* 42 (2007): 915–28.

Keller, Catherine. *The Face of the Deep: A Theology of Becoming.* London and New York: Routledge, 2003.

Kim, Jaegwon. "'Downward Causation' in Emergentism and Nonreductive Physicalism." In *Emergence or Reduction? Essays on the Prospects of Nonreductive Physicalism*, edited by Ansgar Beckermann, Hans Flohr, and Jaegwon Kim, 119–38. Berlin and New York: de Gruyter, 1992.

———. "Emergence: Core Ideas and Issues." *Synthese* 151 (2006): 547–59.

———. "Making Sense of Emergence." *Philosophical Studies: An International Journal for Philosophy in the Analytic Tradition* 95 (August 1999): 3–36.

———. "The Myth of Nonreductive Materialism." *Proceedings and Addresses of the American Philosophical Association* 63 (1989): 31–47.

———. *Philosophy of Mind.* 3rd ed. Boulder, CO: Westview Press, 2011.

———. "Supervenience as a Philosophical Concept." *Metaphilosophy* 21 (1990): 1–27.

Koons, Robert. "Staunch vs. Faint-Hearted Hylomorphism: Toward an Aristotelian Account of Composition." *Res Philosophica* 91 (2014): 151–77.

Koons, Robert C., and Timothy Pickavance. *The Atlas of Reality: A Comprehensive Guide to Metaphysics.* Oxford: John Wiley & Sons, 2017.

———. *Metaphysics: The Fundamentals.* Oxford: Wiley-Blackwell, 2015.
Koslicki, Kathrin. "Aristotle's Mereology and the Status of Form." *Journal of Philosophy* 103 (2006): 715–36.
———. *The Structure of Objects.* Oxford: Oxford University Press, 2010.
Krause, Karl Christian Friedrich. *Der Begriff der Philosophie.* Leipzig: Otto Schulze, 1893.
———. *Der zur Gewissheit der Gotteserkenntnis als des höchsten Wissenschaftsprinzips emporleitende Theil der Philosophie.* Prague: Tempsky, 1869.
———. *Vorlesungen über das System der Philosophie.* Göttingen: Dieterich'sche Buchhandlung, 1828.
Lataster, Raphael. "The Attractiveness of Panentheism—a Reply to Benedikt Paul Göcke." *Sophia* 53, no. 3 (2014): 389–95.
———. "Theists Misrepresenting Panentheism—Another Reply to Benedikt Paul Göcke." *Sophia* 54, no. 1 (2015): 93–98.
Lauer, Quentin. *Hegel's Concept of God.* Albany: State University of New York Press, 1982.
Laughlin, Robert B. *A Different Universe: Reinventing Physics from the Bottom Down.* New York: Basic Books, 2005.
Levine, Michael P. *Pantheism: A Non-Theistic Concept of Deity.* London and New York: Routledge, 1994.
Lewes, George Henry. *Problems of Life and Mind.* Vol. 2. London: Kegan Paul, Trench, Trübner, & Co., 1875.
Losee, John. *Theories of Causality: From Antiquity to the Present.* New Brunswick, NJ, and London: Transaction Publishers, 2011.
Louth, Andrew. "The Cosmic Vision of Saint Maximos the Confessor." In Clayton and Peacocke, *In Whom We Live and Move*, 184–96.
Lowe, Edward Jonathan. *The Four-Category Ontology: A Metaphysical Foundation for Natural Science.* New York: Oxford University Press, 2006.
Mander, William. "Pantheism." *Stanford Encyclopedia of Philosophy.* Spring 2020 ed. Edited by Edward N. Zalta. https://plato.stanford.edu/archives/spr2020/entries/pantheism/.
Marenbon, John. "The Medievals." In Beebee, Hitchcock, and Menzies, *Oxford Handbook of Causation*, 40–54.
Martin, Charles Burton. *The Mind in Nature.* Oxford and New York: Clarendon, 2008.
Mason, Richard. *The God of Spinoza: A Philosophical Study.* Cambridge: Cambridge University Press, 1997.
Mawson, Tim. "God's Body." *Heythrop Journal* 47, no. 2 (2006): 171–81.
Mayr, Erasmus. "Powers and Downward Causation." In *Philosophical and Scientific Perspectives on Downward Causation*, edited by Michele Paolini Paoletti and Francesco Orilia, 76–91. New York: Routledge, 2017.
McClure, Matthew Thompson. *The Early Philosophers of Greece.* New York: Appleton-Century-Crofts, 1935.

McLaughlin, Brian, and Karen Bennett. "Supervenience." *Stanford Encyclopedia of Philosophy*. Winter 2018 ed. Edited by Edward N. Zalta. https://plato.stanford.edu/archives/win2018/entries/supervenience/.

McWhorter, Matthew R. "Aquinas on God's Relation to the World." *New Blackfriars* 94, no. 1049 (2013): 3–19.

Mellor, David Hugh. "Counting Corners Correctly." *Analysis* 42 (1982): 96–97.

———. "In Defense of Dispositions." *Philosophical Perspectives* 12 (1974): 283–312.

Melsen, Andreas Gerardus Maria van. *The Philosophy of Nature*. Pittsburgh: Duquesne University Press, 1961.

Mill, John Stuart. *A System of Logic*. New York and London: Harper & Brothers, 1846.

Molnar, George. *Powers: A Study in Metaphysics*. Edited by Stephen Mumford. New York: Oxford University Press, 2003.

Moreno, Alvaro, and Jon Umerez. "Downward Causation at the Core of Living Organization." In Andersen et al., *Downward Causation*, 99–116.

Morgan, Conwy Lloyd. *Emergent Evolution*. London: Williams & Norgate, 1923.

Morowitz, Harold J. *The Emergence of Everything: How the World Became Complex*. Oxford: Oxford University Press, 2002.

———. "Emergence of Transcendence." In *From Complexity to Life: On the Emergence of Life and Meaning*, edited by Niels Henrik Gregersen, 177–86. New York: Oxford University Press, 2003.

Müller, Klaus. *Glauben, Fragen, Denken*. Vol. 3, *Selbstbeziehung und Gottesfrage*. Münster: Aschendorf, 2010.

Mullins, Richard T. "The Difficulty with Demarcating Panentheism." *Sophia* 55, no. 3 (2016): 325–46.

Mumford, Stephen Dean. "Causal Powers and Capacities." In Beebee, Hitchcock, and Menzies, *Oxford Handbook of Causation*, 265–78.

———. *Dispositions*. Oxford and New York: Oxford University Press, 1998.

———. "The Power of Power." In *Powers and Capacities in Philosophy: The New Aristotelianism*, edited by Ruth Groff and John Greco, 9–24. New York: Routledge, 2013.

Mumford, Stephen, and Rani Lill Anjum. "Causal Dispositionalism." In *Properties, Powers, and Structures: Issues in the Metaphysics of Realism*, edited by Alexander Bird, B. D. Ellis, and Howard Sankey, 101–18. New York: Routledge, 2012.

———. *Getting Causes from Powers*. Oxford and New York: Oxford University Press, 2011.

———. "A Powerful Theory of Causation." In *The Metaphysics of Powers: Their Grounding and Their Manifestations*, edited by Anna Marmodoro, 143–59. New York: Routledge, 2010.

Murphy, Nancey. "Divine Action in the Natural Order: Buridan's Ass and Schrödinger's Cat." In Russell, Murphy, and Peacocke, *Chaos and Complexity*, 325–57.

———. "Emergence and Mental Causation." In Clayton and Davies, *Re-Emergence of Emergence*, 227–43.

———. "Reductionism: How Did We Fall into It and Can We Emerge from It?" In Murphy and Stoeger, *Evolution and Emergence*, 19–39.

Murphy, Nancey, and William R. Stoeger. *Evolution and Emergence: Systems, Organisms, Persons*. Oxford: Oxford University Press, 2007.

Nagel, Ernst. *The Structure of Science: Problems in the Logic of Scientific Explanation*. New York: Harcourt, Brace and World, 1961.

Nahm, Milton C. *Selections from Early Greek Philosophy*. Englewood Cliffs, NJ: Prentice-Hall, 1964.

Nesteruk, Alexei V. "The Universe as Hypostatic Inherence in the Logos of God: Panentheism in the Eastern Orthodox Perspective." In Clayton and Peacocke, *In Whom We Live and Move*, 169–83.

O'Connor, Timothy. "Philosophical Implications of Emergence." In *The Routledge Companion to Religion and Science*, edited by James W. Haag, Gregory R. Peterson, and Michael L. Spezio, 206–12. New York: Routledge, 2012.

Oderberg, David S. *Real Essentialism*. New York: Routledge, 2007.

Oomen, Palmyre M. F. "God's Power and Almightiness in Whitehead's Thought." *Open Theology* 1 (2015): 277–92.

Paoletti, Michele Paolini, and Francesco Orilia. "Downward Causation: An Opinionated Introduction." In *Philosophical and Scientific Perspectives on Downward Causation*, edited by Michele Paolini Paoletti and Francesco Orilia, 1–21. New York: Routledge, 2017.

Peacocke, Arthur. "Articulating God's Presence in and to the World Unveiled by the Sciences." In Clayton and Peacocke, *In Whom We Live and Move*, 137–54.

———. "Biological Evolution—a Positive Theological Appraisal." In *Evolutionary and Molecular Biology: Scientific Perspectives on Divine Action*, edited by Robert J. Russell, William R. Stoeger, and Francisco José Ayala, 357–76. Vatican: Vatican Observatory; Berkeley, CA: Center for Theology and the Natural Sciences, 1998.

———. "Emergence, Mind, and Divine Action: The Hierarchy of the Sciences in Relation to the Human Mind-Brain-Body." In Clayton and Davies, *Re-Emergence of Emergence*, 257–78.

———. "Emergent Realities with Causal Efficacy: Some Philosophical and Theological Applications." In Murphy and Stoeger, *Evolution and Emergence*, 267–83.

———. "God's Action in the Real World." *Zygon* 26 (1991): 455–76.

———. "God's Interaction with the World." *Studies in Science and Theology* 3 (1995): 135–52.

———. "God's Interaction with the World: The Implications of Deterministic 'Chaos' and of Interconnected and Independent Complexity." In Russell, Murphy, and Peacocke, *Chaos and Complexity*, 263–87.

———. *Paths from Science towards God: The End of All Our Exploring.* Oxford: Oneworld Publications, 2001.

———. *Theology for a Scientific Age: Being and Becoming—Natural, Divine, and Human.* Minneapolis: Fortress, 1993.

Peterson, Gregory R. "Species of Emergence." *Zygon* 41 (2006): 689–712.

———. "Whither Panentheism?" *Zygon* 36, no. 3 (2001): 395–405.

Place, Ullin T. "Dispositions as Intentional States." In *Dispositions: A Debate*, by David Malet Armstrong, Charles Burton Martin, and Ullin T. Place, 19–32. London: Routledge, 1996.

Plato. *Theaetetus.* In *Dialogues*, translated by Benjamin Jowett, 4:328–39. London: Oxford University Press, 1931.

———. *Timaeus.* In *Plato*, vol. 9, translated by R. G. Bury, 16–253. Cambridge, MA: Harvard University Press, 1929.

Plotinus. *Enneads.* 7 vols. Translated by A. H. Armstrong. Cambridge, MA: Harvard University Press, 1966–88.

Polkinghorne, John C. *Belief in God in an Age of Science.* New Haven: Yale University Press, 1998.

———. *Faith, Science and Understanding.* New Haven: Yale University Press, 2001.

———. "Kenotic Creation and Divine Action." In *The Work of Love: Creation as Kenosis*, edited by J. C. Polkinghorne, 90–106. Grand Rapids, MI / Cambridge, UK: Eerdmans, 2001.

———. "The Laws of Nature and the Laws of Physics." In *Quantum Cosmology and the Laws of Nature: Scientific Perspectives on Divine Action*, edited by Robert J. Russell, Nancey C. Murphy, and C. J. Isham, 429–40. Vatican: Vatican Observatory; Berkeley, CA: Center for Theology and the Natural Sciences, 1999.

———. "The Metaphysics of Divine Action." In Russell, Murphy, and Peacocke, *Chaos and Complexity*, 147–56.

———. *Science and Christian Belief: Theological Reflections of a Bottom-Up Thinker.* London: SPCK, 1994.

———. *Science and Theology: An Introduction.* Minneapolis: Fortress, 1998.

———. *Science and the Trinity: The Christian Encounter with Reality.* New Haven: Yale University Press, 2004.

———. *Theology in the Context of Science.* New Haven and London: Yale University Press, 2009.

———, ed. *The Work of Love: Creation as Kenosis.* Grand Rapids, MI / Cambridge, UK: Eerdmans, 2001.

Popper, Karl R. *The Logic of Scientific Discovery.* New York: Basic Books, 1959.

Popper, Karl R., and John Carew Eccles. *The Self and Its Brain.* Berlin/Heidelberg/London/New York: Springer, 1977.

Prabhu, Joseph. "Hegel's Secular Theology." *Sophia* 49, no. 2 (2010): 217–29.

Putnam, Hilary. "The Nature of Mental States." In *Mind, Language, and Reality: Philosophical Papers*, 2:429–40. Cambridge: Cambridge University Press, 1975.

Quine, Willard V. O. *The Roots of Reference*. La Salle, IL: Open Court, 1974.

Rea, Michael C. "Hylomorphism Reconditioned." *Philosophical Perspectives* 25 (2011): 341–58.

Reck, Andrew J. "Substance, Subject and Dialectic." *Tulane Studies in Philosophy* 9 (1960): 109–33.

Reeves, Gene. "God and Creativity." In *Explorations in Whitehead's Philosophy*, edited by Lewis S. Ford and George Louis Kline, 239–51. New York: Fordham University Press, 1983.

Rotenstreich, Nathan. *From Substance to Subject: Studies in Hegel*. The Hague: Martinus Nijhoff, 1974.

Russell, Robert J. *Cosmology from Alpha to Omega: The Creative Mutual Interaction of Theology and Science*. Minneapolis: Fortress, 2008.

Russell, Robert J., Philip Clayton, Kirk Wegter-McNelly, and John Polkinghorne, eds. *Quantum Mechanics: Scientific Perspectives on Divine Action*. Vatican: Vatican Observatory; Berkeley, CA: Center for Theology and the Natural Sciences, 2001.

Russell, Robert J., Nancey C. Murphy, and Arthur Robert Peacocke, eds. *Chaos and Complexity: Scientific Perspectives on Divine Action*. Vatican: Vatican Observatory; Berkeley, CA: Center for Theology and the Natural Sciences, 2000.

Saunders, Nicholas. *Divine Action and Modern Science*. Cambridge and New York: Cambridge University Press, 2002.

Schelling, Friedrich Wilhelm Joseph von. *System of Transcendental Idealism*. Translated by Peter Heath. Charlottesville: University Press of Virginia, 1978.

Schleiermacher, Friedrich. *The Christian Faith*. Translated by H. R. Mackintosh and J. S. Stewart. Edinburgh: T&T Clark, 1928.

Scott, Alwyn C. *The Nonlinear Universe*. Berlin, Heidelberg, New York: Springer, 2007.

Shanley, Brian J. *The Thomist Tradition*. Dordrecht: Kluwer Academic, 2002.

Shapin, Steven. *The Scientific Revolution*. Chicago: University of Chicago Press, 1998.

Sherman, Jeremy, and Terrence W. Deacon. "Teleology for the Perplexed: How Matter Began to Matter." *Zygon* 42 (2007): 873–901.

Shmueli, Efraim. "Hegel's Interpretation of Spinoza's Concept of Substance." *International Journal for Philosophy of Religion* 1 (1970): 176–91.

Shoemaker, Sydney. "Causal and Metaphysical Necessity." *Pacific Philosophical Quarterly* 79 (1998): 59–77.

———. "Causality and Properties." In *Time and Cause: Essays Presented to Richard Taylor*, edited by Richard Taylor and Peter van Inwagen, 109–35. Dordrecht and Boston: Reidel, 1980.

Silberstein, Michael. "In Defence of Ontological Emergence and Mental Causation." In Clayton and Davies, *Re-Emergence of Emergence*, 203–26.

Silva, Ignacio. "A Cause among Causes? God Acting in the Natural World." *European Journal for Philosophy of Religion* 7, no. 4 (2015): 99–114.
———. "Divine Action and Thomism: Why Thomas Aquinas's Thought Is Attractive Today." *Acta Philosophica* 25 (2016): 65–84.
———. "John Polkinghorne on Divine Action: A Coherent Theological Evolution." *Science and Christian Belief* 24, no. 1 (2012): 19–30.
———. "Providence, Contingency, and the Perfection of the Universe." *Philosophy, Theology and the Sciences* 2 (2015): 137–57.
———. "Revisiting Aquinas on Providence and Rising to the Challenge of Divine Action in Nature." *Journal of Religion* 94 (2014): 277–91.
———. "Thomas Aquinas Holds Fast: Objections to Aquinas within Today's Debate on Divine Action." *Heythrop Journal* 48 (2011): 1–10.
———. "Thomas Aquinas on Natural Contingency and Providence." In *Abraham's Dice: Chance and Providence in the Monotheistic Traditions*, edited by Karl W. Giberson, 158–74. New York: Oxford University Press, 2016.
Smart, John Jamieson Carswell. "Physicalism and Emergence." *Neuroscience* 6 (1981): 109–13.
Sosa, Ernest, and Michael Tooley, eds. *Causation*. Oxford and New York: Oxford University Press, 1993.
Sperry, Roger Wolcott. "Discussion: Macro- versus Micro-Determinism." *Philosophy of Science* 53 (1986): 265–70.
Sprigge, Timothy L. S. *The God of Metaphysics*. Oxford: Clarendon, 2006.
Stephan, Achim. "Emergence—a Systematic View on Its Historical Facets." In *Emergence or Reduction? Essays on the Prospects of Nonreductive Physicalism*, edited by Ansgar Beckermann, Hans Flohr, and Jaegwon Kim, 25–48. Berlin and New York: de Gruyter, 1992.
Stoeger, William R. "Describing God's Action in the World in Light of Scientific Knowledge of Reality." In Russell, Murphy, and Peacocke, *Chaos and Complexity*, 239–61.
Swinburne, Richard. *The Coherence of Theism*. Oxford: Oxford University Press, 1977.
Tabaczek, Mariusz. *Emergence: Towards a New Metaphysics and Philosophy of Science*. Notre Dame, IN: University of Notre Dame Press, 2019.
———. "Hegel and Whitehead: In Search for Sources of Contemporary Versions of Panentheism in the Science–Theology Dialogue." *Theology and Science* 11 (2013): 143–61.
———. "Pantheism and Panentheism." In *T&T Clark Handbook of the Doctrine of Creation*, edited by Jason Goroncy. London and New York: Bloomsbury T&T Clark, forthcoming.
Temple, William. *Nature, Man, and God*. London: Macmillan, 1934.
te Velde, Rudi A. *Aquinas on God: The "Divine Science" of the Summa Theologiae*. Aldershot: Ashgate, 2006.

———. *Participation and Substantiality in Thomas Aquinas*. Leiden, New York, Cologne: Brill, 1995.

Thomas, Owen C. "Problems in Panentheism." In *The Oxford Handbook of Religion and Science*, edited by Philip Clayton and Zachary Simpson, 652–64. Oxford: Oxford University Press, 2006.

Towne, Edgar A. "The Variety of Panentheisms." *Zygon* 40, no. 3 (2005): 779–86.

Tracy, David. *Blessed Rage for Order: The New Pluralism in Theology*. New York: Seabury Press, 1975.

Tracy, Thomas F. "Creation, Providence, and Quantum Chance." In *Quantum Mechanics: Scientific Perspectives on Divine Action*, edited by Robert J. Russell, Philip Clayton, Kirk Wegter-McNelly, and John Polkinghorne, 235–58. Vatican: Vatican Observatory; Berkeley, CA: Center for Theology and the Natural Sciences, 2001.

———. "Special Divine Action and the Laws of Nature." In *Scientific Perspectives on Divine Action: Twenty Years of Challenge and Progress*, edited by Robert J. Russell, Nancey C. Murphy, and William R. Stoeger, 249–83. Vatican: Vatican Observatory; Berkeley, CA: Center for Theology and the Natural Sciences, 2008.

Vlastos, Gregory. "Organic Categories in Whitehead." *Journal of Philosophy* 34 (1937): 253–62.

Wallace, William A. *Causality and Scientific Explanation*. 2 vols. Ann Arbor: University of Michigan Press, 1972–74.

Ware, Kallistos. "God Immanent yet Transcendent: The Divine Energies according to Saint Gregory Palamas." In Clayton and Peacocke, *In Whom We Live and Move*, 157–68.

Weeks, Sophie. "The Role of Mechanics in Francis Bacon's Great Instauration." In *Philosophies of Technology: Francis Bacon and His Contemporaries*, edited by Claus Zittel, Gisela Engel, Romano Nanni, and Nicole C. Karafyllis, 131–96. Leiden, Boston: Brill, 2008.

Weinandy, Thomas Gerard. *Does God Change? The Word's Becoming in the Incarnation*. Still River, MA: St. Bede's Publications, 1984.

Whitehead, Alfred North. *Process and Reality*. New York: Free Press, 1979.

———. *Religion in the Making*. New York: Macmillan, 1926.

———. *Science and the Modern World*. New York: New American Library, 1950. First published 1925.

Whittemore, Robert C. "Hegel as Panentheist." *Tulane Studies in Philosophy* 9 (1960): 134–64.

———. "The Meeting of East and West in Neglected Vedanta." *Dialogue & Alliance* 2 (1988): 33–47.

Wilkens, Steve, and Alan G. Padgett. *Christianity and Western Thought*. Vol. 2, *Faith and Reason in the Nineteenth Century*. Downers Grove, IL: InterVarsity Press, 2000.

Williams, Daniel Day. *The Spirit and the Forms of Love*. New York: Harper & Row, 1968.
Wippel, John F. *The Metaphysical Thought of Thomas Aquinas*. Washington, DC: Catholic University of America Press, 2000.
———. "Thomas Aquinas on Creatures as Causes of Esse." *International Philosophical Quarterly* 40 (2000): 197–213.
———. "Thomas Aquinas on the Distinction and Derivation of the Many from the One: A Dialectic between Being and Nonbeing." *Review of Metaphysics* 38 (1985): 563–90.
Wojtysiak, Jacek. "Panenteizm." In *Filozofia Boga: Część II, Odkrywanie Boga* [Philosophy of God: Part 2, Discerning God], ed. Stanisław Janeczek and Anna Starościc, 506–13. Lublin: Wydawnictwo KUL, 2017.
Wong, Hong Yu. "The Secret Lives of Emergents." In *Emergence in Science and Philosophy*, edited by Antonella Corradini and Timothy O'Connor, 7–24. New York: Routledge, 2010.

INDEX

absence
    Aquinas on nonbeing, 10, 107, 219–28, 234, 297n.102, 298n.105, 299nn.114–15
    Aristotle on, 101–2
    Deacon on, 5, 6, 10, 38, 47, 48–49, 50–51, 53, 55–56, 57, 58, 59, 61, 98, 99–102, 107, 184, 186, 188, 189, 217, 218, 219, 221, 225, 227–28, 234, 250n.70, 251n.83, 251n.85, 251n.90, 261n.88
    as enabling condition vs. cause, 88
accidental change
    Aristotle on, 67, 71, 254n.12
    vs. substantial change, 67, 71, 101, 201, 254n.12, 256n.43
accidental form, 226, 227, 256n.43, 289n.28, 299n.114
    Aquinas on, 162–63, 195, 201, 286n.6
    Aristotle on, 65, 66–67, 71, 79, 94–95, 254n.12
acetyl coenzyme A, 76, 77
agnosticism, 3
Akhenaton, 110
Al-Ash'arī, on occasionalism, 198
Alexander, Samuel, 238
    on emergence (EM), 25, 29, 30, 32
    on God as becoming, 148–49, 181–82, 274n.129
    on laws ruling emergents, 32
    on levels of organization, 31
    Space, Time and Deity, 31
    on theistic naturalism, 148–49, 153
    on universe as body of God, 149, 182
Al-Ghazālī, on occasionalism, 198
analytic philosophy, 25, 248n.39
    and metaphysics, 6, 9, 13, 26, 27, 81, 84, 87, 93, 260n.82
    See also dispositional/powers metaphysics
Anaxagoras, 63, 253n.6
Anaximander, 63
Anaximenes, 63
Anderson, Alan, 238
Anjum, Rani
    on dispositions/powers, 85, 87
    on temporal simultaneity of causal relations, 88
Anselm, St., 114
anthropomorphism, 152
apophatic theology, 147, 180, 274n.123, 284n.63, 294n.69
Aquinas, Thomas
    on actuality, 166, 175, 190, 192, 226, 234, 254n.22, 287n.9, 299n.114
    on angels as spiritual substances, 287n.9, 298n.105
    vs. Aristotle, 109–10, 189, 191, 192, 194, 209, 211, 212, 229, 230, 285n.3, 288n.16, 289n.27, 292n.48, 295n.87

Aquinas, Thomas (*cont.*)
  vs. Augustine, 288n.15
  on causes *per se* vs. *per accidens*, 208, 212, 225–26, 227, 230, 294n.78, 296n.93, 299nn.114–15
  on chance, 191, 205, 211–13, 230, 234, 296n.93, 296n.95
  on contingency, 211–12
  vs. Deacon, 107, 219, 221, 225–28, 234
  on efficient causation, 74, 175, 194, 196, 198, 199–209, 212, 229, 230, 287n.12, 289n.27, 292n.51, 293n.57, 293n.65, 299n.114
  on *esse*/act of being, 189–90, 191, 226–27, 229, 285n.3, 287nn.9–10, 298n.105, 299n.115
  on essence (*essentia*), 189–90, 191, 226–27, 229, 299n.115
  on evil as privation of good, 208, 230, 289n.28
  on final causation, 74–75, 107, 194, 196, 205, 209–11, 212, 218, 229, 230, 291n.35, 294n.82, 295n.83, 295n.85, 299n.114
  on formal causation, 74–75, 107, 175, 192–97, 199, 205, 212, 218, 229, 230, 290n.33
  on Fourth Way, 289n.27
  on God as Creator, 7, 107, 189, 190, 191, 192, 195, 196–97, 199, 200, 202, 204, 211–12, 229, 234, 281n.36, 286n.4, 287n.12, 295n.87
  on God as final end (*telos*), 107, 189, 209–10, 294n.82, 295n.83, 295n.85
  on God as self-subsistent being (*ipsum esse subsistens*), 189, 190, 224, 285n.2
  on God as source of being (*esse*), 189–90, 192, 205, 207, 230, 234
  on God's act of understanding, 221–25
  on God's essence, 190, 195, 222, 229, 285n.2, 289nn.27–28, 298n.110
  on God's existence, 233, 235, 289n.27
  on God's goodness, 209–10, 286n.5, 294n.81, 295n.85, 295n.87
  on God's ideas as exemplar causes, 107, 189, 192–97, 199–200, 207, 219, 221, 227, 234, 288nn.15–17, 288n.23, 289nn.27–28, 290nn.33–34, 291n.37, 291n.39, 294n.81, 297n.100
  on God's immanence, 162–64, 165–66, 174, 175, 191, 229, 230, 239–40
  on God's nature/essence, 189–90
  on God's perfection, 285nn.2–3, 289n.27, 295n.87
  on God's primary and principal causality, 2–3, 7, 107, 189, 190, 207, 208–9, 211–12, 228–29, 230, 234, 242n.2
  on God's transcendence, 7, 163–64, 165, 191, 200, 229, 230, 240
  hylomorphism of, 175, 189, 225, 287n.9
  on interrelatedness of causes, 74, 97, 196
  on material causation, 74, 192, 205, 212, 230, 287n.12
  on miracles, 208, 209
  on natural teleology, 194
  on nonbeing, 10, 107, 219–28, 234, 297n.102, 298n.105, 299nn.114–15
  on occasionalism, 199
  on one and many, 219–21, 223, 225, 227, 234
  on potentiality, 70, 225–27, 285n.3, 287n.9, 299nn.114–15
  on primary and secondary causation, 2–3, 7, 72, 107, 189,

200, 201–5, 207, 208–9, 211–12, 228–29, 230, 234, 242n.2, 292n.48, 293n.57
on primary matter, 70, 97, 175, 191, 192, 195, 199, 226–27, 229, 254n.22, 287n.9, 287n.11–12, 298n.105, 299n.114
on principal and instrumental causation, 2–3, 7, 107, 189, 200, 201–5, 208, 209, 211–12, 234, 291n.45, 292n.51
on privation, 208, 230, 289n.28, 299n.114
on providence, 190, 212–13, 286n.5
vs. Pseudo-Dionysius, 113–14, 284n.63
on secondary (proximate) matter, 226, 227, 299n.114
on secondary and instrumental causation of creatures, 107, 189
on soul/body metaphor, 163, 174, 175, 239–40, 279n.23, 283n.59
on substances and accidents, 162–63, 195, 201, 286n.6
on substantial form, 70, 75, 97, 189, 191, 192, 195, 199, 204, 226, 227, 254n.22, 298n.105, 299nn.114–15
virtual (*virtute*) presence of elements in mixed substances, 75, 97
Aristotle
on absence, 101–2
on accidental change, 67, 71, 254n.12
on accidental form, 65, 66–67, 71, 79, 94–95, 254n.12
on active vs. passive potency, 66, 75, 89, 91, 93
on actuality, 65–66, 69, 70, 72, 89–90, 93, 100, 175, 254n.22, 285n.3
vs. Aquinas, 109–10, 189, 191, 192, 194, 209, 211, 212, 229, 230, 285n.3, 288n.16, 289n.27, 292n.48, 295n.87
on atomism, 64
on being and change, 65-67
on causation, 2, 5–6, 7, 9, 13, 45, 46, 47, 55, 56–57, 59–60, 61–62, 63, 64–65, 67–75, 79–80, 81, 89, 93–97, 104, 107, 184, 189, 191, 192, 198, 199, 212, 214, 216, 229, 230, 233, 255n.25, 285n.70
on chance, 9, 78–80, 96–97, 191, 192, 211, 212, 257nn.50–52, 259n.71
on change and stability, 63, 65–67
Deacon on, 55, 57, 59–60
on efficient causation, 5, 46, 55, 71–73, 74–75, 79, 95–96, 97, 103, 175, 255n.30, 261n.92
on essence, 67, 68, 69, 70, 91, 244n.9, 255n.25
on final causation, 5, 55, 70, 71, 72–73, 74–75, 79, 97, 209, 255n.31, 256nn.32–33, 256n.42, 257n.51, 259n.75, 261n.92
on first principle, 141
on formal causation, 5, 55, 67–68, 69–71, 73, 74–75, 79, 94–95, 97, 99–100, 175, 188, 189, 255nn.24–25, 261n.92
vs. Heraclitus, 62, 63, 65, 66–67, 253n.5
hylomorphism of, 6, 74, 78, 90–93, 94–95, 99–100, 175, 189, 252n.2, 254n.19, 260nn.83–84, 261n.86
on interrelatedness of causes, 73, 74–75, 97
on material causation, 5, 46, 55, 59, 67–69, 74, 98, 99–100
on necessity, 80
on nonrational vs. rational active powers, 66
vs. Parmenides, 62, 63, 64, 65, 66

Aristotle (*cont.*)
  on *per se* vs. *per accidens*/incidental causes, 79–80, 96, 102, 212, 257n.50, 259n.71
  on philosophy and wonder, 15, 243n.1
  on potentiality, 65–66, 68, 69, 71–72, 74, 89–91, 93, 100, 254n.22, 285n.3
  on primary and secondary causation, 72, 200, 292n.48
  on primary matter, 68–69, 70, 80, 91, 93, 94–95, 97, 100, 175, 226, 252n.2, 254nn.18–19, 255n.30
  on privation of form, 101–2
  on secondary (proximate) matter, 68
  on substances and accidents, 65, 66–67, 68
  on substantial change, 67, 69, 70–71, 97, 100, 254n.12, 255n.30
  on substantial form, 6, 67, 68, 69–70, 72, 74, 79, 91, 93, 94–95, 97, 99–100, 103, 252n.2, 254n.19, 260n.84, 261n.92, 285n.3
  on transcendental attributes of being, 178
  on universals, 248n.39
  on Unmoved Mover, 198, 285n.3, 295n.87
Armstrong, David, 27, 260n.80
atheism, 3, 122, 171, 179
atomism, 2, 24, 198
  of Democritus, 64, 65, 75, 84
ATP (adenosine triphosphate), 76, 77
attractors, 45, 51
Augustine, St., 193, 208, 277n.6
  on chance, 213
  on divine ideas, 288n.15
autocatalytic reactions, 51
autogenesis, 52–54, 102, 103, 251nn.82–83, 251n.86
Averroës, on concurrentism, 198–99

Bacigalupi, Augustus, 55
Bacon, Francis
  empirical methodology of, 22
  on final causality, 54
Baltzly, Dirk, on Plato and panentheism, 263n.6
Barbour, Ian, 238, 282n.44, 295n.91
  on dipolar God, 136, 167
  on Whitehead and God's transcendence, 170–71
Barth, Karl, on God's immanence, 279n.23
Bénard cells, 17, 18, 51
Berdyaev, Nicolai, 238
Berger, Peter, 238
Bergson, Henri, 131, 145, 238
Bethune-Baker, James, 238
Biernacki, Loriliai: *Panentheism across the World's Traditions*, 268n.50
big bang theory, 135, 152, 183
biology, 25, 28, 135, 145
  laws of, 17
  molecular biology, 11, 12, 26
  systems approach in, 12, 26–27, 41, 58, 184, 233, 245n.15
  See also evolution
Birch, Charles, 238
Bird, Alexander, on powers, 6, 85, 258n.60, 259n.76
Boethius, 144, 290n.34
  De Trinitate, 219, 223
Boff, Leonardo, 238
  on panentheism, 133
Böhme, Jakob, 114, 120, 237
Bonhoeffer, Dietrich, 238
Borg, Marcus, 238

Born, Max, 24
Bostock, David, 73, 255n.31, 256n.32
Bracken, Joseph, 238
  on God as regnant subsociety, 161, 170
  on God-world relationship, 157, 161, 170, 278n.17, 283n.60
  on "in" metaphor, 157, 161–62
  on Trinitarian field theory, 161–62, 278n.15
Branick, Vincent, on divine understanding and nonbeing, 221, 222, 223, 224, 225, 299n.115
Bridgman, Percy, operationalism of, 24
Brierley, Michael W.
  on "in" metaphor, 156–58
  on panentheism, 106, 132, 156–58, 237, 262n.4
British empiricism, 25, 28
Broad, Charlie Dunbar
  on emergence (EM), 25, 30, 32
  on trans-ordinal laws, 32
broken symmetry, 17, 20, 27–28
Bruno, Giordano, 237
Buber, Martin, 238
Bulgakov, Sergei, 238
Bultmann, Rudolf, 238
  on divine action, 3
  on God as Creator, 242n.4
Bunge, Mario, 27
Burtt, Edwin A.
  on God and final causality, 2
  on God as Efficient Cause, 2

Caird, John and Edward, 238
Campbell, Donald Thomas, on downward causation (DC), 35, 36
Capra, Fritjof, on emergence, 11
Carroll, William, on divine action, 204, 293n.65

Cartwright, Nancy, on powers and Aristotle's metaphysics, 89
Case-Winters, Anna, 238
Cashman, Tyrone, 128, 251n.83, 261n.88, 284n.63
causal closure of the physical/causal inheritance principle, 34, 41, 138, 148, 247n.32
  and DC-based EM, 36, 39, 42, 43, 250n.70
  and SUP-based EM, 39
  and whole-part constraints-based EM, 37–38
causal joint conundrum, 169, 216
  Peacocke on, 3, 150, 183, 184, 187, 205, 229–30
causal pleiotropy, 88
causal polygeny, 6, 88
causal preemption
  early preemption, 87
  late preemption, 87
  simultaneous preemption, 87
causal reciprocity, 88
causal reductionism, 26, 42–44, 55, 94
chance
  Aquinas on, 191, 205, 211–13, 230, 234, 296n.93, 296n.95
  Aristotle on, 9, 78–80, 96–97, 191, 192, 211, 212, 257nn.50–52, 259n.71
  and divine action, 138, 181, 191, 192, 211–13
  and downward causation (DC), 96–97
  and necessity, 78–80, 137, 148
chaos, 4, 136, 42, 273n.108, 295n.91
chemistry, 17, 25, 28–29, 41
Christian, William, on Whitehead, 168
citric acid cycle, 76, 77

classical theism, 165, 231
  divine immanence in, 162–63, 164, 239
  vs. panentheism, 4, 105–6, 109–10, 111, 113–14, 123–25, 131, 133, 134, 135, 138, 151, 156, 159, 162–63, 164, 166–67, 191, 262n.4, 270n.71, 281n.36, 284n.61, 285n.70, 285n.1 (chap. 5)
  vs. pantheism, 276n.5
  vs. process theism, 131, 156
  See also Aquinas, Thomas
Clayton, Philip, 3, 167, 295n.91
  on divine action and double agency, 206, 207
  emergentist panentheism of, 4, 8, 133, 140–42, 151, 152, 157–58, 160, 161, 162, 164–65, 174–76, 182, 185, 228, 231, 235, 238, 272n.101, 278n.12
  on Fichte, 267n.37
  on God as Creator, 141, 151, 269n.61, 281n.36
  on God as dipolar, 172–73
  on God as immanent, 162
  on God as One Ultimate Reality, 160
  on God as personal, 141, 151
  on God's transcendence, 169
  on "in" metaphor, 157–58, 160, 161, 162
  on panentheistic analogy (PA), 175–76
  on Peacocke, 151–52
  on reductionism and emergence, 11
  on Schelling, 120, 141
  on soul/body metaphor, 174–76
Cobb, John
  on actual occasions, 282n.44
  on God's immanence, 162
  on "in" metaphor, 162
  on panentheism, 127, 238, 270n.71
Coleridge, Samuel Taylor, 238

concurrence, 3, 198–200, 209
Cone, James, 238
contingency, 80, 84, 88, 94, 146, 213, 257n.52
  Aquinas on, 211–12
  Deacon on, 146
  God and contingent entities, 131, 159, 170, 190, 215, 216, 219, 220, 231, 262n.4, 270n.61, 279n.22, 292n.48, 295n.87
  Hegel on, 124, 126, 141, 168
convection cells, 17, 18
Cooper, Barton, on Whitehead, 169, 270n.71
Cooper, John, 109
  on Clayton, 272n.101
  on Fichte, 115, 116
  on Plotinus, 112, 113
  on Whitehead, 128
Copleston, Frederick, 120
  on Fichte, Schelling, and Hegel, 121
counterfactual view of causation (CVC), 81–82
  vs. DVC, 87, 89
  and possible worlds modality, 83, 89
Cowdell, Scott, 238
Craig, William Lane, 163–64
creation
  creatio continua, 190, 281n.36, 286n.4
  creatio ex nihilo, 112, 130, 142, 170, 172, 178, 180, 190, 199, 201, 262n.4, 269n.61, 278n.14, 281n.36, 282n.42, 286n.4, 297n.98
  See also God: as Creator
Cudworth, Ralph, 237
Culp, John
  on Bracken, 161
  on panentheism, 132
  on Whitehead, 169
Curley, Edwin: *Spinoza's Metaphysics*, 264n.12

Davies, Paul
  on downward causation (DC), 35, 36
  on God and laws of nature, 171–72, 181
  on God as Creator, 136–37, 148, 171, 181, 281n.36
  on God's transcendence, 171
  on God-world relationship, 157, 171–72, 187
  on modified uniformitarianism, 137, 148, 181
  on theistic naturalism, 148, 185
Dawkins, Richard, reductionist secularism of, 267n.42
Deacon, Terrence
  on absences (constraints), 5, 6, 10, 38, 47, 48–49, 50–51, 53, 55–56, 57, 58, 59, 61, 98, 99–102, 107, 184, 186, 188, 189, 217, 218, 219, 221, 225, 227–28, 234, 250n.70, 251n.83, 251n.85, 251n.90, 261n.88
  on abstract formal properties/abstract ideal forms, 48
  on apophatic tradition, 147, 180, 274n.123, 284n.63
  vs. Aquinas, 107, 219, 221, 225–28, 234
  on Aristotelian causes, 55, 57, 59–60
  on autogenesis, 52–54, 102, 103, 251nn.82–83, 251n.86
  on causal nonreductionism, 54, 55
  on consciousness, 47
  on definition of emergence (EM), 28
  on downward causation (DC), 35, 36, 48
  on dynamical depth model of EM, 5–6, 7–8, 9, 10, 13, 46, 47–60, 61, 98–104, 178, 179, 184, 186, 188–89, 217–19, 221, 227–28, 234, 250n.70, 251n.83, 251n.85, 260n.85
  on efficient causation, 55, 56, 57, 59, 100, 103, 218, 261n.88
  on eliminative reductionism, 5–6, 54–55
  on emergent properties, 49
  on ententional phenomena, 47, 56, 57–58, 103, 250n.61
  on entropy, 49
  on extrinsic vs. intrinsic constraints, 49, 50, 51–52
  on final causation, 5–6, 54, 57, 59, 61, 98, 102–3, 188, 234
  on formal causation, 5, 55–57, 59, 61, 98, 99–100, 103, 179, 188, 234, 251n.90, 261n.87
  vs. Gillett, 38
  on higher order attractors, 52
  on homeodynamics, 50, 51, 59, 217, 219
  on human self-creation, 147, 274n.120
  on life's origin, 47, 48, 54, 56, 102–3
  on matter, 99–100, 260n.85
  on mind-like attributes, 128
  on morphodynamics, 50–51, 52, 54, 59, 217, 219
  on orthograde and contragrade changes, 49, 50, 51, 55–56, 98–99, 100, 103, 179, 217, 218–19
  on probability space, 55, 56, 59, 98, 99
  on reductionism, 48, 54–55
  on religious naturalism, 146–48, 179, 180
  on selfhood, 51, 54, 251n.86
  on self-organizing systems, 47–48, 50–51
  on spontaneity of changes, 98–99, 100, 103, 179
  on stages of dynamical depth, 50–54
  on teleodynamics, 51–54, 56–58, 59, 103–4, 177, 189, 217, 219, 227, 251n.86

Deane-Drummond, C., 157
deism, 3, 141, 180–81, 185, 205, 207, 208
Democritus, atomism of, 64, 65, 75, 84
Descartes, René
  on final causality, 54
  on mechanism, 23, 244n.9
  theory of dense aether, 21
determinism, 80, 257n.52
dialectics, 118, 120–23
  between being and nonbeing, 219–21, 222, 225
  Hegel on, 121–23, 124, 126, 134, 173, 225, 283n.51
  in Neoplatonism, 112, 113, 114, 116
  Vlastos on, 173, 283n.51
diamond lattice crystal structure, 76–77
dispositional/powers metaphysics
  and Aristotelian metaphysics, 6–7, 9, 13, 89–93, 99, 188
  vs. bundle theory of substance, 92, 93
  dispositional properties vs. categorical properties, 91–92, 93, 260n.80
  dispositions as irreducible, 85, 87, 94, 95–96, 103
  dispositions as natural-kind specific, 6, 93, 94, 96, 99, 103, 215, 227, 259n.71
  and downward causation (DC), 94, 95–96, 188–89, 215
  and dynamical depth version of EM, 98–99, 179, 188
  and emergence (EM), 6–7, 94, 95–96, 107, 188–89, 215, 217–18, 219, 233–34
  and essentialism, 6–7, 85, 86–87, 92–93, 94, 95–96
  and hylomorphism, 90–93, 260n.83

  identity theory regarding dispositional and categorical properties, 91, 92
  manifestations of dispositions, 85–87, 88, 89–90, 92–93, 94, 96, 103, 207, 259n.71
  Mumford on, 85, 87, 96, 258n.59, 258n.63, 259n.76
  and neutral (dual-aspect) monism, 91–92
  ontological dualism regarding dispositional and categorical properties, 92, 260n.80
  and pan-categoricalism, 91, 92, 259n.77
  as pan-dispositionalism, 91, 92, 259n.76
  powers as properties, 85–87, 91
  powers as universals, 85
  probability in, 259n.71
  relationship to formal causation, 6–7, 188–89, 227, 228
  role of teleology in, 6–7, 90, 93, 188–89, 217–18
dispositional view of causation (DVC), 6, 9, 13, 84–93, 94, 261n.90
  causal necessity as suppositional in, 88, 103
  vs. CVC, 87, 89
  and hylomorphism, 91
  vs. MVC, 87, 89
  vs. ProbVC, 87, 88
  vs. ProcVC, 87
  role of specific conditions in, 85, 88
  vs. RVC, 87
  vs. SVC, 89
Dodds, Michael J.
  on Aquinas, 165, 201, 204–5
  on chance, 213
  on divine action, 204–5, 209, 285n.71
  on God and causality, 1, 213

on God as first final cause, 210–11
on God's immanence, 165–66
on Heraclitus and Whitehead, 252n.3
on substantial form, 70, 105, 287n.10
Dombrowski, Daniel, 238
Doolan, Gregory
  on divine exemplars, 197, 222, 224, 288n.16, 289nn.27–28, 290n.34, 291n.39, 297n.100
  on divine self-knowledge, 194
  on divine understanding and nonbeing, 221, 222, 223, 224, 225
Dorner, Isaak, 237
double agency theory, 205–7, 208, 294n.69
downward causation (DC)
  and Aristotelian metaphysics, 93–97, 214, 216
  and causal reductionism, 26, 42–44, 94
  and chance, 96–97
  Deacon on, 35, 36, 48
  definition, 34–35
  and dispositional/powers metaphysics, 94, 95–96, 188–89, 215
  and efficient causation, 43–44, 95–96, 184, 228
  emergence (EM) based on, 4, 5, 6, 7–8, 9, 10, 13, 28, 31, 34–37, 39, 40, 41, 42–44, 48, 59, 93–97, 107, 136, 184, 185, 188–89, 214, 216, 217, 219, 228, 231, 233–34, 250n.70
  Emmeche et al. on, 5, 36, 44–45, 251n.90
  and hylomorphism, 94–95, 96
  and interrelatedness of causes, 97
  Kim on, 5, 36–37, 42, 248n.34
  medium DC (MDC), 44, 45

Murphy on (DC), 35, 36, 43
Peacocke on, 35, 36, 137, 248n.38
strong DC (SDC), 44, 45
and systems approach in life sciences, 3
and teleology, 96, 188–89
van Gulick on, 35, 36, 43, 247n.32
weak DC (WDC), 44–45
what is acted upon in, 36
what is causal in, 35–36, 43, 247n.32, 248n.34
Drees, Willem, on theistic naturalism, 148, 180–81
dualism, 30, 31, 44
Dudley, John, 257n.52

Eastern Orthodox Christianity, 143, 280n.31, 284n.63
Eckhart, Meister, 114, 141, 237
eddies in water, 17, 18, 51, 95
Edwards, Denis, 238
Edwards, Jonathan, 237
efficient causation, 22, 39, 42, 45, 64, 78, 214
  Aquinas on, 74, 175, 194, 196, 198, 199–209, 212, 229, 230, 289n.27, 292n.51, 293n.57, 293n.65, 299n.114
  Aristotle on, 5, 46, 55, 71–73, 74–75, 79, 95–96, 97, 103, 175, 255n.30, 261n.92
  Deacon on, 55, 56, 57, 59, 100, 103, 218, 261n.88
  double agency of God and creatures, 200, 205–7, 208, 294n.69
  and downward causation (DC), 43–44, 95–96, 184, 228
  vs. formal causation, 175
  in science, 23–24, 80
  See also primary and secondary causation; principal and instrumental causation

elasticity, 16
El-Hani, Charbel Niño, 5
   on matter and form, 46
eliminativism, 5, 9, 31, 42, 54–55, 144, 188. *See also* reductionism
Ellis, Brian
   on causal powers, 87
   on dispositions/powers, 6, 260n.80
   on teleonomic goals, 46
Ellis, George, on downward causation (DC), 35, 36
emergence (EM)
   DC-based version, 4, 5, 6, 7–8, 9, 10, 13, 28, 31, 34–37, 39, 40, 41, 42–44, 48, 59, 93–97, 107, 136, 184, 185, 188–89, 214, 216, 217, 219, 228, 231, 233–34, 250n.70
   definitions of, 27–28
   and dispositional/powers metaphysics, 6–7, 94, 95–96, 107, 188–89, 215, 217–18, 219, 233–34
   dynamical depth version, 5–6, 7–8, 9, 10, 13, 46, 47–60, 61, 93, 98–104, 137, 178, 179, 185, 186, 188–89, 217–19, 221, 227–31, 234, 250n.70, 251n85, 260n.85
   epistemological (weak) EM, 9, 20, 28, 33–34, 151
   God's goodness communicated by, 214, 215–16, 218–19, 228, 234
   God's perfection communicated by, 214–16, 218–19, 228, 234
   irreducibility of emergents, 28, 32, 33, 34, 35, 36, 37, 38, 40–41, 42, 56, 94, 95–96, 97, 100, 145, 188, 214, 216
   levels of organization of parts in emergent wholes, 28, 31–33, 41–42, 43, 44–45, 48, 189, 217, 227, 234
   as mereological, 5, 6, 28, 30–31, 46, 59, 107, 184, 189, 217, 228, 234
   as mystery, 147, 148, 152–53
   necessity and insufficiency of parts for existence of emergent whole, 28, 30–31
   nondeducibility of emergents, 28, 33
   nonpredictability of emergents, 28, 32–33
   ontological (strong) EM, 9, 20, 28, 33–46, 47, 48, 49, 151, 188–89, 214, 233
   origin of emergentism, 25–27, 245n.12
   relationship to broken symmetry, 17, 20, 27–28
   relationship to causal closure of the physical, 36, 37–38, 39, 42, 43, 250n.70
   relationship to new laws of nature characteristic of emergent wholes, 28, 32
   relationship to nonadditivity of causes, 28–29
   relationship to novelty of emergent processes, entities, and properties, 28, 29–30, 34, 35, 36, 37, 41, 42, 43
   relationship to ontological/physical monism, 28, 30, 31, 34, 36, 37–38, 39, 41–42, 44, 137–38, 146, 151–52, 160, 249n.45
   and religious naturalism, 146–48, 177–80
   scientific vs. philosophical aspects, 41–42
   and serendipitous creativity, 152–53
   supervenience (SUP)-based version, 28, 39–41, 61
   and systems approach in life sciences, 3, 41

and theistic naturalism, 146, 148–49,
    180–81
whole-part constraints-based
    version, 28, 31, 36, 37–39, 40, 41,
    49, 61, 248n.38
See also emergentist panentheism;
    emergent phenomena
emergentist panentheism, 7, 8, 9–10,
    109
  critique of, 10, 107, 155, 181–86,
    187–88, 189, 229–31, 233–34
  definition of, 4
  vs. Thomistic interpretation of
    divine action in/through EM, 107,
    228–31
  See also Clayton, Philip; Peacocke,
    Arthur
emergent phenomena, 13–20, 50, 94–
    95, 233, 235
  as communicating God's goodness,
    214, 215–16, 218–19, 228, 234
  as communicating God's perfection,
    214–16, 218–19, 228, 234
  complexity of, 137, 138, 189, 213–14,
    217, 219, 227, 234
  emergent entities (substances), 28,
    29, 39
  emergent laws, 32, 33, 38, 40
  emergent properties (qualities), 28,
    29–30, 38, 39, 40, 49, 94–95
  first-order emergents, 9, 15–16
  human mind, 17, 27, 35, 95, 176–77,
    260n.84
  novelty of, 28, 29–30, 34, 35, 36, 37,
    41, 42, 43
  qualitative difference of emergents,
    28, 30, 31, 36
  relationship to broken symmetry, 17,
    20, 27–28
  second-order emergents, 9, 16–17,
    18, 51

third-order emergents, 9, 17, 19
wonder of, 15–20, 233, 235
Emerson, Ralph Waldo, 238
Emmeche, Claus, et al.
  on downward causation (DC), 5, 36,
    44–45, 251n.90
  on efficient causation, 44
  on emergent relations as
    nonhomomorphic, 31–32
  on extended typology of causes, 45
  on formal causation, 44, 45, 251n.90
  on levels of complexity, 31–32, 44–45
  on organizational principles, 45
Empedocles, 63–64, 253n.7
entropy, 49–51, 53–54, 99
epiphenomenalism, 35, 100
Eriugena, John Scotus, 114, 141, 237,
    242n.6
eschatological panentheism, 144, 156,
    173, 274n.114
essentialism
  of Aristotle, 67, 68, 69, 70, 91,
    244n.9, 255n.25
  and dispositional/powers
    metaphysics, 6–7, 85, 86–87,
    92–93, 94, 95–96
evolution, 17, 51, 54, 135, 139, 140,
    148–49, 151, 152, 153

Farrer, Austin, 3
  on divine action and double agency,
    206–7
  on downward causation (DC), 35, 36
Fechner, Gustav, 238
Feser, Edward, 259n.75
Fichte, Johann, 114–16, 126, 237
  on Absolute Ego, 115–16, 119, 122,
    264n.16, 267n.37
  on creation, 269n.61
  Hegel on, 121, 122
  Schelling on, 118–19

Fiddes, Paul, 238
  on divine suffering, 144, 272n.94
  on panentheism, 133, 134
fideism, 205, 207, 208
final causation, 22, 44, 63, 217
  Aquinas on, 74–75, 107, 194, 196, 205, 209–11, 212, 218, 229, 230, 291n.35, 294n.82, 295n.83, 295n.85, 299n.114
  Aristotle on, 5, 46, 55, 70, 71, 72–73, 74–75
  Deacon on, 5–6, 54, 57, 59, 61, 98, 102–3, 188, 234
  God as final end (*telos*), 107, 189, 209–11, 213, 294n.82, 295n.83, 295n.85
  relationship to divine action, 129, 209–11, 218–19, 228
  relationship to efficient causation, 78, 79, 227
  See also teleology
Fine, Kit, on hylomorphism, 261n.86
flocking birds, 17, 19
formal causation, 22, 64, 80, 217
  Aquinas on, 74–75, 107, 175, 192–97, 199, 205, 212, 218, 229, 230, 290n.33
  Aristotle on, 5, 46, 55, 67–68, 69–71, 73, 74–75, 79, 94–95, 97, 99–100, 175, 188, 189, 255nn.24–25, 261n.92
  Deacon on, 5, 55–57, 59, 61, 98, 99–100, 103, 179, 188, 234, 251n.90, 261n.87
  Emmeche et al. on, 44, 45, 251n.90
  relationship to dispositional/powers metaphysics, 6–7, 188–89, 227, 228
  relationship to divine action, 105, 192–97, 218–19, 228
  relationship to downward causation (DC), 251n.90

relationship to efficient causation, 78, 79, 175
  See also hylomorphism
fortune, 211, 212, 229, 230, 257n.51
Fox, Matthew, 238
friction, 16

Gadamer, Hans-Georg, 238
Galileo Galilei, on science and mathematics, 257n.53
Garrison, Jim, 238
Gasser, Georg, on God's omnipresence, 279n.22
Gaukroger, Stephen
  on Aristotelian essentialism, 244n.9
  on mechanism, 244n.9
  on natural philosophy, 244n.6
  on science, 1
Gelpi, Donald, 238
General Divine Action (GDA) vs. Special Divine Action (SDA), 150, 185, 275n.138, 279n.22, 293n.63, 295n.91, 297n.96
genetics, 135
geological polygons, 51
German idealism, 113, 114–26, 237
  and God's transcendence, 167, 168, 170, 172
  and panentheism, 105, 109, 120–21, 123–26, 167, 173, 242n.6, 268n.50, 269n.65, 272n.101
  relationship to Spinozism, 114, 115, 116, 120, 121–22, 123, 124, 126
Gerrida (*Gerridae*), 15, 16
Gillett, Carl
  vs. Deacon, 38
  on natural laws, 38
  on whole-part constraints-based version of emergence (EM), 37–38
Gilson, Étienne, 8, 295n.87
Göcke, Benedikt Paul

on "in" metaphor, 158
on Krause, 117–18, 119, 164, 265nn.21–22, 266n.23, 269n.61, 271nn.75–76, 277n.8
on panentheism vs. classical theism, 262n.4
God
  as the Absolute, 115–16, 117–18, 121, 123, 124, 125, 126, 131–32, 164, 265n.21, 268n.50
  chance and divine action, 137, 181, 191, 192, 211–13
  as creativity, 152–53, 182, 276n.150
  as Creator, 1, 2–3, 4, 7, 105, 123–24, 132, 136, 137, 138, 139, 140, 141, 142–45, 148–49, 152–53, 159, 160, 165–66, 168, 170–71, 172, 176–77, 178, 180–81, 189, 190, 192, 194, 195, 196–97, 199, 200, 201, 202, 204, 206, 207, 211–12, 213, 215, 220, 221, 229, 230, 231, 234, 242n.4, 262n.4, 264n.16, 269n.61, 278n.14, 281n.36, 282n.42, 286n.4, 287n.12, 295n.87, 297n.98
  as dipolar, 110–11, 118, 126–32, 135, 136, 137, 141, 144, 151, 157, 167, 168, 172–73, 174, 185, 231, 271n.86, 273n.112, 274n.129, 280n.31
  divine ideas as exemplar causes/extrinsic forms, 107, 189, 192–97, 199–200, 207, 215, 219, 221, 222, 223, 225, 227, 234, 287n.12, 288nn.15–17, 288n.23, 289nn.27–28, 290nn.33–34, 291n.37, 291n.39, 294n.81, 297n.100
  essence (*essentia*) of, 120–21, 124, 132, 134, 138, 141, 143, 162, 166, 173, 174, 177, 180, 189, 190, 191, 195, 197, 214, 215, 219, 220, 221, 222, 223, 224–25, 229, 237, 271n.86, 285n.2, 289nn.27–28, 297nn.98–100, 298n.110
  as eternal, 4, 7, 121, 131, 132, 134, 137, 139, 142, 143, 144–45, 167, 171, 172, 234, 273n.112
  existence (*esse*) of, 131, 132, 214, 220, 222, 231, 233, 235, 285nn.1–2, 289n.27, 297nn.98–99
  as final end (*telos*), 2, 107, 189, 209–11, 213, 234, 287n.12, 294n.82, 295n.83, 295n.85
  freedom of, 134, 141, 142, 214, 223–24, 279n.23
  General Divine Action (GDA) vs. Special Divine Action (SDA), 150, 185, 275n.138, 279n.22, 293n.63, 295n.91, 297n.98
  goodness of, 174, 209–11, 214, 215–16, 218–19, 223, 228, 230, 234, 286n.5, 294n.81, 295n.85, 295n.87
  immanence of, 106, 111, 115, 121, 125, 126, 135, 137, 138, 143–44, 149, 155–56, 162–67, 168, 174, 175–76, 191, 216, 218–19, 228, 229, 230, 231, 234, 239–40, 262n.2, 276n.5, 279nn.22–23, 281n.36
  as immutable, 4, 7, 121, 123, 131, 132, 134, 167, 190, 191, 231, 234, 262n.4
  as impassible, 4, 7, 134, 167, 172, 234
  as impersonal, 114, 115
  as infinite, 4, 7, 115, 123, 124, 138, 148, 149, 159, 160, 182, 187, 191, 192, 200, 226, 234, 340
  as interventionist, 136, 142, 145–46, 150, 198, 205–6, 295n.91
  as loving, 134, 136
  and natural teleology, 91, 194–95, 209–11, 213, 225, 234

God (*cont.*)
nonbeing and divine understanding, 221–25
as omnipotent, 4, 7, 132, 134, 139, 140, 142, 167, 170, 172, 231, 234, 273n.112
omnipresence of, 162–63, 279n.22
as omniscient, 4, 7, 132, 134, 138, 139, 142, 143–44, 167, 172, 231, 234, 273n.112
participation in being of, 183–84, 190, 192, 214, 223, 225, 227, 285n.1, 286n.7, 297n.100
perfection of, 131, 168, 197, 203, 210–11, 214–16, 218–19, 220, 221–22, 223, 224, 225, 227, 228, 230, 231, 234, 285nn.2–3, 289n.27, 295n.87, 297n.98
as personal, 121, 126, 139, 141, 150–51, 152, 179, 268n.50
primary and principal causation of, 2–3, 7, 107, 189, 200, 201–5, 207, 208–9, 211–12, 213, 215, 228–29, 230, 234, 242n.2, 281n.36, 291n.45, 292n.51, 293n.57, 293n.63
providence of, 190, 203, 212–13, 275n.138, 286n.5
relationship to formal causation, 2, 192–97
relationship to laws of nature, 4, 171–72, 181, 198, 275n.138
relationship with the world, 2–3, 4, 7–8, 106, 107, 111, 114, 115, 117, 119, 120–21, 122, 124–25, 126, 128–31, 132–35, 136, 137, 138–39, 140, 141, 142–44, 145–46, 149–52, 155, 156–66, 173–77, 182–83, 184, 185, 187–88, 189–91, 198, 205–6, 215–16, 229, 230, 237, 265n.21, 266n.23, 271n.76, 276n.1, 276n.5, 277n.6, 278n.17, 279n.23, 280n.31, 281n.36, 295n.91
self-knowledge of, 194, 289n.28
self-limitation of, 4, 134, 138, 139, 140, 142, 143, 144, 167, 172, 173, 230, 273n.112
as self-subsistent being (*ipsum esse subsistens*), 189, 190, 224, 285n.2
simplicity of, 123, 131, 220, 223, 298n.105
as source of being (*esse*), 189–90, 192, 199, 201–2, 203, 204, 205, 207, 213, 214, 215, 219, 226, 227, 230, 231, 234, 285n.1, 286n.4, 288n.14, 293n.58, 297n.99
as source of intrinsic forms, 192–93, 196, 197, 199
as source of primary matter, 2, 192, 199, 207, 213, 229, 287n.12
as temporal, 137, 139, 143–45, 173, 273n.112
transcendence of, 7, 111, 121, 124–25, 129–30, 137, 140, 143, 149, 155–56, 158, 160, 163–64, 165, 166–73, 175–76, 177, 179, 180, 181–82, 191, 200, 216, 219, 230, 231, 234, 240, 262n.2, 264n.16, 276n.5
as Trinity, 114, 125, 142, 144, 161, 163, 278n.15, 280n.31
as univocal cause, 2–3, 7, 166, 184–85, 187, 190–91, 216, 230, 285nn.70–71
See also classical theism; emergentist panentheism; panentheism; process theism; religious naturalism; theistic naturalism
Goodenough, Ursula
on creation, 177–78
on emergent teleology, 177–78
on irreducible complexity, 146

on religious naturalism, 146, 153, 177–79, 274n.123
Graham, Daniel W., 253n.5
Green, Thomas Hill, 238
Gregersen, Niels Henrik
  on Alexander's theistic naturalism, 149
  on Aquinas, 162–63, 279n.23
  on divine immanence in classical theism, 162–63, 164, 239
  on evil and process theism, 174
  on panentheism, 8, 164, 262n.2
  on soul/body metaphor and God-world relationship, 177
Griffin, David
  on Christology, 135
  on God and evil, 174
  on mind/brain dependency, 278n.12
  on panentheism, 127, 133, 134, 238, 269n.61, 270n.71, 278n.12
  on Whitehead, 127
Guilherme, Alexandre, 266n.26
Gunton, Colin, 161
Gutiérrez, Gustavo, 238

H_2O molecules, 15, 16
Haag, James
  on emergentist panentheism and monism, 151
  on Peacocke, 151
Hartshorne, Charles
  on Akhenaton, 110
  on God as dipolar, 110–11, 118, 131, 135, 167, 170
  on God as inclusive of reality, 161
  on God's transcendence, 169–70
  on *Hymn to Time*, 110
  panentheism of, 105, 109, 127, 131–32, 134, 135, 140, 238, 270n.71
  on surrelativism of God, 131, 136, 167, 170
  on Taoism, 111

Hegel, G. W. F., 114, 116, 237, 267n.37
  on Absolute Idea, 121, 123, 124–25, 141, 173, 268n.50
  vs. Aristotle, 141
  on contingency, 124, 126, 141, 168
  on dialectics, 121–23, 124, 126, 134, 173, 225, 283n.51
  on Fichte, 121, 122
  on God, 123–25, 268n.50
  on Nature, 124, 168
  on negativity, 123
  panentheism of, 123–26, 134, 141, 242n.6, 268n.50, 269n.65
  on Schelling, 121, 122
  on Spinoza, 121–22, 123, 267n.32
  on substance, 123, 124
Heidegger, Martin, 131, 238
Heil, John, on natural intentionality, 90
Heim, Karl, 238
Heraclitus, 62, 63, 65, 66–67, 127, 252n.3, 253n.5
Herder, Gottfried, 237
Higgs bosons, 16
Hocking, William, 238
Hodgson, Peter, 238
horizontal explanation, 244n.7, 244n.9
Hulswit, Menno, on downward causation (DC), 5, 43
human agency, 73, 80, 95, 175, 257nn.51–52, 274n.120
  and dispositionalism, 86
  relationship to divine action, 120–21, 138, 197, 206
Hume, David
  bundle theory of substance, 92, 93, 260n.83
  on causation, 24, 81, 87, 91, 94, 246n.26, 257n.56, 258n.57
Humphreys, Paul, on supervenience (SUP), 40

Hutchings, Patrick, 268n.50
hylomorphism, 214, 260n.82
   of Aquinas, 175, 189, 225, 287n.9
   of Aristotle, 6, 74, 78, 90–93, 94–95, 99–100, 175, 189, 252n.2, 254n.19, 260nn.83–84, 261n.86
   and dispositionalism, 90–93, 260n.83
   and downward causation (DC), 94–95, 96
   and dynamical depth version of emergence, 99–100
   mereological and structural version, 99–100, 261n.86

indeterminism, 32–33, 246n.26, 272n.93, 273n.108, 295n.91
Inge, William Ralph, 238
Intelligent Design movement, 145
intentionality, 85, 90, 103
Internet, 17
Irwin, Terrence, on form, 70
isomerization, 85–86

James, William, 131, 238
Jaworski, William, on hylomorphism, 261n.86
Jensen, Alexander, 169
Jesus Christ, 135, 145, 272n.94
Johnston, Mark, on hylomorphism, 261n.86
Juarrero, Alicia
   on context-sensitive constraints, 38
   on downward causation (DC), 35, 36
Julian of Norwich, 237

Kant, Immanuel, 114–15, 120
   on *noumena* (things-in-themselves), 8
   on transcendental knowledge, 178
Kaufman, Gordon
   on emergentist panentheism, 152–53, 182, 187

   on faith, 153
   on God as creativity, 152–53, 182, 276n.150
   on modalities of creativity, 152–53
   on serendipitous creativity, 152–53, 182
Kim, Jaegwon
   on causal exclusion, 36, 37, 247n.33
   on downward causation (DC), 5, 36–37, 42, 248n.34
   on functional model of reduction, 33
   and mind-brain debate, 27
   on nonreductionism (NP), 43
   on overdetermination, 36, 37, 42, 44
   on reductionism, 33, 40–41, 42–43
   on self-causation/self-determination, 36
   on supervenience (SUP), 37, 40–41
Knight, Christopher, 238
Koons, Robert
   on hylomorphism, 261n.86
   on laws of nature, 246n.26
Koslicki, Kathrin, on hylomorphism, 261n.86
Koutroufinis, Spyridon
   on constraints, 49
   on dynamical depth, 47, 51
   on morphodynamic EM, 51
   on selfhood, 51
Krause, Karl
   on *creatio ex nihilo*, 269n.61
   Göcke on, 117–18, 119, 164, 265nn.21–22, 266n.23, 269n.61, 271nn.75–76, 277n.8
   on God as *Orwesen*/the Absolute, 116, 117–18, 164, 265n.21, 266nn.23–24, 269n.61, 271n.75
   on God as *Urwesen*, 117–18, 266n.23, 271n.75
   on "in" metaphor, 159

panentheism of, 114, 116–18, 119, 120, 126, 159, 237, 265n.21, 266n.23, 269n.61, 271nn.75–76, 277n.8
   on science, 116–17, 118
   on time, 271n.75
   on transcendental categories, 116–17
Krebs cycle, 76, 77
Kripke, Saul, 27
Küng, Hans, 238
Køppe, Simo. *See* Emmeche, Claus, et al.

laminar flows, 50
Lao-tzu, on absence, 58
laser light, 51
Lataster, Raphael, on panentheism, 158, 262n.4, 277n.6
Laughlin, Robert, on broken symmetry, 17, 20, 27–28
Law, William, 237
laws of nature
   Davies on, 171–72, 181
   Koons on, 246n.26
   laws of biology, 17
   laws of chemistry, 17, 28–29
   laws of physics, 17, 21–22, 28–29
   laws of thermodynamics, 4, 48, 99
   relationship of God to, 4, 171–72, 181, 198, 275n.138
Lequier, Jules, 238
Lessing, Gotthold, 237
Levine, Michael P., 282n.42
Lewes, George Henry
   on emergent effects, 25, 29
   on emergent entities (substances), 29
   on resultant effects, 29
Lewis, David, 27
life sciences
   paradigm shift in, 3–4
   systems approach in, 3, 12, 26–27, 41, 58

Locke, John, and substratum theory of particulars, 260n.83
logical positivism, 24, 25–26, 27
Lossky, Vladimir, 284n.63
Lotze, Herman, 238
Louth, Andrew, 280n.31
Lowe, E. J., 260n.80

MacGregor, Gedds, 238
Mach, Ernst, phenomenalism of, 24
Macquarrie, John, 238
   on Christology, 135
   on panentheism, 133, 134
Maimonides, Moses, on concurrentism, 198–99
Mander, William, on pantheism, 242n.6
manipulability view of causation (MVC), 82, 84
   vs. DVC, 87, 89
Mason, Richard, on Spinoza and panentheism, 264n.12
material causation
   Aquinas on, 74, 192, 205, 212, 230, 287n.12
   Aristotle on, 5, 46, 55, 59, 67–69, 74, 98, 99–100
   Bacon on, 22
   See also primary matter
mathematics, 26, 44, 64, 244n.7
   Galileo on, 257n.52
   in science, 2, 12, 22, 23, 24, 80, 95, 257n.53
Mawson, T. J., 284n.61
Maximus the Confessor, 284n.63
McDaniel, Jay, 238
McFague, Sallie, 238
   on Christology, 135
   on panentheism, 133, 134
mechanics, 20–21, 22, 23, 24, 28–29, 244n.7
mechanism, 2, 23–25

Mechtild of Magdeburg, 237
medium downward causation (MDC), 44, 45
Mellor, David Hugh, 259n.76
Melsen, Andrew G. van (Andreas Gerardus Maria van Melsen)
  on finality in natural science, 256n.36
  on species-individual structure of matter, 252n.2
metaphysics
  and analytic philosophy, 26
  as first philosophy, 78
  concerning matter, 20, 21, 22, 23, 24
  relationship to empirical inquiry, 78
Mill, John Stuart
  on emergent entities (substances), 29
  on heteropathic vs. homopathic effects and laws, 29, 32
  on physical vs. chemical causes and transition laws, 29
  on reductionism, 25
mind/brain dependency, 39, 45, 278n.12, 283n.59
mind/brain studies, 27
Molnar, George
  on causation and powers, 87
  on dispositions/powers, 6, 87, 258n.59, 260n.80
  on physical intentionality, 90
Moltmann, Jürgen
  on panentheism, 134–35, 157, 238
  on *zimsum*, 138, 157
Montefiore, Hugh, 238
More, Henry, 237
Moreno, Alvaro, 5
  on matter and form, 46
Morgan, Conwy Lloyd, on emergence (EM), 25, 29
Morowitz, Harold, 149, 157, 182
Müller, Klaus, 164
Mullins, Richard T., 268n.50, 284n.61

Mumford, Alexander
  on dispositions/powers, 85, 87, 96, 258n.59, 258n.63, 259n.76
  on neutral monism, 260n.78
  on temporal simultaneity of causal relations, 88
Mumford, Stephen, 6
Murphy, Nancey, 295n.91
  on divine action and double agency, 205, 206, 207, 208
  on downward causation (DC), 35, 36, 43
Mutakallims, 198

Nagel, Ernest, on reductionism, 26–27, 33
natural geometric patterns, 16–17
natural philosophy. *See* physics
natural selection, 54
necessity
  and chance, 78–80, 137, 148
  of first cause, 211–12
  of God's existence, 131
  of Hegel's Absolute, 141
  as suppositional, 80, 88, 103, 211–12
Neoplatonism, 109, 111, 112–14, 116, 196, 237, 268n.50
Nesteruk, Alexei, 280n.31
neuroscience, 11, 27, 176, 283n.60
Newton, Isaac, 21–22, 38, 120, 210
Nicolas of Cusa, 114, 141, 237, 242n.6
nonadditivity of causes, 28–29
nonreductionist physicalism (NP), 12, 30, 31, 42–43, 137–38, 145, 188, 249n.45
normativity, 85

occasionalism, 22, 198, 199, 205, 207, 208
O'Connor, Timothy, on Deacon, 250n.70

Oderberg, David
  on dispositionalism, 6, 90
  on panpsychism, 259n.75
Ogden, Schubert, 238
Oomen, Palmyre, on Whitehead, 169
Oord, Thomas Jay, 157
Oppenheimer, Helen, 238
overdetermination, 87, 209
  Kim on, 36, 37, 42, 44

Padgett, Alan G., on Fichte, 264n.16
Pailin, David, 238
  on panentheism, 135
panentheism
  in ancient religious traditions,
    110–11
  and Christology, 135
  vs. classical theism, 4, 105–6, 109–
    10, 111, 113–14, 123–24, 131,
    133, 134, 135, 138, 151, 156, 159,
    162–63, 164, 166–67, 191, 262n.4,
    281n.36, 284n.61, 285n.1
  critical evaluation of, 155–77
  definition of, 105–6, 262n.2
  and dialectics, 112, 113, 114, 116,
    118, 120–23, 134
  and emanationism, 112–13
  as eschatological, 144, 156, 173,
    274n.114
  as expressivist, 118, 173
  and German idealism, 105, 109,
    120–21, 123–26, 134, 141, 167,
    173, 242n.6, 268n.50, 269n.65,
    272n.101
  and God's immanence, 133, 135,
    137, 138, 143, 144, 149, 155–56,
    162–67, 191
  and God's transcendence, 133, 137,
    143, 149, 155–56, 158, 160, 167–
    73, 191
  of Hartshorne, 105, 109, 127, 131–32,
    134, 135, 140, 238, 270n.71
  of Hegel, 123–26, 134, 141, 242n.6,
    268n.50, 269n.65
  historical development of, 105–6,
    109, 110–32
  and indeterminacy, 133
  "in" metaphor of, 106, 132–35, 138,
    140, 149–50, 155, 156–62, 164,
    182, 185–86, 187–88, 230, 242n.6,
    276n.1, 276n.5, 277n.8
  of Krause, 114, 116–18, 118, 119,
    120, 126, 159, 237, 265n.21,
    266n.23, 269n.61, 271nn.75–76,
    277n.8
  and Neoplatonism, 109, 111, 112–14,
    141, 143
  panentheistic analogy (PA), 175–77
  vs. pantheism, 7, 106, 114, 124–25,
    138, 140, 158, 160, 181–82, 231,
    242n.6, 262n.2, 263n.6, 264n.12,
    276n.1, 276n.5, 277n.6, 278n.12,
    282n.42, 284n.61
  and participation, 113
  and Plato, 109–10, 111, 263n.6
  as processual, 105, 118, 210
  role of soul (mind)/body metaphor
    in, 132–33, 138–39, 142, 149, 155,
    174–77, 283n.54, 284n.61
  and theodicy, 134, 155, 173–74
  of Whitehead, 127, 128–31, 135,
    136, 140, 149, 182, 238, 270n.65,
    270n.71
  See also emergentist panentheism;
    Krause, Karl
Pannenberg, Wolfhart, 238
panpsychism, 6, 116, 127–28, 259n.75
pantheism, 105–6, 153, 156, 284n.62
  vs. classical theism, 276n.5
  vs. panentheism, 7, 106, 114, 124–
    25, 138, 140, 158, 160, 181–82,
    231, 242n.6, 262n.2, 263n.6,
    264n.12, 276n.1, 276n.5, 277n.6,
    278n.12, 282n.42, 284n.61

pantheism (*cont.*)
   of Spinoza, 114, 116, 124, 242n.6, 276n.1
Parmenides, on being and nonbeing, 62, 63, 64, 65, 66, 220, 253n.4
Paul, St., 157
Peacocke, Arthur
   on causal joint conundrum, 3, 150, 183, 184, 187, 205, 229–30
   on Christology, 135
   vs. Deacon, 184
   on double agency and divine action, 205–6
   on downward causation (DC), 35, 36, 137, 248n.38
   emergentist panentheism of, 4, 8, 9–10, 106, 133, 135, 137–40, 146, 149–52, 157, 159–60, 161, 172, 176, 182–86, 187–88, 216, 228–31, 235, 238
   on God and evil, 174
   on God as Creator, 281n.36
   on God as personal, 139, 150–51
   on God's action, 140, 146, 150–51, 183, 184, 216, 229, 230, 285n.70, 295n.91
   on God's knowledge, 272n.93
   on God's self-limitation, 4, 138, 139, 167, 172, 272n.93
   on God's suffering, 138, 272n.94
   on God-world relationship, 138–39, 150, 176, 182–83, 184, 185, 187–88, 216, 229–30
   on "in" metaphor, 138, 159–60, 161, 172, 187–88
   on monism, 137–38, 160
   on ontological interface between God and world, 150, 182–83, 185, 187
   on soul (mind)/body metaphor, 138–39, 176–77
   on special divine action, 185
   on whole-part/top-down model of divine action, 4, 142, 150, 151, 152, 183, 184, 230, 295n.91
Peirce, Charles Sanders, 131, 238
Pemberton, John, on powers and Aristotle's metaphysics, 89
Pereira, Antonio Marcos, 5, 46
Peterson, Gregory, on mind and body, 283n.59
Pfleiderer, Otto, 238
phase-space physical theory, 44–45, 188
philosophy of mind, 27, 37, 116, 176, 283n.60
physicalism, 2, 36, 37, 38, 39, 42–43, 247n.32, 249n.45, 250n.70
physics, 20–22, 23, 24, 25, 26, 28–29, 41
   big bang theory, 135, 152–53, 183
   See also quantum mechanics
Pickavance, Timothy, on laws of nature, 246n.26
Pittenger, Norman, 238
   on Christology, 135
   on panentheism, 133, 135
Place, Ullin, on dispositions as intentional states, 90
Plato
   on the Good, 112
   on Heraclitus, 253n.5
   on Ideas (Forms), 64, 69, 111, 112, 192–93, 194, 248n.39, 288n.15, 289n.27
   on the Living Being, 111
   and panentheism, 109–10, 111, 263n.6
   on philosophy and wonder, 243n.1
   vs. Plotinus, 112–13
   on the World-Soul, 111–12, 113

Plotinus, 111, 237
  on emanationism, 112–13
  on the One, 112, 113
  on participation, 113
  vs. Plato, 112–13
  on World-Soul, 112, 113
Poincaré, Henri, conventionalism of, 24
Polanyi, Michael, 35
Polkinghorne, John, 3, 285n.70
  on chaos and divine action, 142, 273n.108, 295n.91
  on divine kenosis, 142–43, 144, 167, 173, 273n.108, 273n.112
  on divine suffering, 144, 173
  on double agency theory and divine action, 205, 206, 207, 208, 294n.69
  on eschatological fulfillment, 144, 274n.114
  on God as Creator, 142–44
  on God as dipolar, 144, 273n.112
  on God as Trinity, 142, 144
  on incarnation, 143
  on panentheism, 142–45, 173, 238, 274n.114
  on theodicy, 206, 208
Popper, Karl R., 27, 259n.76
Prabhu, Joseph, on Hegel, 267n.42, 268n.50
prediction, 12, 24
primary and secondary causation, 181, 214, 218–19
  Aquinas on, 2–3, 7, 72, 107, 189, 200, 201–5, 207, 208–9, 211–12, 228–29, 234, 242n.2, 292n.48, 293n.57
  Aristotle on, 72, 200, 292n.48
  primary causation of God, 2–3, 7, 107, 181, 189, 200, 201–5, 207, 208–9, 211–12, 213, 215, 228–29, 230, 234, 281n.36, 293n.57, 293n.63
primary matter, 6, 23, 67, 74, 75, 197, 228, 289n.28, 297n.98
  Aquinas on, 70, 97, 175, 191, 192, 195, 199, 226–27, 229, 254n.22, 287n.9, 287nn.11–12, 298n.105, 299n.114
  Aristotle on, 68–69, 70, 80, 91, 93, 94–95, 97, 100, 175, 226, 252n.2, 254nn.18–19, 255n.30
  God as source of, 2, 192, 199, 207, 213, 229, 287n.12
  as potentiality, 191, 192, 199, 207, 214, 219, 226, 254n.22, 287n.12
  vs. secondary (proximate) matter, 68, 78, 226, 227
  See also hylomorphism
principal and instrumental causation, 218–19, 242n.2
  Aquinas on, 2–3, 7, 107, 189, 200, 201–5, 208, 209, 211–12, 234, 291n.45, 292n.51
  principal causation of God, 2–3, 7, 107, 189, 200, 201–5, 208, 209, 211–12, 213, 215, 228–29, 230, 234, 281n.36, 291n.45, 292n.51
Pringle-Pattison, Andrew Seth, 238
probability view of causation (ProbVC), 82, 83
  vs. DVC, 87
problem of evil, 134, 155, 173–74, 206, 208, 230–31
process theism, 173, 174, 185, 278n.14, 278n.17
  vs. classical theism, 131, 156
  dipolar God of, 126–32, 141, 144, 167, 172, 273n.112
  God's transcendence in, 168–69
  See also Hartshorne, Charles; Whitehead, Alfred North

process view of causation (ProcVC), 82, 84
  vs. DVC, 87
Proclus, 111, 113, 237
  on dialectics, 113
  on the One, 113
Pseudo-Dionysius, 113–14, 237, 284n.63
Putnam, Hilary, on multiple realizability of psychological phenomena, 30

qualitative aspect of reality, 8–9, 11–12, 30, 64, 95
quantitative aspect of reality, 2, 8–9, 12, 30, 64, 99
quantum mechanics, 4, 15, 17, 135, 136, 260n.80
  chance in, 24, 79
  Copenhagen interpretation, 24, 261n.85, 295n.91
Quine, Willard, 27, 259n.77

Rahner, Karl, 238
Rahner's Rule, 144
Rea, Michael C., 259n.76
realist critique of knowledge, 8
Reck, Andrew, 123
reductionism, 9, 30, 244n.9
  regarding causation, 5–6, 42, 43–44, 45, 55, 56, 64–65, 94
  Deacon on, 48, 54–55
  intertheoretical and bridge-laws-based reduction, 26–27, 33
  Kim on, 33, 40–41, 42–43
  Nagel on, 26–27, 33
  in science, 2, 3, 11–12, 20, 23–25, 26–27, 35, 44, 55, 56, 245n.10
Reese, William
  on Akhenaton, 110
  on *Hymn to Time*, 110
  on panentheism, 109, 134
  on Taoism, 111
  on theistic naturalism, 185
Reeves, Gene, 282n.44
regulatory view of causation (RVC), 81, 83
  vs. DVC, 87
relativity, 4, 135
religious naturalism, 150, 176
  Deacon on, 146–48, 179, 180
  and emergence (EM), 146–48, 177–80
  Goodenough on, 146–47, 153, 177–79, 274n.123
  Sherman on, 179, 180, 274n.123
  and teleology, 147, 177, 178, 179–80
Renouvier, Charles, 238
Robinson, John, 238
  on panentheism and Christology, 135
Ruether, Rosemary, 238
Russell, Bertrand
  on causal lines, 82
  on law of cosmic laziness, 256n.36
Russell, Robert John
  on big bang cosmology, 4, 183
  on noninterventionist objective divine action (NIODA), 295n.91
  on Peacocke's emergentist panentheism, 4
  on topology of universe, 183

Saunders, Nicholas, 295n.91
  on General Divine Action (GDA) vs. Special Divine Action (SDA), 275n.138
  on laws of nature, 246n.26
Schelling, Friedrich, 116, 118–21, 126, 237, 266n.26, 266n.29
  on the Absolute as Nature, 119–20
  on Fichte, 118–19
  on God's choice to create, 141
  on God's essence, 120–21

on Nature, 119–20, 122
panentheism of, 120–21, 272n.101
Schleiermacher, Friedrich, 237
  on the absolutely supernatural, 3
Schoonenberg, Piet, 238
Schweitzer, Albert, 238
science
  dialogue with theology, 4, 8, 9,
    106, 109, 127, 135–45, 164–65,
    170, 177, 205, 214, 233, 235, 237,
    285n.71
  efficient causation in, 23–24, 80
  and God of the gaps, 2
  Krause on, 116–17, 118
  mathematics in, 2, 12, 22, 23, 24, 80,
    95, 257n.53
  paradigm shift in, 3–4
  philosophy of, 8, 20, 26, 27, 28, 35,
    37, 116, 216, 244n.5
  reductionism in, 2, 3, 11–12, 20,
    23–25, 26–27, 35, 44, 55, 56,
    245n.10
  revolution of seventeenth century,
    20, 243n.4
  vs. *scientia*, 245n.10
  vs. theology, 1–4, 114
Scott, Alwyn
  on Aristotelian causes, 46
  on downward causation (DC),
    248n.34
Scotus, John Duns, 114, 292n.48
Segundo, Jan Luis, 238
self-organizing (dissipative) systems,
  16–17
self-organizing processes, 36
serendipitous creativity, 152–53
Shanley, Brian, on Aquinas and God's
  immanence, 166
Sherman, Jeremy
  on emergence and teleology, 147–48
  on human self-creation, 147,
    274n.120

on religious naturalism, 179, 180,
  274n.123
Shoemaker, Sidney, 259n.76
Silberstein, Michael, on causation, 5,
  46
Silva, Ignacio
  on divine action, 202–4, 207, 208,
    273n.108, 285n.70, 293n.59
  on divine *kenosis*, 142
singularist view of causation (SVC),
  82, 83–84
  vs. DVC, 89
Smart, John Jamieson Carswell, 27
Smith, John, 237
snowflakes, 17, 18, 51
soul (mind)/body metaphor
  as analogy of God-world
    relationship, 132–33, 138–39, 142,
    149, 174–77, 283n.60, 284n.61
  Aquinas on, 163, 174, 175, 239–40,
    279n.23, 283n.59
  Clayton on, 174–76
  Peacocke on, 138–39, 176–77
  role in panentheism, 132–33, 138–
    39, 142, 149, 155, 174–77, 283n.54,
    284n.61
Sperry, Roger Wolcott
  on downward causation (DC), 36
  and mind/brain debate, 27
Spinoza, Benedict de
  on final causality, 54
  and German idealism, 114, 115,
    116, 120, 121–22, 123, 124, 126,
    266n.26
  on God, 126, 264n.12
  and panentheism, 242n.6, 264n.12
  pantheism of, 114, 116, 124, 242n.6,
    276n.1
  on substance, 115, 123, 124, 126,
    264n.12, 276n.1
spiral phylotaxis, 51
Srivastava, Alok, 55

stereoisomerization, 85, 86
Stewart, Claude, 238
Stjernfelt, Frederic. *See* Emmeche, Claus, et al.
stochastic systems, 50
Stoeger, William R.
   on divine action, 105, 204, 293n.63
   on God as Creator, 105, 204
strong downward causation (SDC), 44, 45, 151
structural isomerization, 85, 86
substance
   vs. accidents, 65, 66–67, 71, 79, 162–63, 195, 254n.12, 286n.6
   Aristotle on, 65, 66–67, 68, 70, 254n.12, 254n.19, 255n.25
   primary vs. secondary substance, 255n.27
   Spinoza on, 115, 123, 124, 126, 264n.12, 276n.1
substantial change, 252n.2, 255n.27
   vs. accidental change, 67, 71, 101, 201, 254n.12, 256n.43
   Aristotle on, 67, 69, 70–71, 97, 100, 254n.12, 255n.30
substantial form, 22, 23, 73, 76, 78, 81, 196, 218, 219, 225, 228, 289n.28
   Aquinas on, 70, 75, 97, 189, 191, 192, 195, 199, 204, 226, 227, 254n.22, 298n.105, 299nn.114–15
   Aristotle on, 6, 67, 68, 69–70, 72, 74, 79, 91, 93, 94–95, 97, 99–100, 103, 252n.2, 260n.84, 261n.92, 285n.3
   Dodds on, 70, 105, 287n.10
   God as Creator of, 107, 189, 207, 213, 234
   soul as, 297n.98
   See also hylomorphism
supervenience (SUP), 48, 100, 176
   definition, 39

   emergence (EM) based on, 28, 39–41, 61
   Kim on, 37, 40–41
   relationship to mind/brain dependency, 39
surface tension, 15–16, 50
swarm behavior, 17, 19
Swinburne, Richard, 284n.61

Taoism, 58, 111
Teilhard de Chardin, Pierre, 238
teleology, 22, 23, 45, 46, 63, 80, 81, 136, 276n.1
   Aquinas on God as final end (*telos*), 107, 189, 209–10, 294n.82, 295n.83, 295n.85
   Deacon on teleodynamics, 51–54, 56–58, 59, 103–4, 177, 189, 217, 219, 227, 251n.86
   and downward causation (DC), 96, 188–89
   as emergent, 17, 51–54, 56–58, 59, 98, 103–4, 177, 178, 179, 189, 215–16, 217, 219, 227, 251n.86
   God and natural teleology, 91, 194–95, 209–11, 213, 225, 234
   and religious naturalism, 146–48, 177, 178, 179–80
   role in dispositional/powers metaphysics, 6–7, 90, 93, 188–89, 217–18
   and theistic naturalism, 148
   See also final causation
temperature, 16
Temple, William, 238
tensile strength, 16
te Velde, Rudi, 203–4, 293n.58
Thales, 63
theistic naturalism, 150, 176, 180–82, 187
   Alexander on, 148–49, 153

Davies on, 148, 185
Drees on, 148, 180–81
and emergence (EM), 146, 148–49, 180–81
and teleology, 148
theodicy. *See* problem of evil
theology
  analogical predication in, 2
  apophatic theology, 147, 180, 274n.123, 284n.63, 294n.69
  deism, 3, 141, 180–81, 185, 205, 207, 208
  dialogue with science, 4, 8, 9, 106, 109, 127, 135–45, 164–65, 170, 177, 205, 214, 233, 235, 237, 285n.71
  natural theology vs. theology of nature, 8
  as nonreductive, 4
  vs. science, 1–4, 114
  via eminentiae in, 113, 180
  via negativa in, 113, 180
  via positiva in, 113, 180
  See also classical theism; panentheism; process theism
thermodynamics
  laws of, 4, 48, 99
  thermodynamic equilibrium, 47–48, 50, 51
Thomas, Owen, 161, 278n.14, 279n.23
Tillich, Paul, 238
Towne, Edgar, on panentheism, 280n.31
Tracy, David, 163–64
Tracy, Thomas F., on divine action, 203, 206, 295n.91
Troeltsch, Ernst, 238

Umerez, Jon, 5, 46
underdetermination, 87
Upanishads, 110

van Gulick, Robert
  on downward causation (DC), 35, 36, 43, 247n.32
  on selective activation, 43
Vedantic traditions, 157
vertical explanation, 244n.7, 244n.9
virtual presence, 75–78, 97, 100
viscosity, 16, 50
vitalism, 25, 145, 152
Vlastos, Gregory, on dialectics, 173, 283n.51

Ward, James, 238
Ware, Kallistos, 238, 280n.31
  on panentheism, 133
water crystals, 17, 18
water erosion formations, 16
water striders, 15
Watts, Alan, 238
weak downward causation (WDC), 44–45, 151
Weiss, Paul, 238
Whichcote, Benjamin, 237
Whitehead, Alfred North, 6, 105, 109, 269n.61
  on the Absolute as creativity, 130, 131, 168, 169, 171
  on actual occasions, 127–28, 129, 130, 136, 141, 161, 168, 169, 171, 174, 278n.14, 282n.44
  on Alexander, 274n.129
  on concrescence, 127, 129, 168, 169
  on creativity, 171, 282n.44
  dialectics of, 283n.51
  on God, 128–31, 157, 160–61, 167, 168–69, 170–71, 174, 274n.129, 282n.44
  vs. Hegel, 270n.65
  panentheism of, 127, 128–31, 135, 140, 149, 182, 238, 270n.65, 270n.71

panentheism of (*cont.*)
   on prehension, 127, 128, 129, 130, 157, 161, 168, 278n.14
   on process view of reality/philosophy of organism, 127–28, 140, 149, 157, 160–61, 167, 168–69, 170, 174, 252n.3
   on universals, 127–28
Whittemore, Robert, 124–25
Wilkens, Steve, on Fichte, 264n.16
Will, James, 238
Williams, Daniel Day, 144
Wippel, John, 202, 292n.51, 298n.105
   on Aquinas on nonbeing, 299n.115
   on divine understanding and nonbeing, 221, 222, 223, 224–25, 297n.102
Wong, Hong Yu
   on emergent vs. resultant properties, 40
   on supervenience (SUP), 40

zimsum, 138, 157

Mariusz Tabaczek, O.P., is a friar preacher, professor of theology, and member of the Thomistic Institute at the Pontifical University of Saint Thomas Aquinas in Rome. He is the author of *Emergence: Towards A New Metaphysics and Philosophy of Science* (University of Notre Dame Press, 2019).

CPSIA information can be obtained
at www.ICGtesting.com
Printed in the USA
LVHW081919110521
687116LV00002B/218